The Psocoptera of tropical South-east Asia

This book is dedicated to the memory of Ian W.B. Thornton (1926-2002).

FAUNA MALESIANA HANDBOOKS

Series editor A.J. de Winter
Project editor G.O. Keijl

Fauna Malesiana
National Museum of Natural History Naturalis
P.O. Box 9517
2300 RA Leiden
The Netherlands

Fauna Malesiana received financial support from the Royal Netherlands Academy of Arts and Sciences, and the National Museum of Natural History Naturalis, Leiden.

FAUNA MALESIANA HANDBOOK 6

The Psocoptera of tropical South-east Asia

T.R. New & C. Lienhard

BRILL
LEIDEN · BOSTON

2007

Library of Congress Cataloging-in-Publication Data
The Library of Congress Cataloging-in-Publication Data is also available.

This book is printed on acid-free paper.

ISSN 1388-3895
ISBN-10 90 04 14902 3
ISBN-13 978 90 04 14902 1

Cover design by John Stelck (Biskaje), Leiden
Typesetting Fritschy opmaak & redactie, Leiden

Printed in The Netherlands.

Contents

Preface

This handbook is a preliminary guide to one of the lesser-known insect orders, Psocoptera, in tropical South-east Asia, a region of considerable interest to entomologists and biogeographers alike. The rich regional fauna of these insects, the barklice and booklice, exemplifies the complex patterns of transition between the Oriental and Australasian realms reported for many other faunal groups.
This synthesis is possible only because specialist collecting in many parts of South-east Asia in the last few decades has provided much new information on psocid diversity and distributions, and on the regional peculiarities of the psocopteran fauna. Many gaps in our knowledge remain, and many taxa are still to be collected and described. However, 725 species (belonging to 25 families and 130 genera) have been reported or described from the region, and it is perhaps only by bringing together the available published information at this stage, despite its inevitable shortcomings, that workers in the region may be encouraged to take up the study of these insects and contribute to constructing a more definitive picture. As with many other insect groups, many of the earliest described species of psocids can be recognised only tentatively, and most types of psocids described from tropical south east Asia are housed far from the region and are thus difficult for local students to study. By providing keys to families and genera (based on published information available to us up to the end of 2005), accompanied by illustrations to facilitate recognition of most of the genera represented, together with a reasonably complete bibliography to the most important references on the order, we hope to provide a taxonomic foundation on which future workers can build. Many of the illustrations are based on published diagrams, but all have been redrawn to a more uniform format for ease of use.

Chapters 1–6 include introductory material to the Psocoptera, with keys to families and genera and present an illustrated systematic overview of the regional fauna. Chapter 7 is a checklist of all species known from the region until late 2005, with references to their original descriptions. The reference section encompasses both parts of the volume.

All species reported from the region until the end of the year 2000 are listed, together with their original taxonomic citations, by Lienhard & Smithers (2002), and their World Catalogue and Bibliography is an indispensable working tool for psocidologists everywhere. Additions and corrections to this compilation are published regularly in 'Psocid News' (on line presently at:http://insect3.agr.hokudai.ac.jp/psoco-web/psoco-net/index.html), up to now as Lienhard 2003a, b, 2004c, 2005, 2006, and individual country-based species lists extracted from the World Catalogue are also available on line (presently at http://www.ville-ge.ch/musinfo/mhng/page-e/ps-coun.htm).

We appreciate greatly the support of the Fauna Malesiana Foundation in the preparation of this handbook, and value especially the continued encouragement of Dr Ton de Winter, and the patient and thorough editorial treatment from Mr Guido Keijl.

T.R. New and C. Lienhard
Melbourne and Geneva, May 2006

T.R. New
La Trobe University
Victoria 3086
Australia

C. Lienhard
Muséum d'histoire naturelle de la Ville de Genève
CH-1211 Geneva 6
Switzerland

Acknowledgments

by T.R. New

Any synthesis such as this must draw heavily on the studies of other specialists, and most psocidologists active during the last 35 years have helped me – in many cases more than they may realise – by exchanges of ideas, reprints and specimens and their ready responses to my numerous queries over that time. I owe a particular debt of gratitude to the late Professor Ian Thornton, who introduced me to the regional fauna during our field trips to Papua New Guinea in the early 1970s, whose studies have been of vital importance in contributing to this synthesis and whose friendship during the whole of my career in psocidology has been an important and lasting influence. This book is dedicated to his memory. I greatly appreciate the friendship also of Dr Courtenay Smithers (Australian Museum, Sydney), Dr Charles Lienhard (Museum of Natural History, Geneva), Dr S.S. Lee (University of Malaya, Kuala Lumpur), Dr Edward L. Mockford (Illinois State University, Normal), Dr Bryan Turner (University of London), Dr Kazunori Yoshizawa (Hokkaido University, Japan) and two of my former psocidological students, Dr Endang Sri Kentjonowati (Universitas Udayana, Bali) and Dr Evan Schmidt. All have freely shared their knowledge and experiences with me and none should be held responsible for any misunderstandings I have unwittingly perpetuated here! In particular, Charles Lienhard provided many perceptive comments in reviewing a draft and proofs of the whole work. I also appreciate greatly the help of many museum curators and collection managers who have generously allowed me access to their collections of psocids for study or loaned types or other material; of particular importance for this work have been repeated visits to the Natural History Museum, London (Dr David Hollis, Dr Peter Barnard), together with the unrivalled resources of the entomology library there (Ms Julie Harvey, Ms Pam Gilbert), and the Bishop Museum, Honolulu (Dr Gordon Nishida, Dr Neal Evenhuis).

Chapter 1

Introduction

by T.R. New

Tropical South-east Asia is a region of immense interest to entomologists and biogeographers alike, and much of the debate over the naturalness of biogeographical divisions has centred on this complex region which is viewed widely as a crucible of evolution and diversification for many taxa. This book is an account of one of the lesser known insect orders, the Psocoptera, in the region and encompasses an identification guide to the taxa recorded and discussion of their biological features and distributional relationships. In general, Psocoptera are diverse in South-east Asia with 725 species described, and manifest high levels of regional endemism. However, in common with many other groups of small and inconspicuous invertebrates they are still quite poorly known, and one major purpose of this handbook is to provide a template for further study, and to indicate the need for additional faunal studies and collections from the region.

About 13% of the described world Psocoptera species and almost 30% of the genera have been reported from the region shown in fig. 1, from Thailand and nearby mainland areas through peninsular Malaysia, and the complex archipelagos of the Philippines and Indonesia to New Guinea and nearby islands. These insects thus parallel many other biota in helping to demonstrate the transitions between the Oriental and Australasian faunas that have for so long intrigued biologists. No similar synthesis of Psocoptera has been made, although New (1977h) summarised the information then available on the Oriental fauna, emphasising the poor knowledge of the group. Since then, considerable study has occurred of collections made by specialists of Psocoptera from Malaysia, Indonesia and Papua New Guinea, together with parts of the eastern Palaearctic and Australia. Despite the country basis of most of these studies, they furnish material for a much more comprehensive faunistic and biogeographical appraisal than possible hitherto. This account treats the fauna to generic level, and provides keys to families and genera within each family. It is thus not as detailed as recent works on the Psocoptera of North America (Mockford, 1993a) or the western Palaearctic (Lienhard, 1998a), both areas for which most of the species have been described. Collections from South-east Asia contain numerous undescribed taxa, but the notes given here should provide a useful starting place for students in the region.

Historical perspective

The Psocoptera are among the diverse suite of insect orders placed formerly within the Linnaean 'Neuroptera', with the three first-described species (Linnaeus, 1758) being placed in the genera *Hemerobius* and *Termes*. The genus *Psocus* was raised by Latreille (1794) to hold all species of the group then known, with the insects then treated variously as 'Neuroptera' or 'Pseudoneuroptera'. The latter group encompasses the great variety of 'Neuroptera s.l.' that have an incomplete metamorphosis. Curtis (1837) added '*Caecilius*' as a second division within the group. The various subsequent group names, listed by Lienhard (1998a), include Psocoptera, Psocida, Psocides, Psocidae, Psocina, all based on the name *Psocus*, before Burmeister (1839) designated the name 'Corrodentia' to include psocids, termites, embiopterans and Neuroptera: Coniopterygidae. Several later workers until the early twentieth century accepted the psocids as a 'suborder' of Corrodentia, which Enderlein (1903d) named 'Copeognatha', referring to the form of the lacinia. This group was raised to ordinal status, as Psocoptera, by Shipley (1904) and this arrangement has remained largely unchallenged, other than for continuing debate over its monophyly, since that time.

Classification within the order has also been complex and some groupings of families, and the limits of some families, continue to be reappraised. Following Enderlein's (1903d) oversimplified and artificial system of grouping psocids by the number of their tarsal segments, several other workers also proposed groupings within Psocoptera, incorporating a range of different characters, and sometimes changing the scope of the order somewhat. For example, Banks (1916) gave the psocids superfamily status (Psocoidea) and added Zoraptera as a second group within the Corrodentia. The Psocoidea were distributed among six families, mainly using leg and antennal characters, although Banks subsequently used wing venation extensively in systematic appraisal. Karny (1930) in his division of the group incorporated fossil taxa into four main categories. Modern intra-ordinal classification stems from an arrangement by Pearman (1936a), which has been modified somewhat by later workers, most importantly Roesler (1944) and Badonnel (1951), before Smithers (1972c) produced a comprehensive synthesis and diagnoses of all genera and higher categories then recognised in an account that remains invaluable to all students of Psocoptera.

Pearman's (1936a) approach was revolutionary in that he, for the first time, incorporated genitalic features and a wide range of other structures to group psocids into 26 families, leaving a residue of genera which could not be allocated satisfactorily at this level. The families were not diagnosed formally, but 'typified' by genera included in each, and were grouped into seven main 'family groups', with an eighth group (Homilopsocidea) erected to include the nine families not placed in the other groups. Both Roesler and Badonnel retained Pearman's concept of family groups, and Roesler was the first worker to use the three currently recognised suborders. Badonnel's classification followed this subordinal arrangement to provide the foundation of classification used by nearly all later workers. The development of this arrangement is summarised in table 1 (p. 7) and is discussed more fully by Smithers (1972c). The positions of many anomalous genera, and others poorly known at the time, have been clarified progressively during the last 30-40 years, but much of this has been achieved without distorting greatly the arrangement outlined above. A few genera are still puzzling, but tables 1 and 2 (p. 9) indicate the widely accepted arrangement of higher classification within the Psocoptera, together with its development, and this broad arrangement is followed here, notwithstanding some more recent changes. In particular, the constitution of segregates within the Psocomorpha continues to be debated. Thus, Yoshizawa (2002; based on extensive cladistic analysis) recognised six infraorders, two of which were newly raised for single anomalous families (Archipsocetae for Archipsocidae, Hemipsocetae for Hemipsocidae). The other four infraorders (all previously established: Epipsocetae, Caeciliusetae, Homilopsocidea, Psocetae) were each considered to be monophyletic. However, the relationships of families within Homilopsocidea remain somewhat uncertain. Yoshizawa (2002) suggested the arrangement of four superfamilies, namely Elipsocoidea (Elipsocidae), Lachesilloidea (Lachesillidae, Ectopsocidae), Pseudocaeciliodea (Pseudocaeciliidae, Trichopsocidae, Calopsocidae) and Peripsocoidea (Peripsocidae, Bryopsocidae, Philotarsidae, Mesopsocidae). Schmidt & New (2004) erected two further small families (Lesneiidae, Sabulopsocidae) in this series. Both are related to Elipsocidae, and neither occurs in South-east Asia. Molecular studies by Johnson & Mockford (2003) also revealed the Homilopsocidea not to be monophyletic and supported the removal of Archipsocidae from this series. In contrast, they found no evidence to support removal of Hemipsocidae from its traditional place within the Psocetae.

Categories within each family found in South-east Asia are included in context below, together with key references to more recent appraisals and revisions of wider interest for each. Smithers' (1972c) account has been the template against which most genera described more recently have been matched, and was followed by a revised set of keys to families and genera by Smithers (1990a). In that synopsis, the number of families was listed as 36, collectively containing some 300 genera. In the recent *World Catalogue and Bibliography of Psocoptera* (Lienhard & Smithers, 2002), 41 families (including two wholly fossil families, Archaeatropidae and Empheriidae) are recognised. These authors, in a compendium which sets new standards for documentation of the order, list 4408 species described up to the end of the year 2000. They noted, also, the dramatic increase in taxonomic knowledge during the last half century or so, with the 1950 total of described species being only 1242 – so that the number of named species has more than tripled during that time. An important recent increase to the species number was the publication of a comprehensive monograph on Chinese Psocoptera (Li Fasheng, 2002; all species also listed by Lienhard, 2003b). At present about 5550 species and 460 genera of Psocoptera are known worldwide (Lienhard in litt.). The above-mentioned catalogue by Lienhard & Smithers (2002) and the supplements to the catalogue published by Lienhard (2003a, b, 2004c, 2005, 2006) include listing for all species recorded from every country and region. A separate species checklist for South-east Asia is provided as chapter 7 of this book; comments on diversity of each family and genus are included in chapter 5.

Regional studies on South-east Asian Psocoptera

General knowledge of the Psocoptera, as with many other insects, commenced with studies on European fauna, followed by that of North America, but only relatively small parts of the fauna of either region had been studied effectively much before the end of the nineteenth century. Knowledge of the fauna of other parts of the world lagged considerably; Hagen (1861b) could list only about a dozen species from South America, for example. A number of species from the Oriental region were indeed described in the nineteenth century, by workers such as Hagen, Kolbe and McLachlan (none of whom had worked in the region), and these are mainly descriptions of single or few species, in isolation from the rest of the fauna. Enderlein's (1903a) monograph marked the beginning of the modern era of regional studies, and included a list of 115 Oriental-Australian species and descriptions of 28 new Oriental taxa, in addition to keys to families and genera, and extensive information from earlier descriptions. It was followed by additional papers by Enderlein (1907a, 1926b) dealing with Psocoptera from Java, the last of these including a list of 60 species and descriptions of 26 new species from the island.

Most other workers in the early twentieth century confined their studies to species descriptions (Karny, Ribaga, Navás, Pearman) as such opportunities arose, but Banks (from 1913 on) prepared a series of papers that included keys and dealt with various groups of 'neuropteroid insects' (including Psocoptera) from the Philippines. All these studies were based on material collected by visiting scientists, and the references are cited fully by Smithers (1965d) and Lienhard & Smithers (2002).

The first resident entomologist in the region to work on Psocoptera, E. Jacobson, discussed the web-spinning behaviour of several species (Jacobson, 1924) in the first biological studies on psocids from Indonesia. Jacobson (1870-1944) was born in Germany but came to Java as a young man and became Director of the Bogor Zoological Museum. His work was followed by Soehardjan's (1958) first major study on the Indonesian Psocoptera, in which he listed all described species from the archipelago, and summarised the content of the earlier accounts noted above. Additional taxa described by Soehardjan & Hamann (1959) brought the total of then known Indonesian psocids to 126 species, giving, for the first time, an indication of the remarkable richness of the fauna.

Considerable increase in knowledge of the region's Psocoptera has occurred since the mid twentieth century, building on studies by Thornton on the Hong Kong fauna and synopses of some major groups in the Pacific region (Pseudocaeciliidae: Lee & Thornton, 1967; Peripsocidae: Thornton & Wong, 1968). These were followed by a series of studies on the Psocoptera of New Guinea and the Melanesian arc, co-authored by Smithers & Thornton (1973-1990), as parts of a wide-ranging study of the biogeography of psocids in the western Pacific, and descriptions of many further taxa from Indonesia (below), Malaysia (mainly by New & Lee, 1991a-e, 1992a-c), and more fragmentary accounts from many other countries in the region. The papers are specified in context later, but they, together with more recent studies on Indonesian taxa (Vaughan et al., 1989, 1991; Endang & Thornton, 1992; Thornton & Browning, 1992; Endang et al., 2002; Endang & New, 2005) have collectively provided the basis for the present synopsis. In parallel, considerably increased knowledge of the Australian fauna (with a catalogue of the known taxa produced by Smithers, 1996a) and that of mainland China (the latter in a long series of descriptive papers by Li Fasheng, culminating in the above-mentioned two volume treatise on the Chinese fauna: Li Fasheng, 2002) have helped biogeographical interpretations and inferences on distributional affinities and peculiarities, together with discussion of putative endemicity. Numerous taxa remain to be discovered and described, but the purpose of this account is to provide a sound template for this to occur, as the first major synopsis of the order from the region since New's (1977h) rather broader treatment of Oriental taxa. The major references on taxonomy of each family are cited in context. It should be noted here that many of the genera raised by Li Fasheng for Chinese taxa (passim, see Li Fasheng, 2002; all taxa are also listed by Lienhard, 2003b.) can not be appraised fully at present. The Chinese psocid fauna now includes more than 1500 species in 170 genera, 64 of which were described by Li Fasheng (2002). Many of the genera appear to differ from previously described taxa in only small features, and those relevant to the present account are noted in context without presumption of formal appraisal or synonymy. Similarly, Li Fasheng's assessment of psocopteran higher classification contains groupings that differ from those used by earlier workers. Some nomenclatoral amendments concerning the Chinese Psocoptera have already been published by Lienhard (2003c).

The study of Psocoptera within the region has been difficult to initiate and foster. The complex taxonomic literature is sometimes elusive, and much of the older classic work is in a variety of European languages. Many of the species descriptions and diagnoses are inadequate for unambiguous recognition without direct examination of the type material. With few exceptions, such types are deposited far distant from South-east Asia, reflecting species descriptions by the above-listed workers in Europe or North America. Much older type material is damaged or lost – or, at least, has proved very difficult to trace. Localities given for some types are imprecise and, as Barlow (1988) noted for Lepidoptera 'material collected before 1900 from South-east Asia was frequently inaccurately labelled'. The problem for Psocoptera is considerably less than for Lepidoptera, simply because rather few psocids were collected at that time!

TABLE 1
Some earlier classic taxonomic arrangements of Psocoptera and the similarities between these. Names of family groeps in **bold**

[CONTINUED] ▸

Pearman (1936a)	Roesler (1944)	Badonnel (1951)
PSOCOPTERA	COPEOGNATHA	PSOCOPTERA
	TROGIOMORPHA	TROGIOMORPHA
Atropetae	**Atropetae**	**Atropetae**
Lepidopsocidae	Lepidopsocidae	Lepidopsocidae
Atropidae		Trogiidae
Psoquillidae	Trogiidae	Psoquillidae
Psocatropetae	**Psocatropetae**	**Psocatropetae**
Psocatropidae		Psyllipsocidae
Scoliopsyllopsidae	Psyllipsocidae	Prionoglaridae
	TROCTOMORPHA	TROCTOMORPHA
Amphientometae	**Amphientometae**	**Amphientometae**
Amphientomidae	Amphientomidae	Amphientomidae
	Plaumanniidae	Plaumanniidae
Nanopsocetae	**Nanopsocetae**	**Nanopsocetae**
Liposcelidae	Liposcelidae	Liposcelidae
Pachytroctidae	Pachytroctidae	Pachytroctidae
	EUPSOCIDA	PSOCOMORPHA
Caecilietae	**Caecilietae**	**Caecilietae**
Calopsocidae	Neurosemidae	Calopsocidae

TABLE 1 [CONTINUED]

Pearman (1936a)	Roesler (1944)	Badonnel (1951)
	[EUPSOCIDA]	[PSOCOMORPHA]
Caeciliidae		Caeciliidae
Amphipsocidae		Amphipsocidae
Stenopsocidae	Polypsocidae	Stenopsocidae
Polypsocidae		Polypsocidae
Epipsocetae	**Epipsocetae**	**Epipsocetae**
Epipsocidae	Epipsocidae	Epipsocidae
	Ptiloneuridae	Ptiloneuridae
	Callistopteridae	Callistopteridae
	Psilopsocidae	Psilopsocidae
Psocetae	**Psocetae**	**Psocetae**
Myopsocidae	Myopsocidae	Myopsocidae
Thyrsophoridae		Thyrsophoridae
Psocidae	Psocidae	Psocidae
Homilopsocidea		**Homilopsocidea**
Hemipsocidae		Hemipsocidae
Elipsocidae		Elipsocidae
Philotarsidae		Philotarsidae
Mesopsocidae	Mesopsocidae	Mesopsocidae
Pterodelidae		Lachesillidae
Peripsocidae		Peripsocidae
Pseudocaeciliidae	Pseudocaeciliidae	Pseudocaeciliidae
Trichopsocidae		Trichopsocidae
Archipsocidae		Archipsocidae

TABLE 2
Recent taxonomic arrangement of Psocoptera to subfamily level, essentially following Smithers (1990a) and Lienhard & Smithers (2002), with some more recent changes incorporated. Names of family groups in **bold**. [CONTINUED] ▸

PSOCOPTERA		
TROGIOMORPHA	**Atropetae**	Lepidopsocidae (subfamilies: Thylacellinae, Perientominae, Lepidopsocinae, Echinopsocinae, Lepolepidinae, Parasoinae)
		Psoquillidae
		Trogiidae (subfamilies: Empheriinae, Trogiinae)
	Psocathropetae	Prionoglarididae (subfamilies: Prionoglaridinae, Speleketorinae)
		Psyllipsocidae
TROCTOMORPHA	**Amphientometae**	Amphientomidae (subfamilies: Amphientominae, Tineomorphinae)
		Compsocidae
		Electrentomidae
		Musapsocidae
		Protroctopsocidae
		Troctopsocidae
	Nanopsocetae	Liposcelididae (subfamilies: Embidopsocinae, Liposcelidinae)
		Pachytroctidae (subfamilies: Tapinellinae, Pachytroctinae)
		Sphaeropsocidae
PSOCOMORPHA	**Epipsocetae**	Cladiopsocidae
		Dolabellopsocidae
		Epipsocidae (subfamilies: Gojinae, Epipsocinae)
		Neurostigmatidae
		Ptiloneuridae
	Caeciliusetae	
	Asiopsocoidea	Asiopsocidae

PSOCOPTERA		
[PSOCOMORPHA]	Caeciliusoidea	Amphipsocidae (subfamilies: Amphipsocinae, Dasydemellinae, Calocaeciliinae)
		Caeciliusidae (subfamilies: Aphyopsocinae, Caeciliusinae, Fuelleborniellinae, Paracaeciliinae)
		Stenopsocidae
	Homilopsocidea	Archipsocidae (subfamilies: Archipsocinae, Pararchipsocinae)
		Bryopsocidae
		Calopsocidae (subfamilies: Callistopterinae, Calopsocinae)
		Ectopsocidae
		Elipsocidae (subfamilies: Elipsocinae, Pseudopsocinae, Propsocinae, Nepiomorphinae)
		Lachesillidae (subfamilies: Eolachesillinae, Lachesillinae)
		Lesneiidae
		Mesopsocidae
		Peripsocidae
		Philotarsidae (subfamilies: Philotarsinae, Aaroniellinae)
		Pseudocaeciliidae (subfamilies: Pseudocaeciliinae, Zelandopsocinae, Electropsocinae)
		Sabulopsocidae
		Trichopsocidae
	Psocetae	Hemipsocidae
		Myopsocidae
		Psilopsocidae
		Psocidae (subfamilies: Amphigerontiinae, Psocinae, Thyrsophorinae)

Chapter 2

Introducing Psocoptera

by T.R. New

Psocoptera (with alternative ordinal names of Corrodentia and Copeognatha, as above), the barklice, booklice or simply 'psocids', are a small order of hemimetabolous terrestrial insects, mostly only a few mm in length. As noted above, about 5550 species have been described globally, distributed among some 460 genera, with most species (and many genera) having limited geographical distributions. The majority is associated with vegetation, but some species are typically found in domestic environments and can be harmful to e.g. stored food, particularly flour, grains and dried fruit.

Systematic position

Psocoptera are included in Paraneoptera, with some authorities recognising a clade Psocodea, containing Psocoptera and the true lice (Phthiraptera), which are clearly closely related and within which Psocoptera are 'non-parasitic Psocodea' (Yoshizawa & Saigusa, 2001). The Psocoptera are treated in this book as a distinct order. The order has a long evolutionary history, with fossils from the Permian period onward. Although the identity of many early fossils purported to be Psocoptera remains uncertain, the rich array of species available in amber from the Cretaceous and more recent Oligocene periods are often clearly related to modern forms. The older fossils exhibit a number of more anomalous features, rendering their identity uncertain, and are discussed by Carpenter (1992). Many of the extant taxa appear to be of considerable antiquity; thus, Larsen (1978) suggested, from studies on the European amber fossil taxa, that the order may have undergone little change since the Mesozoic, and Mockford (1969b) noted that the psocid fauna of northern Mexico may not have changed greatly over the last 27 million years.

Relationships of orders within the Paraneoptera are still debated, although Psocoptera are likely to be a basal group within the series (Yoshizawa & Saigusa, 2003). Monophyly of the Psocoptera also remains undecided, although that of Psocodea is indicated strongly by the specialised form of the hypopharynx. However, Lyal's (1985) suggestion that the booklice (the family Liposcelididae) may constitute the sister group to Phthiraptera indicates the possibility that the Psocoptera may be paraphyletic (see also Smithers, 1991b). However, there are also convincing pointers to maintaining them as monophyletic, and Lienhard (1998a, following Seeger, 1979) noted the following features as apomorphies for the order:

- egg with very thin chorion, lacking micropyles and aeropyles,
- the dorsal orientation of the embryo in the egg, and absence of rotation of the embryo within the egg,

- the unique folding pattern of the legs in the embryo,
- the behaviour of the embryo (pronymph) when it leaves the egg.

However, more recent analyses based on mitochondrial DNA (Yoshizawa & Johnson, 2003) and sequences of the marker 18 Sr DNA gene for 145 taxa distributed across both orders (Johnson et al., 2004) endorse paraphyly, in that the psocopteran infraorder Nanopsocetae (p. 81) was included clearly within the Phthiraptera, with Liposcelididae an apparent sister taxon to the louse order Amblycera.

Synapomorphies of Psocoptera and Phthiraptera are (cf. Lienhard, 1998a):

- the unique form of the hypopharynx, with the ability to absorb water vapour from the atmosphere,
- antennal flagella with mechanism for rupture through 'lines of weakness' (Seeger, 1975),
- polytrophic ovarioles,
- accessory tubules to the axoneme of the sperm tail elliptical in cross section (shown by Dallai & Afzelius, 1991).

Definition and diagnosis

In general, the superficial appearance of winged psocids from vegetation resembles that of many small Homoptera, especially some Psylloidea; many psyllids are of similar size to many psocids, generally hold their wings in a similar position, and have relatively long antennae. They have two tarsal segments, but are differentiated easily by having suctorial mouthparts, differing wing venation and a generally 'tougher' appearance. Nymphs of psyllids, unlike those of psocids, tend to be strongly dorsoventrally flattened. The small 'dustywings' (Neuroptera: Coniopterygidae), likewise, may be confused with psocids, but have moniliform antennae and very different wing venation. Psocids are otherwise unlikely to be confused with other insect groups, although certain superficialities may occasionally cause delay in recognition. Thus, some psychodid flies may be confused with hairy-winged psocids. Conversely, a few psocids may be confused with small beetles. Species of Sphaeropsocidae (not found in South-east Asia) and a number of regional psocid genera with elytriform fore wings (*Calocaecilius*, p. 109) are indeed considered to mimic beetles such as some Coccinellidae.

Psocoptera can be diagnosed as follows (fuller notes on morphology are given below): free living neopterous exopterygote insects belonging to Paraneoptera; usually small (body length ca 1-8 mm measured from the front of the head to the tip of the abdomen); head with large bulbous postclypeus; compound eyes large and globular, sometimes variously reduced; ocelli present or absent; antennae long and filiform, 8 to 50-segmented, often with trichoid sensilla on distal region of flagellomeres (or 'flagellar segments'); mouthparts chewing, mandibles well developed, maxillary palpi 4-segmented; labial palpi short, 1- or 2-segmented; lacinia elongate, slender, forming a characteristic narrow 'maxillary pick'; hypopharynx highly modified, with large basal sitophore sclerite and two distal ovoid lingual sclerites connected to sitophore (= cibarial) sclerite by sclerotised tubes (hypopharyngeal filaments). Pronotum very reduced; small laterocervical sclerites present; mesonotum enlarged, in winged morphs often protruding over pronotum. Wings membranous or, more rarely, scaled or haired; usually two pairs, extending beyond apex of abdomen, but may be variously reduced or absent, with sexual dimorphism in extent of wing development or other polymorphism not infrequent; wings held flat over abdomen or in a steeper 'tent' over the body; venation characteristic, usually simple and with few crossveins; fore wing vein Rs usually 2-branched, M usually 3-branched (figs. 38-45); Cu usually 2-branched, one or two anal veins; often a well developed pterostigma in fore wing; hind wing venation more reduced than that of fore wing. Legs elongate, usually slender; tarsi 2- or 3-segmented; two pretarsal claws, each usually with a basal setiform process and a subbasal pulvillus; hind coxa often with specialised 'rasp' and tympanic area ('coxal organ', 'Pearman's organ'). Abdomen 11-segmented, membranous except for apical segments, which may be more sclerotised; apical segment represented by dorsal epiproct and lateral paraprocts, the latter often with field of small trichobothria, no defined cerci present. Female external genitalia represented by subgenital plate (sternite viii) and three pairs of gonapophyses (sometimes reduced), with spermathecal opening (spermapore) often sclerotised. Male external genitalia comprising hypandrium (sternite ix, or viii + ix fused, simple or variously sclerotised or ornamented), phallosome (usually a pair of elongate parameres, often fused distally to constitute aedeagal arch, lateral branches may be extended as 'external parameres'; sometimes linked to form closed frame), and often ornamentation, such as bosses or hooks, on clunium (fused terminal tergites), epiproct and /or paraprocts.

Morphology

A wide range of structural features are used to identify Psocoptera and, as many of these are used in keys and taxon synopses in this account, the various character states are introduced here to demonstrate their variety and application. Most parts of the body provide taxonomic characters in particular taxa, ranging from gross appearance or shape to fine structural detail. Thus, a few families can be diagnosed broadly on general appearance, and many more on features of wing venation and legs, whereas separation of many genera and species may necessitate examination of genitalic structures, of mouthpart features, details

of body surface sculpture, the form of various antennal segments and sensilla, and other features which can be examined satisfactorily only from good microscopical preparations. Some of these features are discussed here, and others noted in more specific contexts later, to illustrate the major character states used. The general morphology of Psocoptera is discussed in many general entomology texts and in more specialist accounts by Badonnel (1943a), Smithers (1972c), Weidner (1972), Günther (1974a), New (1974h), Mockford (1993a) and Lienhard (1998a), inter alii. More comprehensive references are also included in these works.
The general appearance of the insects in profile is shown in figs. 17-19.

Head (figs. 20-32)

The overall shape and orientation of the head differs somewhat between taxa, with a tendency for prognathy in many subcortical and domestic psocids in the more primitive suborders. The compound eyes are usually large and protruding, with sexual dimorphism (in which the eyes of males are larger and more 'globular') not uncommon. However, eyes can also be strongly reduced and represented by a few ommatidia, particularly in flightless morphs. One known psocid species, the cave-dwelling sphaeropsocid *Sphaeropsocopsis myrtleae* Lienhard & Ashmole, from St Helena, completely lacks eyes (Lienhard & Ashmole, 1999). Ocelli are normally three in number but may be reduced to two or absent, with polymorphism associated with alary polymorphism: wingless and brachypterous psocids usually lack ocelli.
Antennae are invariably slender and filiform, other than for very occasional expansion of the scape and basal flagellum. The number of antennal segments has taxonomic value and can be diagnostic for suborders and some lower groups (see below). Features such as the presence and distribution of placoid sensilla (Archipsocidae, fig. 31), the extent of surface ornamentation (particularly the so-called 'secondary annulations', fig. 32), and the shape of the apical flagellar segment (Philotarsidae) can also have importance. Mouthpart features are of greater value for diagnosis of many groups, with the form of the palpi, mandibles and lacinia all being used as taxonomic characters.

Thorax (figs. 33-37)

With very few exceptions, the appendages rather than the thorax itself provide the major taxonomic interest in this region of the body. Most psocids have the prothorax strongly reduced and the mesonotum often enlarged and expanded dorsally to partially cover the prothorax. In Liposcelididae, however, the pronotum is large and lobed. In winged taxa, the metathorax is invariably smaller than the mesothorax, but both are less developed in apterous morphs of polymorphic taxa than in the corresponding macropterous morphs.

The **legs** (fig. 34) comprise the coxa, trochanter, femur, tibia and tarsus, with associated structures, with the femur and tibia both long and the tarsus consisting of two or three segments, with paired pretarsal claws and pulvilli, the latter occasionally absent. The legs are generally slender, but the hind femur of *Liposcelis* is enlarged and flattened, and occasional other expansions of particular segments occur throughout the order. In some Amphientomidae, Amphipsocidae and Epipsocetae, the fore femora have a row of spines or 'cones'. The tibia is occasionally dilated rather than cylindrical and, in some families, may bear a row of ctenidiobothria (fig. 34): strong short setae arising from an ornamented basal socket. These structures may occur also on the ventral edge of the tarsal segments, and their incidence and number can have taxonomic significance.

The number of tarsal segments can be highly characteristic in particular families and genera and the form of the tarsal claws and pulvilli (figs. 35-36) is often associated with life in particular habitats (see Morphological habitat correlates under Biology). Thus, each claw may be simple, have a strong subapical tooth, or more elaborate ornamentation. Likewise, the pulvillus may be slender and tapered, dilated at the apex, or more conspicuously expanded to constitute a broad 'pad' beneath the claw.

The hind coxae in many families bear 'Pearman's organ' (or the 'coxal organ', fig. 34, p. 176), a small rugose of lined dome ('rasp') and adjacent smooth membranous area ('tympanum'), believed to be stridulatory in function, although this has not been confirmed experimentally. The organ is sometimes poorly developed in neotenic or short-winged individuals in comparison with macropters of the same species. Some Lepidopsocidae have a presumed coxal interlocking mechanism on the mesothoracic legs. This comprises a small projection from the left coxa, which fits a corresponding depression on the right coxa (Menon, 1938b).

Wing form and venation are amongst the most important taxonomic characters of Psocoptera and, when present, are universally used in systematic studies. Some trivial variation in detail is not infrequent, especially with polymorphic morphs, but the basic pattern is widely reliable for taxonomic use. The normal condition is of macroptery, with two pairs of membranous wings, with fore wings being larger than hind wings, and with more elaborate fore wing venation. Wings usually project well beyond the apex of the abdomen when the insect is at rest, but a few taxa have shortened, elytriform or coriaceous wings. In some Lepidopsocidae the hind wing may be strongly reduced. The simple states of macroptery or aptery, with or without intermediate morphs, may itself help to define some taxa, as may the way in which the wings are held at rest. Thus, many of the more supposedly primitive psocids hold the wings horizontally over the abdomen (fig. 17), while most others hold the wings in a more or less steep 'tent' (fig. 19). The latter is often associated with some form of wing-coupling structure, predominantly of a small lobe ('stigmasac', fig. 38) on the ventral side of the fore wing at the base of the pterostigma, and with which the hind wing costa interlinks when at rest (New, 1974e).

Wing venation (figs. 38-39) is generally rather simple, with few crossveins and little multiple forking of the main longitudinal veins of either wing, and the pattern is of major importance in diagnosing many different taxa. The typical pattern can become severely distorted in brachypterous psocids, and individual anomalies occur in macropterous psocids. Otherwise, the characteristic venation ranges from relatively complete venation to reduction to only one or two main veins, or, rarely, to complete lack of venation on small alary lobes (figs. 40-45).

The most commonly used nomenclature for psocid wing venation is illustrated in figures 38-39, and the following notes augment this illustration. In the fore wing, the subcostal vein (Sc) is usually reduced, other than in some Lepidopsocidae, and most commonly ends free in the membrane of the costal cell. More rarely, particularly in some Psocidae, it may reach the costa or the radius. A distal part of this vein (Sc′) constitutes the base of the pterostigma and is only rarely absent. The radius (R) runs parallel to the costa near the base of the wing, and R1 forms the posterior margin of the pterostigma found in many higher psocids. In most taxa the pterostigma is free posteriorly, but it may be linked with the radial sector (Rs) by a crossvein, or have a short spur vein extending from the hind angle. Veins Rs and the media (M) are joined in the central region of the wing, commonly at a point, in a short fused length, or by a crossvein, and separate subsequently. Rs typically forks once before the wing margin, to give R2+3 and R4+5; M has one or two branches to produce two or three veins (M1-M3). The anterior cubital vein (Cu1) is confluent with M basally and separates to diverge toward the hind margin, sometimes simply but often forked, to produce the 'areola postica', a very characteristic feature of many psocids as the cell on the posterior margin formed between Cu1a and Cu1b. This cell is most commonly free, but can be linked to M by a crossvein, or by more extensive fusion (Psocidae) to form the discoidal cell in the central region of the forewing. A second, posterior, cubital vein (Cu2) is simple and commonly evanescent, and only one anal vein (A) is usually present; this may meet Cu2 on the wing margin, to constitute the nodulus. A second anal vein occurs in some groups and may join the first.

Hind wing venation (fig. 39) is typically less complex than in the fore wing, with no pterostigma present and veins M and Cu1 commonly unforked (so that there is no areola postica); anal veins, if distinct, often do not reach the hind margin.

In addition to venation, the presence and distribution of macrosetae on the wing veins, margin and membrane all differ between taxa, often in characteristic ways. Some families are characterised by most of their members having glabrous wings (Myopsocidae, Lachesillidae, many Psocidae). In others, the veins may bear one or two rows of setae, setae may be limited to particular veins or regions (such as the hind wing radial fork margin of Elipsocidae), or the whole wing surface may be covered with hairs (Archipsocidae) or these modified to become flattened scales (Lepidopsocidae, Amphientomidae).

Abdomen (figs. 46-64)

Taxonomic interest in the psocid abdomen emphasises the values of genitalic structures on the posterior segments. At one level, these structures may characterise families and they are very often characteristic of genera and species as well. Genitalia are universally used as specific characters in descriptions. However, pregenital segments also have taxonomic value in some psocids. As examples, 1) the anterior tergal sculpturing of many Liposcelididae can have specific value, 2) the number and distribution of ventral abdominal vesicles (or 'blisters') (fig. 37, p. 176) of Caeciliusidae can help to characterise segregates within the family, and 3) a few genera of Mesopsocidae and Psocidae have raised 'humps' or tubercles which may aid crypsis.

The typical complement of genital and associated terminal structures (epiproct, paraprocts, clunium) of each sex are noted here (see figs. 46-64); the collective epithet 'terminalia' or 'ano-genital complex' is applied to these (Lienhard, 1998a), and the term 'telson' is sometimes used for the epiproct and paraprocts together. These structures show specific modifications (figs. 54-56), and the margin of male tergites ix + x, the clunium (fig. 57), may be elaborately produced or ornamented in some taxa (such as many Psocidae).

Males and females of many Psocoptera can be differentiated most easily on genitalic features. Many species are otherwise not markedly sexually dimorphic, although differences in head and eye size and in wing development sometimes occur. The basic plan of the genitalic structure of each sex is summarised here.

Female genitalia (figs. 46-53)

The ventral subgenital plate (sternite viii) may be slightly rounded and convex, or have a well-defined and sclerotised rounded, tapered or angular median lobe posterior to a broadened 'discal area'; the posterior (apical) lobe (fig. 51) is sometimes referred to as the 'egg guide'. The paired gonapophyses (fig. 47) are each composed of the ventral valve (segment viii) and those of sternite ix differentiated into dorsal valve and external (or outer) valves. In sequence, these are sometimes referred to as first (v1), second (v2) and third (v3) valvulae, respectively, and each may be present, reduced or absent in particular taxa. Typically, the ventral valves are slender and glabrous; dorsal valves are also usually elongate but may be broad and have a well defined apical style, sometimes also bearing short setae; the external valves are usually rounded or otherwise enlarged, short, and with long setae. Gonapophyses are absent in the viviparous *Archipsocopsis*.

The median area of sternite ix contains the opening of the spermatheca (figs. 52-53), and can be well-sclerotised and characteristic to form a distinctive 'sperma-pore plate'. The spermatheca can also be well sclerotised in some taxa.

Male genitalia (figs. 57-64)

The male genitalia comprise the hypandrium (equivalent to the subgenital plate) and the phallosome. The hypandrium (figs. 61-64) is derived from sternite ix, and forms a cavity to hold and support the phallosome, the organ used for copulation. In many psocids, the hypandrium is a simple rounded lobe, but it can become highly ornamented and elaborated and bear characteristic symmetrical or asymmetrical spines, hooks or other apophyses. In Psocidae: Amphigerontiinae it is formed by fusion of sternites viii and ix, so that the male abdominal apex appears 'encapsulated' in a single hard structure. Homologies of some of the accessory sclerites, such as the lateral 'claspers' of some Lachesillidae are not wholly clear, but are assumed widely to be derived from the hypandrium.

The phallosome (figs. 58-60), an open or closed frame, comprises several elements. The lateral parameres are sometimes rodlike and may be entirely free or linked basally (anteriorly) by a more or less sclerotised membrane; their apex may be variously expanded and bear sensilla of various kinds, commonly visible as distinct pores. The aedeagus is a posterior arch, formed usually of the inner branches from the parameres that link and are fused in the midline. Again, some parts of the phallosome are difficult to homologise unambiguously and can become highly elaborated. The third main structure present is the 'endophallus', representing the apical region of the spermatic duct, which can be enclosed between the parameres and there form a more or less expanded area (symmetrical or asymmetrical), often with numerous microtrichia or small 'teeth' constituting a radular structure.

Many variations on these basic genitalic plans are illustrated in the systematic sections. The keys given to suborders and families illustrate the need to examine many different structural features in order to identify Psocoptera. The study of other features, such as genetic complement and molecular characteristics as tools to aid studies on intraordinal relationships and phylogeny is in its infancy, and almost all inferences on relationships between the various families and other groups rely solely, or almost solely, on morphological features alone. However, a number of chromosome counts for psocids have been published, with surveys by Wong & Thornton (1966) and Meinander et al. (1974), revealing male diploid components in the range $2n = 4+1$ to $28+1$, with $16+1$ being the most frequent complement in the taxa studied. The male is thus the heterogametic sex, and Meinander et al. (1974) suggested that the Psocomorpha may be a rather homogeneous group on their chromosome complement.

Eggs

Eggs are simple ovoid structures, lacking a micropyle and, in most advanced psocids for which they are known, lacking the areolate or ridged sculpturing common on the chorion of many Trogiomorpha. Rarely (some South American Epipsocetae; New, 1972a), eggs have pronounced filamentous or capitate projections. The sculpturing of trogiomorph eggs is thought to help protect them against desiccation (Seeger, 1979) and the smooth eggs of many Psocomorpha have some form of protective covering believed to serve the same purpose as well as to aid camouflage – both as general crypsis and as protection from predators and parasitoids.

Whether eggs are single or batched, any covering can be very characteristic and aid in identification of the parent family. Pearman (1928c) noted the categories present in British psocids and these have wide geographical application. Eggs may be bare or encrusted with faecal material (sometimes appearing as bark flakes, fig. 12), and either may then be covered with a dense silken web (fig. 11), overlain by a few silken strands, or naked. The naked eggs of some taxa are initially slightly 'sticky' and can become coated casually with debris, and Broadhead (1947d) reported that female *Embidopsocus* kick particles onto freshly laid eggs.

Hatching is accomplished by use of a specialised oviruptor (figs. 2-8) on the frontal region of the head of the pronymphal cuticle. This structure is also characteristic in form for many taxa and is usually either a) a simple median knife-like blade, or b) has a row of serrations ('teeth') in some foliage-frequenting psocids (in which the egg batches are commonly covered with dense silk) augmented by lateral brachia with additional rows of teeth on each side of the median blade.

Nymphs

Nymphs usually resemble adults in body shape and marking and can sometimes be identified clearly to order and, sometimes, lower levels. Nymphs never have functional ocelli or wings. The first instar (fig. 9) may have fewer antennal segments than adults, and the tarsi are always two-segmented. In winged morphs, the first two instars lack any trace of wingpads, but thereafter these develop progressively with each moult (fig. 10), becoming recognisable articulated wings at the final moult to the adult stage. The usual number of six instars is sometimes reduced to five (*Psyllipsocus*), four or even – very rarely – three instars, with the the reduced numbers associated with alary polymorphism. Many taxa with sexual alary dimorphism differ in sexual instar number, with the apterous sex undergoing one fewer moult than macropterous ones. Typically, apterous or micropterous morphs of Psocomorpha are nymphoid, so that lack of development of some typical adult features is regarded as neoteny. Finally, nymphs can be recognized by the lack of external genitalic structures, whereas these are developed fully in the adult insect.

Biology

Much of the early information on biology of Psocoptera was summarised by New (1987a), in which more comprehensive references to many of the topics noted here are given. The following section is largely a summary of that account, updated where possible to incorporate more recent findings.

Life histories, early stages and reproduction

The basic hemimetabolous life cycle of Psocoptera is 'egg – six nymphal instars – adult', but this scheme has undergone many modifications, for example in the development of viviparity in a few taxa and reduction of instar number. The genus *Archipsocopsis* for instance is believed to be entirely viviparous and females lack gonapophyses. The pattern of voltinism is also variable. Many psocids on vegetation undergo well defined seasonal generations in temperate regions (varying in different species from univoltinism to several largely discrete or variously overlapping generations over the more clement part of the year), but much less information is available on seasonality in tropical psocids. Many temperate region psocids pass winter in some diapausing stage, most commonly as eggs. In contrast, species living in climatically equable environments, such as in stored products or domestic situations, may breed continuously throughout the year.

Eggs

Eggs of different taxa are characteristically laid singly, in small groups or in larger batches. Batches are rarely (some Psocidae) of more than 12 or 16 eggs, reflecting the numbers of ovarioles present in a single female (New, 1970c). However, larger batches can result from communal oviposition, so that egg masses of *Peripsocus nitens* in Chile can sometimes contain more than 300 eggs (New, 1985b), and some *Archipsocus* spp. also produce large batches. In both these instances, females 'guard' the large egg masses.

Nymphs

All instars are active from the time of hatching and feed in a similar manner to the adults.

Reproduction

Most psocids are bisexual, but parthenogenesis (thelytoky) is also widespread in the order and is sometimes obligatory. Whereas many species appear to have a natural sex ratio of close to unity, males of some are unknown or extremely rare. In some species there are clear bisexual and unisexual populations and the extent of parthenogenesis thus varies widely. Broadhead (1954b) recognised five main modes of reproduction in the psocids:

- sex ratio near unity; females oviposit only after mating has occurred;
- sex ratio near unity; females lay infertile eggs until mating has occurred. In a few species the level of fertility declines with time after mating and a second mating may then increase fertility again;
- sex ratio near unity; virgin females lay fertile eggs, so species exhibit facultative thelytoky;
- males very rare; females almost entirely parthenogenetic;
- males absent (more correctly, unknown); obligate thelytoky.

As implied above, more than one of these categories can occur within the same species and the extent of parthenogenesis (category 3, above) can vary considerably. The extent of thelytoky amongst the world's Psocoptera is still unknown, simply because most studies on this topic (as with other aspects of biology) have been concentrated on temperate region species, and there is very little information on members of diverse tropical faunas. Any case in which population samples show a high preponderance of females indicates likelihood of thelytoky. Some recent studies have revealed that species complexes of psocids may comprise a sexual population and one or more parthenogenetic segregates in which apparently obligate thelytoky occurs (Betz, 1983c).

Developmental rates and seasonality

Other than for diapausing stages, which may facilitate overwintering in many temperate region psocids living out of doors, a wide variety of psocids studied have broadly similar patterns of continuous development. Generation time is usually of the order of a few weeks, but can vary considerably with temperature, and adult longevity is also commonly a few weeks to around two months. Thus, egg duration is commonly about one to three weeks and nymphal development at least two weeks (usually within the range of two to six weeks). These times may be sufficiently well synchronised to produce characteristic patterns of voltinism, to in turn produce peaks in seasonal incidence and abundance.

Again, information on tropical psocids is poor. In Jamaica, Turner (1974b) recognised three seasonal patterns of psocid development, but the existence of these or of similar categories has not been confirmed for the South-east Asian fauna. These categories are:

- species with low populations for most of the year and a summer peak;
- species with peaks in both spring and summer, but low populations for much of the remainder of the year;
- species with continuously low population levels and no seasonal peaks.

Mode of existence

Although they can occur in very large numbers, most psocids are regarded as 'solitary' animals. The best-known exceptions to this are the colonial Archipsocidae, a family of small tropical Psocoptera all of which construct large sheet webs on tree trunks and live under these for extended periods to produce large colonies over a sequence of generations. The webs sometimes enclose substantial areas of tree surface and cause comment because of their unsightliness. The functions of such webs are not clearly understood, but these may include protection from predators and climatic extremes (such as those leading to desiccation on exposed surfaces; New, 1973c), in addition to protecting a feeding area from competitors.

A few other species of Psocoptera 'herd graze' (New & Collins, 1987), a habit in which individuals move around together whilst feeding and live in close proximity to each other. This habit occurs in some Myopsocidae and Psocidae from the time of hatching. In North America nymphs of some Psocidae: Cerastipsocini 'form large, dense herds' which disperse after the adults pass the teneral stage (Mockford, 1993a). Communal brooding in Psocoptera is rare (New, 1985b).

Food and feeding

Psocoptera feed by 'grazing'. They are among the few insect groups that specialise in feeding on epiphytic microflora – lichens or algae – on plant surfaces, combined with varied organic debris and fungal spores on foliage or bark. Others show considerable feeding specificity, so that several distinct ecological categories occur. Domestic psocids incorporate farinaceous debris into the variety and some species can also cause primary damage to stored grain by their feeding activities. Psocids are thus grazing herbivores and detritivores, but a few species have been reported to be predators (perhaps only opportunistically) or feed on exudates from plant trichomes. The variety of foods of arboreal psocids has been divided into functional categories by Turner (1975b), as follows:

- primary producer components as the major energy input; these are green algae such as *Pleurococcus*, and the algal components of lichens,
- fungi, including the fungal components of lichens, which are not primary producers but utilise exudates as energy sources,
- substrate material, such as bark flakes; these may be eaten in small amounts, with rarer records of ingestion of pollen in substantial quantities as opportunistically available on the substrate.

Much of the habitat limitation of particular arboreal psocids noted below appears to be attributable to feeding 'preferences' and specialisation and the relative availability of different foods on living or dead foliage and bark. Food supply may also influence patterns of seasonal abundance. For example, food supplies on bark may be viewed as essentially 'permanent' as available throughout the year, whereas that on deciduous foliage increases from zero to a maximum over a period of a few months, the life of a single leaf. In Britain, the abundance of some foliage-frequenting psocids thus increases markedly toward late summer and early autumn preceding leaf fall. These species are then secondary litter frequenters (p. 25) and do not re-enter trees until food has increased on new foliage in spring (New, 1970b). Much remains to be discovered about the diets of tropical psocids and examination of gut contents of specimens being prepared for taxonomic study (p. 35) has been the main source of inferences on this. Many such specimens contain a large faecal pellet or other gut content, and teasing this apart on a slide can reveal the major constituents of recent meals.

One important outcome of apparent feeding specificity is that some tree species support many psocid species, whilst others support few or none.

Habitat

Psocoptera occupy a wide variety of habitats. All are terrestrial and many taxa are associated primarily with vegetation, some showing specialisations to particular kinds of vegetation or vegetation formation. Mockford (1993a) noted six major 'habitat series' in the order, each of which can be related broadly to particular taxonomic segregates. The 'habitat series' are living foliage, dead foliage, ground litter, bark of trees (divided into open bark and subcortical), rock surfaces, and human habitats (more broadly, domestic habitats, including stored products). The last category includes species of considerable economic importance, particularly as pests in grains, flour and dried fruits in storage. Many of this group, typified by the 'booklouse', *Liposcelis*, have been transported widely by human activities and some species are virtually cosmopolitan, with their area of origin unknown. As an example, Broadhead (1961) noted that one such species, *Psoquilla marginepunctata*, had then been recorded from parts of South America, Hawaii, Malaysia, Britain, continental Europe, and parts of Africa, to which Mockford (1993a) added North America (Florida). This species has been taken under bark in remote parts of Brazil and may well be a South American native, but most other species have not had their possible origins traced in that way and their distributions are so wide that origins are unlikely ever to be determined with certainty. Because stored products are usually maintained in well-buffered environments such as silos and warehouses, the psocids present are not susceptible to vagaries of climate and seasonal change and can breed continuously to build up very large populations.

In contrast to this, most Psocoptera found in more natural outside habitats tend not to be as widespread as those distributed through commerce, and the variety of habitats reflects the predominant feeding habit of the insects, namely grazing on epiphytic microflora (such as algae, fungal spores and lichens) growing on plant and other surfaces. For insects specialising on such foodstuffs or on more general organic debris, it is a small transition to thriving in domestic environments. Some psocids are known from habitats such as bird nests, mammal burrows and caves and, indeed, some have never been found elsewhere.

Each of the main habitat groups noted above tend to support characteristic psocids, with corresponding adaptations in their life cycles to match seasonal availability of foods. Some such species can become very abundant in particular habitats, with strong seasonal peaks. These patterns of seasonality have been elucidated mainly from studies in northern temperate regions with strongly seasonal climates; very little is known about developmental patterns and seasonality of psocid abundance in the tropics. Other than some preliminary observations from South America (see New, 1973j), very little detailed information is available on this theme, although scattered relevant information is provided also by Turner (1974c), Turner & Broadhead (1974), Broadhead (1983), Broadhead & Wolda (1985) and Wolda & Broadhead (1985).

Despite this lack of basic information, inferences and parallels with temperate region fauna are sufficiently strong to be able to categorise the 'preferred' habitats of many taxa, at least at family and often at generic level. Psocoptera are a relatively minor constituent of the leaf litter fauna, but three main ecological categories are found in that broad habitat (New, 1969a; García Aldrete, 1990a). New (1987a) described these as

- primary litter frequenters – species that occur only in leaf litter, undertake their whole life cycle there and generally do not frequent other habitats;
- secondary litter frequenters – species that undergo generations in the litter and elsewhere on a regular seasonal basis. These are typified by foliage frequenting species that lay overwintering eggs on deciduous leaves that drop to the ground. The spring generation is passed in the litter, but resulting adults fly onto trees, where the next generation(s) is/are passed;
- casual litter frequenters – species that do not normally breed in the litter but can survive there and commonly occur as a result of being removed or dropped from vegetation. These may return to other habitats, or remain in the litter until death.

Primary litter frequenters can exhibit morphological features associated with life in litter (García Aldrete, 1990a).

These features include a high level of alary polymorphism (see below), with brachyptery being common and associated commonly with a general neotenic appearance. Because of the relatively greater representation of primary litter frequenters in Trogiomorpha and Troctomorpha than in Psocomorpha, Mockford (in García Aldrete, 1990a) suggested that ground litter may be a primitive habitat for psocids. On the other hand, many of the litter psocids are related to those associated with dead foliage (and, indeed, themselves found also in this habitat) and this may represent another, separate origin of the litter-frequenting habit.

The great majority of psocid species living out of doors occur on vegetation, predominantly trees and shrubs. Most arboreal psocids are associated clearly with one or other of the three main habitats noted above, namely bark, dead foliage or living foliage. The latter is the smallest category and perhaps also the most specialised. Superimposed on this broad habitat differentiation is some evidence of host-plant specificity, perhaps mediated through features of the epiphytic microflora (New, 1970b). At one level, some taxa are associated primarily with low-growing plants such as grasses or heaths and this habit appears to have been adopted independently from a number of different groups of arboreal psocids, so that closely related taxa may occupy one or other stratum. Species of *Caecilius* associated with palms are apparently more closely related to congenors living on grasses than to other species (Mockford, 1965b). Within the tree layer, comparative collections on different species have sometimes revealed considerable specificity, either on major plant groups or at a lower taxonomic level. Thus, as examples:

- in Brazil, several species of *Lachesilla* frequenting dead foliage are common on *Syagrus* palms and not found on a large range of other trees in cerrado (New, 1971f);
- in the northern hemisphere, conifers and broadleaved trees each support abundant psocid species not found on the other group,
- within an assemblage of broadleaved trees, particular foliage-frequenting psocids can occur in markedly different numbers on different tree species growing close together.

These differences may reflect the amounts and kinds of potential food growing on different plants, together with the form of leaf surfaces (smooth, waxy, hairy, et cetera) and the mobility of leaves during winds. New (1970b) found very few psocids on trees such as *Salix*, in which the foliage 'trembles' even in low winds, and suggested that this might reflect the ease with which the insects can be dislodged.

Even in relation to evergreen foliage, and with exceptions such as *Eucalyptus* species with deciduous bark, bark may be regarded as a more 'permanent' habitat for psocids. In some regions, psocids are the most abundant animals in bark-surface communities (Broadhead, 1958a) and corticolous (bark-frequenting) psocids are by far the largest ecological category of Psocoptera. A further segregate, best represented in Trogiomorpha and Troctomorpha, is being subcortical, so that long-dead trees can continue to support large numbers of psocids as their bark loosens. Very large populations of Liposcelididae: Embidopsocinae, for example, can develop under bark of dead standing trees left after fires (New, 1973j).

Morphological habitat correlates

Many psocids from South-east Asia can be allocated tentatively to one or other of the above habitat groups on the various morphological features associated with, and which may be adaptations to, these. Thus, many subcortical psocids show extreme dorsoventral flattening, and the wings are held flat over the abdomen. Many also show alary polymorphism, a feature which Hamilton (1978) noted as often associated with the subcortical environment, where large colonies of many insects can be built up. Corticolous psocids are often very well camouflaged and some have intricate wing patterning suggesting resemblance to lichens or other bark ornamentation. Highly unusually, a British species

(*Mesopsocus unipunctatus*), with apterous females, appears to have adopted true industrial melanism in urban areas on larch trees whose bark is free of lichen (Popescu et al., 1978). Some other Mesopsocidae and Psocidae have dorsal abdominal projections that Mockford (1965b) likened to small thorns or swellings or leaf scars. A few corticolous psocids shelter beneath silken webs.

An important correlated structural feature of many bark-frequenting psocids is that the pretarsal claws have a subapical tooth and are subtended by narrow pulvilli. In contrast, many foliage-frequenting psocids have smooth claws and the pulvillus is expanded (figs. 35-36). This is thought to help them hold on to smooth or flat leaf surfaces. Many foliage psocids also have eversible abdominal 'blisters' (vesicles) (Turner, 1974b), parts of the ventral intersegmental membranes between segments v-vi and vi-vii, believed to serve the same purpose.

Dispersal

In general, many psocids have been regarded as 'reluctant fliers', despite the fact that a period of flight activity is an integral feature of the life cycle of taxa such as secondary litter frequenters, in order for adults to re-enter trees. Many psocids do indeed seem to fly rather little and are commonly observed to run rather than fly to escape. However, some have been taken in nets in the upper air (Hardy & Milne, 1938), from plane nets (Glick, 1939) and from nets suspended from ships (Thornton, 1964, 1985a; Thornton & Harrell, 1965) and are well represented in catches of suction traps or Malaise traps (as examples New, 1969d (Britain), 1975a (Australia), 1984 (Brazil)). In most cases, it is unknown whether the psocids captured were flying actively or drifting in the aerial plankton. The latter is widely regarded as the predominant mode (or, at least, the more likely mode) of long-distance dispersal. It is, though, clear that psocids can be transported viably over considerable distances. Species captured from boat nets in the western Pacific have sometimes been considerable distances from the nearest land and the numerous taxa known from the Krakatau islands, Indonesia, are most likely to have arrived from Java or Sumatra, each some 40 km away to each side of the Sunda Strait (Thornton, 1985b; Thornton et al., 1988; Vaughan et al., 1989). This widespread tendency for psocids to disperse aerially has not clouded zoogeographical analysis to any major extent because most dispersal seems, very broadly, to be intraregional. However, the very wide potential for dispersal of some psocids to attain a broad tropical distribution led Mockford (1974a) to term these 'tropical waifs'.

Other modes of dispersal occur. One, the possibility of phoresy on vertebrates, was noted by Mockford (1967a) in recording psocids from plumage of birds in Korea and the Philippines, so that transport on migrating hosts could provide for long-distance dispersal. Perhaps some taxa found in bird nests may be especially prone to this form of dispersal. Thornton (1980) suggested that burrow-nesting birds could act as dispersal agents for some ground-dwelling flightless psocids. Psocids have also been associated with rodents and, more rarely, other mammals. Collectively, most psocid taxa found with birds or mammals are of genera known to have a broad habitat range, rather than being specialised arboreal taxa.

Artificial introduction by human agency may have been the principal mode of dispersal of some apterous or flightless domestic psocids, particularly of Troctomorpha and Trogiomorpha. Thus, Broadhead's (1954a) analysis of 230 quarantine interceptions from ship holds or stored products in Britain yielded 30 species, several of them still unknown from elsewhere in Britain. Unlike the 'outdoor' psocids discussed above, the origins of many such species of domestic psocids are not known, and some are known only from highly disjunct records from trade interceptions. Few such species appear to have spread subsequently into more natural habitats.

Alary polymorphism

Many Psocoptera are flightless and either lack wings completely or have them variously reduced in all individuals or as a form of sexual dimorphism or alary polymorphism. The general categories recognised are macroptery (fully winged), brachyptery (short winged, venation present to varying extents), microptery (wings strongly reduced to tiny lobes and lacking most venation) and aptery (wings absent). Polymorphic taxa occur in all three suborders. There can be well-defined sexual dimorphism (thus, *Mesopsocus* females are generally micropterous and males macropterous; *Embidopsocus* may have apterous males and macropterous or apterous females). In most taxa such morphs are definable clearly, but in a few species (such as the widespread *Lachesilla pedicularia*) individuals show all gradations from macroptery to extreme microptery and it is not clear whether such instances reflect a pattern of continuous variation.

Mockford (1957a, 1965e) recognised three categories of alary polymorphism in relation to sex and noted that short winged individuals may be 'nymphoid' in manifesting features typified by psocopteran nymphs, including lack of ocelli, relatively simple thoracic segments and (sometimes) lack of ctenidiobothria on the legs. These categories are:

- males nymphoid, females nymphoid or macropterous,
- males macropterous, females nymphoid,
- males macropterous, females nymphoid or brachypterous.

Aptery and microptery are particularly characteristic of some edaphic or domestic psocids, and others associated with low habitats.

Sound production

Among other biological features of Psocoptera, sound production has attracted intermittent attention. The noise produced by *Trogium pulsatorium* by tapping the end of its abdomen on the substrate in buildings earned it the name of 'death watch' in Britain (Pearman, 1928b; Lienhard, 1982a). *Lepinotus* females make a similar noise. The 'Pearman's organ' on the hind legs of many taxa is also thought to be involved in sound production, but there has been little experimental work on its function. For the North American *Trichadenotecnum alexanderae* (Psocidae), Betz (1983c) suggested that the male's sideways movements during courtship might involve stridulation through coxal organs.

Natural enemies

Many predators, among which birds, appear to take psocids only as part of a wider prey spectrum. The wide variety of arthropod predators recorded attacking psocids include larval Neuroptera (Coniopterygidae, Hemerobiidae, Chrysopidae), Heteroptera (Anthocoridae, Reduviidae), Coleoptera, Dermaptera, Lepidoptera (predatory *Eupithecia* in the Hawaiian Islands: Montgomery, 1982), mites, spiders and opilionids, stem-nesting sphecid wasps that specialise on psocid prey (Danks, 1970), and ants. Invasive taxa such as Argentine ants (*Linepithima humile*) may take psocids on trees (Gurney, 1939, on *Archipsocus* in Louisiana, USA). This list is far from complete, but indicates the wide susceptibility of psocids to many generalist predators and that some natural enemies are more selective.

Birds are presumed to be the major vertebrate predators of psocids but, again, there is little quantitative information available. Popescu's (1979) study in Yorkshire, Britain, implied that feeding by great tits (*Parus major*) on larch trees could be involved in the maintenance of industrial melanism in *Mesopsocus*.

The parasitoids may be very specific to Psocoptera and, even, to particular species of hosts. These natural enemies have been reported by many authors, but there have been few specific studies on their impacts on psocid populations and levels of attack, with the most thorough quantitative study being on an egg parasitoid wasp in Britain (Broadhead & Wapshere, 1966b; Broadhead & Cheke, 1975). In addition to the above groups, psocids can harbour gregarine protozoans and internal nematode worms, but there is no information on whether these are harmful or benign. Likewise, psocids have been reported as attacked by fungi, but the effects are unknown.

Parasitoid wasps

Egg parasitoids

Eggs of Psocoptera in many parts of the world are attacked by minute species of Mymaridae, at only some 0.1-0.2 mm in body length some of the smallest insects known. Indeed, perhaps the smallest recorded insect is the mymarid wasp *Dicopomorpha echmepterygis*, recorded by Mockford (1997a) from an egg of a North American species of *Echmepteryx* (Lepidopsocidae). The genus *Alaptus*, the egg parasitoids most frequently reported from psocids, appears to be virtually cosmopolitan, although species level identifications are not possible for some parts of the world. Few records of these wasps exist from South-east Asia, although New (1987a) reported the incidence of *Alaptus* in Papua New Guinea.

Biological information is derived almost wholly from species found in Britain and North America.

Newly laid psocid eggs are attacked by female wasps, and a single egg laid through the chorion. The whole wasp development occurs within the host egg. Generation times of species attacking foliage-frequenting psocids are in the range of 18-30 days (New, 1969b) and that of univoltine wasps attacking univoltine bark-frequenting psocids is much longer (*Mesopsocus*: Broadhead & Wapshere, 1966b). Some species of *Alaptus* appear to be thelytokous, and several bisexual species examined are amphitokous. Levels of parasitisation are generally quite low: those recorded in Britain are often below 10% (with a maximum of around 20% for foliage-frequenting psocids; New, 1969c), but can be much higher (up to 80.7%) for bark-frequenting psocids with long-lived eggs (Broadhead & Cheke, 1975). Superparasitisation is extremely rare.

Nymphal and adult parasitoids

Euphorine braconid wasps of the genera *Euphoriella* (North America) and *Leiophron* (Europe) are the only parasitoids reported from nymphal and adult psocids and are most closely related to taxa that attack Heteroptera. They have been found only in the northern temperate regions and such parasitisation appears to be very rare (or absent) in other areas where surveys have been undertaken. The pronounced seasonality and univoltinism of some of the wasps can result in one psocid generation being attacked heavily and other(s) virtually free from parasitoids, so that surveys to detect them must extend over considerable periods. Eggs are laid in psocid nymphs, which become grossly distended as the wasp larva develops internally, and the last instar emerges by rupturing the body wall of the late instar nymph or adult host, completing its development in a pupal cocoon outside the host.

Ecological and applied roles

Although psocids can be very abundant on trees, their overall ecological importance in these habitats has generally been assumed to be small (New, 2000) and remains poorly defined. Aspects of the dynamics of epiphytic microflora and their consumers, including Psocoptera, have been studied in Britain by Turner (1975b, 1983a, 1984a), but no parallel investigation has yet been made in Southeast Asia. Speculative roles include (1) influencing levels of photosynthetic efficiency by grazing on honeydew moulds and other covering that would impede light penetration on foliage surfaces, (2) constituting significant proportions of the food of a variety of arboreal predators and so being important constituents of complex arboreal food webs, and (3) being important indicators of habitat quality, such as air pollution in industrial countries (Mey & Tietze, 1979). On rare occasions (Spieksma & Smits, 1975), psocids in buildings have been suggested as an allergen in house dusts, and the putative roles of psocids as vectors of parasites such as fringed tape worm of sheep or as intermediate hosts for cestodes elswhere need confirmation. Likewise, suggestions that they may be vectors of plant diseases have not been confirmed.

The major interest in impacts of Psocoptera has devolved on domestic taxa as pests of stored products, where they range from being minor contaminants (as in many non-industrial situations) to causing substantial economic losses through direct damage and secondary contamination. Until recently, taxa such as the booklice (*Liposcelis*) were regarded as 'secondary pests' in stored products, with them becoming abundant only after major primary damage caused by Coleoptera and other insect groups. However, such psocids are now recognised as causing primary damage and, in Australia, have recently become recognised as significant pests in central storage systems, where attempts to control them effectively are a substantial current focus of interest.

As with many other stored products pests, the abundance and 'performance' of the psocids is influenced strongly by temperature and humidity regimes, with cooler temperatures and dryer conditions slowing or curtailing development, and the converse. Because of differences between species in optimal conditions for development, critical identification of the species involved may be important in formulating control measures in any given situation. The very small size of the insects renders them difficult to control effectively, with Ghani & Sweetman (1951) noting that control of psocids is 'difficult, tedious and expensive'. The species cited most frequently as a pest, often as a result of misidentification, is the cosmopolitan *Liposcelis bostrychophila*, whose biology has been studied in detail (see Turner, 1987, 1994a and Turner & Maude-Roxby, 1988, for background and references).

Collecting and studying Psocoptera

Psocids are collected by beating and sweeping vegetation, by sifting leaf litter and extracting organisms from it in devices such as Tullgren funnels, by direct searching and retrieval of specimens with an aspirator, and by sifting or direct searches in domestic or storage environments.

Most collectors use a beating tray for collecting arboreal Psocoptera. Rather than sweep low vegetation with a coarse-meshed net, some workers prefer to 'shake' it over a tray to reduce direct damage to these very delicate insects. Beating or 'shaking' consists of holding a framed canvas tray or umbrella-like tray (see Lienhard et al., 1987) horizontally below the vegetation and then striking the vegetation sharply with a stick or shaking it to dislodge insects, that then fall onto the tray, from where they can be collected individually. For general collecting, dimensions of the tray are not critical, but a tray diameter or side length of 70-100 centimetres provides a useful and accessible area from which to collect. More quantitative sampling, of course, necessitates standardisation of tray size across all samples to be compared.

Beating can be undertaken only when the vegetation is dry, because soft-bodied psocids are damaged easily if they are wet. In reasonably cool and calm weather, most individuals will remain on the beating tray without flying for at least several minutes, sometimes motionless and well camouflaged so that careful observation may be needed to find them. They can be collected with a standard aspirator (or pooter) and stored in small (labelled) vials of 70-80% ethanol or propanol for later examination and sorting. Adults and nymphs are collected in the same manner. Direct searching of tree trunks, aided by removing patches of loose bark to find subcortical taxa, and surfaces of lichen-coated rocks may reveal many psocids that can be collected directly and which are not susceptible to conventional beating. These can be aspirated or picked up individually with a fine moistened paint brush.

Litter psocids can be extracted by a variety of standard techniques for animals in this habitat. Simple sieving of debris over a white tray may often yield some of the more active taxa, but is laborious in relation to the (usually) small numbers obtained. Tullgren funnels are particularly useful for extracting psocids from substrates such as bird nests. The active principle involved, of creating a heat or desiccation gradient, can be applied in various ways to extract stored products psocids from flour or grain.

Rearing and maintenance

Rearing of psocids is a useful tool to clarify patterns of development and in order to associate nymphs with the corresponding adults. Many psocids are relatively easy to maintain and rear in captivity. Indeed, some stored-products taxa thrive simply in vials of flour or broken grain with an occasional drop of water added and maintenance within a broad temperature range of about 15-25 °C. Rearing of arboreal psocids can be slightly more complex. As Lienhard (1998a) noted, maintenance of arboreal species over several generations is still relatively simple. Living nymphs are transferred carefully to glass vials containing small pieces of bark, dead or living foliage – the substrate from which each was collected – as a source of natural food. When the substrate becomes mouldy or overly dry it is simply replaced, but stoppered vials will usually maintain sufficient humidity for the psocids for at least several days. Most bark-frequenting psocids can not walk on smooth glass, so that the containers may be kept uncovered. In contrast, covering is needed for the more versatile foliage-frequenting species.

Preservation and storage

Psocids are preserved and stored in small vials containing 70-80% alcohol. If kept dry and micro-pinned or glued onto card points or rectangles, as was formerly a much more widespread mode of retention, they shrivel badly and become very brittle. Such specimens commonly lose parts of appendages needed for identification, such as legs and antennae, and may be very difficult to soften for preparation of genitalic structures. However, one disadvantage of alcohol preservation is that the elaborate scale patterns on the wings of many Amphientomidae and Lepidopsocidae, which give them the appearance of tiny and brightly coloured moths, may wash off so that their often diagnostic species-specific appearance is lost. For these, a few individuals preserved dry are invaluable in helping to relate denuded but structurally complete specimens to the 'real appearance' of the insect.

Permanent storage vials, preferably of clear glass for ease of inspection of the contents should contain full label data on locality and habitat of the specimens, together with other standard information. In order to prevent damage to the specimens it is often useful to separate the specimens and label(s) by a cotton plug, and placing labels between the plug and vial stopper. Series of vials can be stored together in larger jars of alcohol which, other than the need to inspect and 'top up' at intervals, are essentially permanent storage units. For specimens dissected for taxonomic study, additional slide storage is needed and histological examination may necessitate use of more specific fixatives for particular purposes.

Laboratory examination

Taxonomic inspection of small insects such as psocids can only rarely be undertaken without some detailed treatment. Low power examination of specimens in vials, or transferred in alcohol to a solid watchglass or other container for examination under a binocular microscope is often sufficient to recognise family and, often, genus, and to separate mixed series into putative species or other groups. However, permanent slide mounts of dissected specimens are usually needed to examine the structural characters, particularly of genitalia, used to separate and diagnose species in families, and to facilitate accurate linear measurements of various structures. For some groups of small psocids (such as Liposcelididae), the whole specimen may be slide-mounted (or for small species, mounted between two coverslips so that both dorsal and ventral structures can be examined easily under high power; Lienhard, 1998a), but specialists otherwise differ somewhat in their preference for mounting all parts of a dissected psocid or for mounting only particular parts and associating these clearly, through cross-reference labelling, to the residual parent body in alcohol. Likewise, most workers dissect the apex of the abdomen to mount the terminalia (that is, the genitalia and associated structures, namely the epiproct, paraprocts and clunium), whereas others prefer to retain the cleared abdominal apex entire in alcohol.

Initial examination of specimens for identification includes description of body and wing colouration and measurements of head and total body length (see below). When describing colour and pattern, note condition of specimens, such as 'freshly killed', or 'after alcohol storage for x years', because preservation usually causes changes. Subsequent treatment usually includes preparing a slide mount of the appendages of one side of the body, namely of antenna, legs (either all, or of the hind leg only) and wings together on the same slide. These are the basis for more accurate linear measurements and, for new taxa, drawing of the wings and description of venation. Lienhard (1998a) also recommended mounting mouthparts, but most workers have not done this as standard procedure and examine them in situ on the head.

The series of appendages rarely, if ever, needs staining or other treatment, and these are mounted after further dehydration and clearing. Clearing can be in xylol or clove oil (preparatory to mounting in Canada balsam) or euparal essence (for Euparal mounts), both of which provide generally good results.

Preparations of teminalia are more complex. The apical half to third (depending on the extent of genitalic modification) of the abdomen is cut off and macerated to remove the muscular tissues, following washing in distilled water to remove alcohol. Maceration involves use of 10-15% sodium hydroxide or potassium hydroxide (be careful: HIGHLY CAUSTIC), which can be heated in a water bath or on a slide warmer plate for more rapid processing, but alternatively used cold in a covered solid watchglass for more gentle cleaning over several hours. Occasional inspection and 'palpation' with fine forceps over this period will ensure that overclearing does not occur. If a large faecal pellet is present, this can be removed carefully with forceps (and may be useful as a separate slide mount to detect diet). Following washing in very dilute acetic acid to neutralise the macerant and /or in several changes of distilled water, the abdomen tip is dehydrated through a series of progressively increased concentration alcohols (usually 30, 50, 70, 90 % and absolute alcohol) before clearing and transfer to a slide for careful dissection and display of the various parts. However, for many pale or lightly sclerotised structures, staining is necessary before these steps are undertaken. Examples of the stains and procedures useful for psocids are

1. acid fuchsin: use 1-5% in 2% acetic acid, stain from water for two hours to overnight, then dehydrate through 70% alcohol and absolute alcohol (5 minutes, 2 x 5-10 minutes),
2. lignin pink: stain in dilute aqueous solution for about three hours, then dehydrate in 50% and pure cellosolve,
3. fast green: 1-2% in 1% acetic acid, stain for 1-3 hours and dehydrate as for acid fuchsin.

Many species will not need staining because they are already well sclerotised and contrasts in texture and limits of structures are clear. After whatever treatment is needed, the abdomen apex of either sex is dissected on the slide, in clearing agent or mountant, as follows:

- female
1 separate the subgenital plate by inserting a fine mounted needle between it and the gonapophyses and cutting/separating it at the lateral margins,
2 separate the gonapophyses of left and right sides medially or, if a complex central spermathecal area is present, cut across tergites VIII and IX on each side above the gonapophyses to retain the two sets together,
3 clear surplus tissue and gut traces from between the paraprocts, remove any faeces, and spread the paraprocts and epiproct, where possible without separating these from the clunium.

- male
1 separate the hypandrium by cutting laterally, as noted for subgenital plate,
2 ease the phallosome away from the hypandrium or more dorsal structures. If very fine, or complex (Lachesillidae), leave with the hypandrium,
3 treat paraprocts, epiproct and clunium as above (steps 1-2).

Transfer all parts to a clean slide with a fine needle and mount with the external surfaces uppermost (except phallosome, which is mounted with the ventral side upward). In the case of particularly thick or complex structures, such as the hypandrium of some psocids, it may be necessary to mount these separately under a raised coverslip to avoid undue distortion or damage by crushing, or to retain them in a microvial in alcohol with the parent body.
In general, observations and notes should be made on the cleared specimens before dissection, so that any damage can be evaluated in relation to the 'true' structure.

Measurements

A series of standard measurements is an integral part of the description of most psocid species and these can provide useful diagnostic features for separating related species. Other than for body length and head dimensions, measurements are made from slide-mounted material, and should be correct to within 1 or 2 microns. Standard measurements are as follows:

1 'body length' is the greatest length from the front of the head in normal orientation to the tip of the abdomen,
2 head measurements (figs. 13-14). The measurements used commonly for psocids are 'IO/D' and 'PO/D', as follows (figs. 13-14). IO/D is the ratio of interocular distance (IO) to eye diameter (D), and two rather different interpretations of this have developed:
    a Pearman's method is the ratio of shortest interocular distance to the greatest horizontal diameter of the eye as seen in facial aspect of the head (Pearman, 1934a), and

b Badonnel's method is the ratio of greatest interocular distance to the length of the longest axis of the eye as seen in dorsal aspect of the head (Ball, 1943; see also Lienhard, 1998a).

PO/D (cf. Mockford & Gurney, 1956) is the ratio of shortest eye axis (PO) to greatest eye axis (D) as seen in dorsal aspect and thus indicates the extent of prominence of the eye,

3 wing lengths (fig. 15) are given as the greatest length from point of attachment to the thorax to the apex, neglecting any marginal hairs extending beyond the membrane,

4 antennal segment lengths (normally of the first and second flagellar segments, f1 and f2) are the greatest lengths omitting the intersegmental membrane,

5 antennal length is the greatest length from base to apex, including intersegmental lengths,

6 leg segment lengths are usually taken as the greatest length of the femur (F) and tibia (T), and from condyle to condyle for tarsal segments (t1, t2, t3) (fig. 16), whence the total tarsal length is the sum of the two or three segments present.

Chapter 3

Biogeography

by T.R. New

Tropical South-east Asia is an immensely complex biogeographical region, of major significance in interpreting patterns of distribution and diversity of the world's biota. It incorporates the region known widely as 'Malesia', the major focus of this book. Malesia is a primarily botanical construct, based on the region including southern Thailand, Malaysia, Indonesia, New Guinea* and the Philippines. To the north-west, the region has traditionally been considered to extend to the Isthmus of Kra, which marks a major floristic division between the floras of continental Asia and Malesia. To the east, various boundaries for the region have been suggested, either including (Kato, 1990, after Van Steenis, 1950) or excluding (Laumonier, 1997) New Britain and other islands to the east and south-east of the island of New Guinea. Malesia thus extends over some 60 degrees of longitude (c. 95-153 °E) and 30 degrees of latitude (c. 18 °N-11 °S), and marks the meeting of the Asian and Australasian biotas.

Malesia

Within the large area described here, including the massive Indonesian archipelago, various biogeographical and floristic subdivisions have been proposed, some of them the topics of ongoing debate. The following notes (after New, 2002) are illustrated by figures 68, 69. For primarily plant-associated insects, including Psocoptera, floristic regions and diversity may be an important distributional correlate, but most biogeographical discussion (other than for major floristic provinces of Malesia depicted by Van Steenis, 1950, and followed by many later workers; fig. 68) have emphasised the region's significance for faunal distributions. Lines of faunal demarcation (fig. 69), including Wallace's line, Weber's line and Lyddeker's line, have resulted from attempts to denote natural faunal breaks or transitions between the Oriental and Australasian biotas in the region. Wallace's line purports to delimit the boundary of Asia (Simpson, 1977), and the complexities of trying to generalise meaningfully over such criteria for numerous and biologically diverse animal groups were emphasised by MacKinnon & MacKinnon (1986), amongst other workers. Wallace's line, as later modified by Huxley (fig. 69), follows the edge of the Sunda shelf. Lydekker's line is defined by the edge of the Sahul shelf (as a true boundary of the Papua-Australian region), and Weber's line indicates the faunal balance between these. The separation of the 'Sundaic subdivision' (Malesia to the west of Wallace's line) and 'Sahul subdivision' (Malesia to the east of Wallace's line) is sometimes recognised, but major boundaries within the region are the subjects of ongoing debate. Inclusion or exclusion of the Philippines in 'Malesia proper' on biogeographical grounds has been debated since their exclusion was suggested by Dickersen et al. (1928).

* In this book New Guinea refers to the entire island (including surrounding islands), Papua (Irian Jaya) refers to the western part and Papua New Guinea refers to the eastern part.

Faunal analysis

Appraisal of the distribution of many insect groups in relation to these conventional divisions is often difficult, as pointed out by Gressitt (1959). He emphasised that Wallace's line is not a particularly distinct boundary for many insect groups. From his study of Psocoptera from Bali and Lombok, the Indonesion islands immediately on each side of Wallace's line, Thornton (1984b) implied that some genera appeared to have their most easterly limits on Bali (*Calocaecilius, Amphipsocus, Psocidus* s.str., *Blastopsocidus*) and a number of purported distributional limits for particular taxa are summarised in the family accounts later in this book. A recent survey of Psocidae from eastern Indonesia (Endang et al., 2002) also implied some attenuation of genera toward the east of the region, and also that some genera may be much better represented in New Guinea than further west in Indonesia. Various areas of endemism for particular Psocoptera within the region can indeed be implied but much of this process must be tentative because of the relatively incomplete extent of collecting that has occurred. As for Neuroptera (New, 2002), appraisal of somewhat fragmentary data limits firmer interpretation, but synthesis, even as far as possible at present, provides some suggestions of distributional patterns as hypotheses for future investigation. Future surveys, however, will be increasingly difficult to pursue and analyse, because much pristine and near-pristine habitat in the region is being, or has already been, changed drastically to accommodate needs of burgeoning human populations, so that the already limited distributions of many putative endemic taxa are being fragmented further, or lost. As for many other insect groups, many of the more intriguing taxa of particular significance in biogeographical interpretation are known from few specimens, and some only from their type localities. At the other extreme, some Psocoptera have been distributed widely in commerce (most of the 'stored products species', for example) and their original geographical affinities are largely supposition or conjecture. A few other species (such as *Lachesilla pedicularia*, p. 122) are very widespread, apparently naturally. Other levels include regional species radiations within widespread genera, such as in *Epipsocopsis* (p. 92) to constitute characteristic regional faunal elements within Psocoptera. As a further example, Mockford (1981) discussed the New World species of *Psococerastis* (p. 155), a genus represented strongly also in Africa and South-east Asia, and suggested the possibility that Old World species may be derived from one or more phylogenetic lines of the New World taxa. He noted the need for greater study of Asian taxa to appraise such suggestions more completely.

Biogeographical syntheses and interpretations of Psocoptera in Melanesia, other than distributional comments in species diagnoses and descriptions of numerous taxa, have been attempted for Philotarsidae (Thornton, 1980, 1981g), Calopsocidae (Thornton & Smithers, 1984), Myopsocidae (Smithers & Thornton,

1974b; but note changes in generic attributions since this account, p. 139), and Psocidae (Smithers & Thornton, 1990; Endang et al., 2002). Smithers & Thornton (1981b) provided a broader appraisal of the role of New Guinea in psocid biogeographical interpretations. Despite somewhat inadequate data for many taxa, New Guinea appears to be an important link between the faunas of the Greater Sundas and the Outer Melanesian Arc, as evidenced by members of the families Pseudocaeciliidae, Peripsocidae, Ectopsocidae, Calopsocidae, Epipsocidae and Myopsocidae. New Guinea also appears to have contributed to northern/eastern Australian fauna of these families, in addition to Stenopsocidae and Psilopsocidae, perhaps not surprisingly in that a land bridge between the two areas existed only some 8000 years ago.

The papers mentioned above all drew on the substantially enhanced perspective gained from recent studies of specialist collections in the region. Calopsocidae is predominantly a South-east Asian family, but all the other families mentioned are much more widespread. Perhaps the most complete appraisal until now is for the Philotarsidae, with Thornton's papers drawing on a succession of revisionary studies of the family for New Guinea, New Caledonia, the Bismarck Archipelago, the Solomon Islands, Australia and the New Zealand region, so that the data incorporated are relatively robust and enable consideration of some 150 species. The family Philotarsidae is particularly diverse in the western Pacific region, but much more diverse in the eastern parts of Malesia than in western Malesia or Malaysia, where they seem to be only poorly represented. The pseudocaeciliid subfamily Zelandopsocinae (formerly included in Philotarsidae, p. 125) is apparently confined to the Australian plate and its fragments; it is absent from the Outer Melanesian Arc. Psocidae, particularly diverse in the region, shows a complex variety of patterns that are gradually becoming clearer (Endang et al., 2002). Several small genera are known only from New Guinea, whilst other more westerly taxa apparently do not occur as far east as these. Interpretation in this family is hampered by imprecise generic definitions – for example, confusion over the boundaries between *Copostigma* and *Ptycta* (p. 163), so that further taxonomic study is a prerequisite for refining biogeographical interpretation in this family.

Table 3 is a geographical summary of Psocoptera in tropical South-east Asia, to indicate likely endemic elements. Distributional notes (with references to each genus) are included in the synoptic notes later in this book (chapter 5). These are almost wholly published data, as summarised in the checklist (chapter 7), but a few additional unpublished country records are also included.

TABLE 3
Psocoptera of tropical South-east Asia: indication of regional endemicity above species level. Families with endemic genera are marked with an asterisk.

[CONTINUED] ▸

FAMILY	COMMENTS
*Lepidopsocidae	many genera widespread; *Echinopsocus* and *Papuapsocus* possible New Guinea endemics
Psoquillidae	all genera widespread
Trogiidae	all genera widespread
*Prionoglarididae	single genus, *Siamoglaris*, apparent endemic
Psyllipsocidae	both genera widespread
*Amphientomidae	well represented; two genera apparent regional endemics: *Stigmatopathus*, *Cymatopsocus*
*Troctopsocidae	*Coleotroctellus* endemic to South-east Asia, but not yet found in Malesia; *Thaipsocus* and *Selenopsocus* found in Malesia
Liposcelididae	two genera (*Embidopsocus*, *Liposcelis*) widespread
Pachytroctidae	genera widepread
Epipsocidae	many endemic species of *Epipsocopsis*
*[Dolabellopsocidae]	*Auroropsocus* from India, China, not yet from South-east Asia
*Caeciliusidae	*Asiocaecilius* widespread from Himalaya to South-east Asia; *Maoripsocopsis*, *Pericaecilius*
*Amphipsocidae	*Tagalopsocus* largely endemic; *Calocaecilius* (as endemic subfamily Calocaeciliinae)
Stenopsocidae	genera widespread
Archipsocidae	genera widespread
*Calopsocidae	seven endemic genera, (*Callistoptera*, *Calosema*, *Cyclopsocus*, *Dendropsocus*, *Nemupsocus*, *Neurosema*, *Torrepsocus*), with the family centred on the region
Ectopsocidae	genera widespread
[Elipsocidae]	regional representation tentative
Lachesillidae	genus widespread
Mesopsocidae	regional representation very low
Peripsocidae	*Peripsocus* widespread; *Diplopsocus* also speciose in region
Philotarsidae	genera widespread

FAMILY	COMMENTS
*Pseudocaeciliidae	*Allopsocus*, *Novopsocus*
Hemipsocidae	genus widespread
*Myopsocidae	*Smithersia*, *Thorntonodes*
Psilopsocidae	*Psilopsocus* apparently particularly well represented, but with few species in region
*Psocidae	*Anomaloblaste*, *Euclismioides*, *Indoblaste*, *Javablaste*, [*Neoblaste* diverse], *Sundapsocus*, *Cycetes* [also Australia], *Elytropsocus*, *Javapsocus*, *Kaindipsocus*, *Thyrsopsocopsis*

Major groups of Psocoptera absent from tropical South-east Asia

This section is a brief synopsis of the major taxa, namely families and sometimes subfamilies, that have not been recorded from the region, to facilitate notice should any be found. Several are characteristically southern temperate or New World groups, whose incidence is highly unlikely, but some others may indeed be discovered in Asia with more assiduous collecting.

Troctomorpha, Amphientometae

Musapsocidae

This small family of electrentomoid psocids (Mockford, 1967b) includes the genera *Musapsocus* Mockford (type species *Musapsocus huastecanus* Mockford) and *Musapsocoides* García Aldrete & Mockford (type species *Musapsocoides nadleri* García Aldrete & Mockford). It is known only from Central and South America and Trinidad (West Indies).

It shares many features with related groups, the Troctopsocidae (p. 79) and Compsocidae (below), but Musapsocidae have 12 antennal segments and the tarsi are 2- or 3-segmented; the other electrentomoid families all have 3-segmented tarsi. Musapsocidae is a New World endemic family and is highly unlikely to be discovered elsewhere. García Aldrete & Mockford (1996) noted that *Musapsocoides* shares several probable apomorphies with some troctopsocid genera. The relationships of the two families may need further examination.

Electrentomidae

Also discussed by Mockford (1967b). The four genera are known respectively from Zimbabwe (*Manicapsocus* Smithers, type species *Manicapsocus alettae* Smithers), Central America (*Epitroctes* Mockford, type species *Epitroctes tuxtlarum* Mockford) and South America (*Nothoentomum* Badonnel, type species *Nothoentomum palpale* Badonnel; *Phallopsocus* Badonnel, type species *Phallopsocus carminatus* Badonnel). The family Electrentomidae thus appears to have a highly disjunct southern distribution and seems to be absent from the entire Palaearctic (other than for Baltic amber fossils) and Oriental regions.

Compsocidae

The two included genera are *Compsocus* Banks (type species *Compsocus elegans* Banks, Central America) and *Electrentomopsis* Mockford (type species *Electrentomopsis variegata* Mockford, Mexico). The family thus appears to be a restricted New World endemic group (Mockford, 1967b).

Protroctopsocidae

A recently separated segregate from Troctopsocidae, of which it was formerly a subfamily, with few species from the Western Palaearctic and one from Central America. See discussion on p. 79 and Lienhard & Mockford (1997).

Troctomorpha, Nanopsocetae

Sphaeropsocidae

These small, generally black, elytriform psocids (in which the hind wing is absent) are found on low vegetation, such as grass tussocks, leaf litter, in caves and under stones in southern South America, Africa, Ascension and St Helena islands and Australia (*Sphaeropsocopsis* Badonnel, type species *Sphaeropsocopsis chilensis* Badonnel) and Chile and Europe (*Badonnelia* Pearman, type species *Badonnelia titei* Pearman). In Europe, *B. titei* Pearman is associated with human-altered habitats and, as other species of the genus are edaphic in Chile (Badonnel, 1963a), may have been introduced. Male *Badonnelia* are apterous. The Sphaeropsocidae appear to be entirely southern and on mid-Atlantic islands and, other than by sporadic introductions from elsewhere, are unlikely to be found in South-east Asia.

Psocomorpha, Epipsocetae

Dolabellopsocidae

Treated on p. 93.

Cladiopsocidae

This small Neotropical family was established by Eertmoed (1973) to contain *Cladiopsocus* Roesler (type species *Dendroneura ramulosa* Enderlein (three tarsal segments)) and *Spurostigma* Eertmoed (type species *Spurostigma dominica* Eertmoed (two tarsal segments)). As in Dolabellopsocidae (p. 93), the fore wing has two anal veins, but Cladiopsocidae have spur veins from the posterior angle of the pterostigma and/or from other fore wing veins.

Neurostigmatidae

A small Neotropical family with few species, characterised in part by having distinct crossveins within the pterostigma.

Ptiloneuridae

This Neotropical family contains the largest and most spectacular Epipsocetae: many species have elaborately-patterned fore wings and additional branches to one or more of the main veins. *Cladiopsocus* was earlier included in this family on having 3-segmented tarsi, the major diagnostic feature separating this family traditionally from Epipsocidae. 'Typical' Ptiloneuridae lack spur veins from the pterostigma or other fore wing veins. The species are associated primarily with dead foliage and bark.

Psocomorpha, Caeciliusetae

Asiopsocidae

This small family is based on *Asiopsocus* Günther (type species *Asiopsocus mongolicus* Günther), occurring in North America, Mexico, Spain, Yemen, Oman, Mongolia, and *Notiopsocus* Banks (type species *Notiopsocus simplex* Banks) occurring in Brazil and Africa. These two genera were placed by earlier authors in Caeciliusidae and Peripsocidae, respectively. The third genus included in the family is *Pronotiopsocus* Mockford (type species *Pronotiopsocus amazonicus* Mockford), from the southern United States of America and Brazil.

The family was raised by Mockford & García Aldrete (1976), who treated it (as superfamily Asiopsocoidea) as a sister group of the other families constituting Caeciliusoidea. A third, neotropical, genus was added by Mockford (1983a), who provided an augmented definition of the family, noted the 'bizarre' disjunct distribution of the family and confirmed relationships with other Caeciliusetae.

Psocomorpha, Homilopsocidea

Archipsocidae: Pararchipsocinae

This neotropical subfamily was described by Badonnel et al. (1984). It differs from Archipsocinae (p. 113) by the hind wing lacking a closed cell, or with this cell being very strongly reduced, and lack of sexual alary dimorphism and neoteny. It was founded on *Pararchipsocus* Badonnel, Mockford & García Aldrete (type species *Pararchipsocus poblanus* Badonnel, Mockford & García Aldrete) and *Pseudarchipsocus* Mockford (type species *Pseudarchipsocus guajiro* Mockford). Pararchipsocinae has since been augmented by *Notarchipsocus* Mockford (type species *Archipsocus macrurus* New). It has not been reported from outside the Neotropical region.

Bryopsocidae

See notes under Pseudocaeciliidae (p. 128) for introductory comments on this family. Mockford (1984d) raised Bryopsocidae to contain *Bryopsocus* Thornton, Wong & Smithers (type species *Austropsocus townsendi* Smithers), formerly placed in Philotarsidae. The genus also contains a species (*B. angulatus* (Smithers)) earlier allocated to *Zelandopsocus*. *Bryopsocus* is intermediate between Philotarsidae and Pseudocaeciliidae, as diagnosed by Mockford (1984d), but also has distinctive features. It is known only from New Zealand and is associated with moss and trees in wet forests.

Elipsocidae

Treated on p. 120.

Lachesillidae: Eolachesillinae

Eolachesillinae is by far the more diverse subfamily of Lachesillidae at the generic level, but it is geographically more restricted than Lachesillinae (p. 121). In discussing the subfamily, Mockford & Sullivan (1986) noted possible confusions with some Elipsocidae, particularly through *Eolachesilla* Badonnel, but placed all other genera in the tribe Graphocaeciliini, represented entirely by New World genera. Eolachesillinae is diagnosed, in part, by the phallosome being a closed frame with some endophallic sclerites present (these absent in Lachesillinae) and by the female paraprocts having a field of short, relatively stout setae along the median ventral margin (these lacking in Lachesillinae).

Lesneiidae

This family was raised recently (Schmidt & New, 2004) to contain the African genus *Lesneia* Badonnel (type species *Lesneia capensis* Badonnel), containing four described species. It was earlier placed as a subfamily (Lesneiinae) of the Elipsocidae.

Sabulopsocidae

This small family is also closely related to Elipsocidae and contains two small genera (*Moapsocus* Schmidt & New, type species *Moapsocus angelicus* Schmidt & New; *Sabulopsocus* Smithers, type species *Sabulopsocus tractuosus* Smithers) from New Zealand (Schmidt & New, 2004).

Trichopsocidae

This small family includes only the genus *Trichopsocus* Kolbe (type species *Caecilius hirtellus* McLachlan), with the disjunct distribution including North and South America, Europe, Africa and Australia including Tasmania. The superficial appearance of *Trichopsocus* is of a pseudocaeciliid, but the hind wing marginal setae behind the wing apex are alternately long and short rather than all being of similar length. The hypandrium and subgenital plate are both simple. The species appear to be foliage-frequenters.

Chapter 4

Keys to suborders and families

by T.R. New

Three distinct suborders of Psocoptera are recognised, among other features based on the number of antennal segments. However, in practice, many specimens may have antennae broken after preservation, so that the number of antennal segments may not be determinable accurately. Because of this, it may be preferable to work directly from the key to families provided (see further). Although hypopharynx features (figs. 65-67) are included in formal diagnoses of suborders, they are rarely used in practical identification. Note also that first instar nymphs (and, occasionally, others), may have fewer antennal segments than their corresponding adults.

General

It is assumed widely that the three suborders, Trogiomorpha, Troctomorpha and Psocomorpha, mark a transition from 'most primitive' (Trogiomorpha) to 'most advanced' (Psocomorpha), with the last by far the most diverse in terms of both number of families and number of species. Within each suborder the various families have been grouped into 'series' or 'groups' (commonly referred to as 'infraorders', as on p. 4) – not paralleling strictly the superfamilies of many other insect orders, but which Smithers (1972c) regarded 'merely as convenient categories' for helping to indicate possible relationships between families. The early classification systems for the order were noted briefly on p. 7, and are discussed more comprehensively by Smithers (1972c). All three suborders are well-represented in South-east Asia and the relevant various 'groups' are defined in context in the systematic sections.
The family key includes all the families of Psocoptera recorded from tropical South-east Asia, or which are likely to be found there. The latter are indicated by an asterisk (*). Because of practical problems of keying through suborders that result from broken specimens (for example antennae are often broken in alcohol, so that simply counting number of antennal segments can be impossible), these categories are ignored in the key, which goes directly to families. Suborder classification, however, forms the basis for the family arrangement in the rest of this account. The key is based on adults and care is needed to distinguish nymphoid adults from immature stages. (See chapter 2 **Nymphs**, under **Morphology**, for differences between adults and nymphs.) Recognition of nymphs to family level is tentative, although attempts to do this for the relatively well known faunas of North America (Mockford, 1987a) and Britain (New, 1977a) have been made, and those keys may be used as guides to the character suites used. In general, claw, pulvillus and antennal characters (allowing for the lower segment numbers of young nymphs), together with body form can help diagnose nymphs, but many can be confidently recognised only when they are taken with corresponding adults. Nymphs of some families of Psocoptera from the region have not been described or diagnosed adequately.

The key is designed primarily for use without need to dissect or slide mount the specimens and thus differs from a number of the keys to lower levels given later, in which close examination of genitalia and other structural features of specimens is an integral part of identification. However, a binocular microscope is necessary and specimens should be examined covered in liquid (such as glycerine or 70% alcohol) to avoid problems of reflection from exposed surfaces. In some instances (such as for checking tarsal segment number and claw/pulvillus form of small specimens), temporary placement in liquid in a cavity slide for higher power examination may be useful. The key, necessarily, considers only a limited range of characters and specimens should be checked carefully against the fuller family diagnoses and illustrations provided later. Illustrations provided for each family include representative wing venation and genitalic structures of many genera, where these are distinctive, and are designed as adjuncts to the key and later taxonomic synopses. Many figures are simplified by omission of unnecessary detail and they are not meant for species level diagnoses except where particular species names are mentioned in accompanying legends. Scale lines are not given, but the text for each family gives generalised 'range measurements' as fore wing or body length. These should be treated as a guide only, as exceptions are likely to occur, but indicate that some groups of psocids are 'large' and others 'smaller' insects.

Alary polymorphism (figs. 70-71) necessitates multiple entries for many families, in which more than one state of macroptery, aptery or some intermediate category (brachyptery or microptery) occurs. In some instances, one or more such states recorded for a given family have not been reported from the region, but all morphs recorded globally for each family are included, to help facilitate recognition should they be found. These morphs are summarised in table 4, as a 'working guide' to key content. This key owes much to those by earlier workers, predominantly that by Smithers (1990a).

TABLE 4
Categories of wing polymorphism or condition included in key to families of Psocoptera. Symbols in parentheses represent morphs not yet reported from tropical South-east Asia. [CONTINUED] ▸

FAMILY	APTEROUS/ STRONGLY MICROPTEROUS; COUPLETS 3-14*	BRACHYPTEROUS/ MICROPTEROUS**; COUPLETS 15-37	MACROPTEROUS; COUPLETS 38-67
Lepidopsocidae	(+)	+	+
Psoquillidae	–	+	+
Trogiidae	+	***	–
Prionoglarididae	–	–	+

TABLE 4 [CONTINUED]

FAMILY	APTEROUS/ STRONGLY MICROPTEROUS; COUPLETS 3-14*	BRACHYPTEROUS/ MICROPTEROUS**; COUPLETS 15-37	MACROPTEROUS; COUPLETS 38-67
Psyllipsocidae	+	+	+
Amphientomidae	(+)	(+)	+
Troctopsocidae	–	+	+
Liposcelididae	+	–	+
Pachytroctidae	+	(+)	+
Epipsocidae	(+)	+	+
Dolabellopsocidae	–	–	(+)
Caeciliusidae	(+)	(+)	+
Amphipsocidae	(+)	(+)	+
Stenopsocidae	–	(+)	+
Archipsocidae	+	+	+
Calopsocidae	–	(+)	+
Ectopsocidae	(+)	(+)	+
Elipsocidae	(+)	(+)	(+)
Lachesillidae	–	(+)	+
Mesopsocidae	(+)	(+)	+
Peripsocidae	–	(+)	+
Philotarsidae	–	(+)	+
Pseudocaeciliidae	–	(+)	+
Hemipsocidae	–	+	+
Myopsocidae	–	–	+
Psilopsocidae	–	–	+
Psocidae	–	(+)	+

* wings rudimentary if present, completely veinless

** some veins usually visible, at least in fore wing

*** in Trogiidae winglets veinless, articulated to mesothorax (often broken), hind wings absent

Key to the suborders

Adults of the three suborders are separable as follows:

1 Antennae 20-50-segmented. Labial palpi 2-segmented, second segment the larger. Tarsi 3-segmented. Filaments of hypopharynx separate throughout their lengths, never fused on midline (fig. 65). Paraproct usually with strong posterior spine (fig. 56). Fore wing lacking pterostigma or with pterostigma membranous ▸ **Trogiomorpha**

– Antennae 9-15-segmented (usually 15-segmented); flagellar segments beyond segments 4 or 5 with secondary annulations (fig. 32). Labial palpi usually 1-segmented, occasionally 2-segmented. Tarsi usually 3-segmented, more rarely 2-segmented. Filaments of hypopharynx fused on midline for part of their length, separate distally (fig. 66). Paraproct without strong posterior spine. Fore wing, when present, without sclerotised pterostigma ▸ **Troctomorpha**

– Antennae with 13 segments at most; flagellar segments lacking secondary annulations but occasionally with reticulate sculpturing on surface. Labial palpi of only one segment. Tarsi 2- or 3-segmented. Filaments of hypopharynx fused on midline for part of their length, separate distally (fig. 67). Paraproct without strong posterior spine. Fore wing, when present, usually with sclerotised pterostigma ▸ **Psocomorpha**

Key to the families

The figures provided for each family should be used to help clarify key characters, if necessary. Families not yet recorded from tropical South-east Asia but likely to be found there are indicated with an asterisk (*).

1 Macropterous: wings extending at least to apex of abdomen, often well beyond this ▸ 38

– Apterous, brachypterous or micropterous: if the last two, wings not extending beyond apex of abdomen ▸ 2

2 Apterous or strongly micropterous (wing rudiments, when present, completely veinless) ▸ 3

– Brachypterous or micropterous, at least traces of wing lobes present together with some venation, particularly in fore wing ▸ 15

3 Tarsi 2-segmented ▸ 4

– Tarsi 3-segmented ▸ 9

4 Antennae 9-segmented (no secondary annulations on flagellum) ▸ **Liposcelididae p. 81**

– Antennae with more than 10 segments ▸ 5

5 Labrum elongate, with two longitudinal sclerotised ridges (fig. 72) ▸ **Epipsocidae p. 90**

– Labrum not conspicuously elongate, lacking longitudinal sclerotised ridges (fig. 73) ▸ 6

6 Pulvillus narrow, rarely absent ▸ ***Elipsocidae p. 120**

– Pulvillus broad (fig. 36) ▸ 7

7 Female: subgenital plate bilobed or with posterior lobe/projection, gonapophyses usually complete (rarely reduced) with external valve setose. Male: phallosome with complex radular sclerites, epiproct and clunium commonly ornamented (apterous individuals very scarce) ▸ **Ectopsocidae p. 118**

– Female: subgenital plate rounded or slightly emarginate at apex, gonapophyses reduced (occasionally absent). Male: phallosome frame simple, lacking complex defined radular sclerites, some rugose sclerification may be present ▸ 8

8 Female: gonapophyses reduced to broad setose external valve and small dorsal valve, or absent. Male: phallosome frame simple, sometimes open, no radular sclerites ▸ **Archipsocidae p. 112**

– Female: gonapophyses reduced, external valve a small lobe, sometimes with 1 or 2 setae. Male: apterous morphs unknown, but should have rugose sclerifications on penial bulb ▸ **Amphipsocidae p. 105/Caeciliusidae (very few) p. 93**

9 Pretarsal claws without subapical tooth. (Maxillary palpi with conical sensillum on second segment: fig. 74) ▸ **Trogiidae p. 68**

– Pretarsal claws with subapical tooth ▸ 10

10 Body sometimes with flattened scales on surface, also on wing rudiments and legs ▸ **Amphientomidae p. 75**

– Body and appendages without flattened scales ▸ 11

11 Paraproct with prominent spine (fig. 56) ▸ **Psyllipsocidae p. 72**

– Paraproct without prominent spine ▸ 12

12 Antennae 13-segmented, without secondary annulations ▸ 13

– Antennae with 14-15 segments, with secondary annulations on distal flagellar segments ▸ 14

13 Antennae long, usually about as long as body (large, ca 3 mm or more, females only [males winged]) ▸ **Mesopsocidae p. 122**
(Mesopsocidae are scarcely recorded from tropical South-east Asia.)
– Antennae short, usually not extending beyond anterior region of abdomen (small, ca 2 mm) ▸ ***Elipsocidae p. 120**

14 Body dorsoventrally compressed; hind femora broad (fig. 18) ▸ **Liposcelididae p. 81**
– Body not dorsoventrally compressed; hind femora normal, slender ▸ **Pachytroctidae p. 85**
(very rarely, Amphientomidae may key here: check family notes and illustrations)

15 Tarsi 2-segmented ▸ 16
– Tarsi 3-segmented ▸ 29

16 Fore wing strongly elytriform, thickened and not translucent ▸ 17
– Fore wing translucent, not strongly thickened ▸ 19

17 Vertex rounded, normal; venation strongly reduced, but closed discoidal cell evident in fore wing ▸ **Psocidae (*Elytropsocus*) p. 164**
– Vertex sharp, may be medially emarginate; venation relatively complete, fore wing venation sometimes extensively reticulate ▸ 18

18 Fore wing with some secondary venation, sometimes extensive ▸ **Calopsocidae p. 114**
– Fore wing without secondary or supernumerary venation ▸ **Amphipsocidae (Calocaeciliinae) p. 109**

19 Labrum elongate, with two longitudinal sclerotised ridges ▸ **Epipsocidae p. 90**
– Labrum normal, without longitudinal sclerotised ridges ▸ 20

20 Venation not discernible (females) ▸ 21
– Venation at least partially distinct ▸ 22

21 Gonapophyses complete, external valve large, with numerous setae ▸ **Pseudocaeciliidae p. 128**
– Gonapophyses reduced, external valve a small bare lobe ▸ **Amphipsocidae p. 105**

22 Fore wing membrane with dense hairs ▸ **Archipsocidae p. 112**
– Fore wing membrane bare ▸ 23

23 Areola postica of fore wing absent ▸ 24
– Areola postica present in fore wing ▸ 25

24 Pterostigma rectangular ▸ **Ectopsocidae p. 118**
– Pterostigma not rectangular, narrowed basally ▸ **Peripsocidae p. 123**

25 Pterostigma and areola postica each connected to other veins by crossveins ▸ **Stenopsocidae p. 110**
– Pterostigma free; areola postica connected to M by crossvein ▸ **Hemipsocidae p. 138**
– Pterostigma and areola postica both free ▸ 26

26 Fore wing veins glabrous or almost glabrous ▸ 27
– Fore wing veins setose ▸ 28

27 Wing veins glabrous. Male: phallosome with median process extending anteriorly. Female: gonapophyses reduced ▸ **Lachesillidae p. 121**
– Wing veins sparsely setose. Male: phallosome rounded anteriorly. Female: gonapophyses complete ▸ ***Elipsocidae p. 120**

28 Pulvillus slender. Male: phallosome usually with distinct radular sclerites. Female: gonapophyses complete, external valve large and setose ▸ **Philotarsidae p. 125**
– Pulvillus broad. Male: phallosome without defined radular sclerites. Female: gonapophyses reduced, external valve a small lobe, usually with a seta ▸ **Caeciliusidae p. 93**

29 Body and wing rudiments with flattened scales ▸ 30
– Body and wing rudiments without flattened scales ▸ 31

30 Paraproct without strong posterior spine; antennae with 15 segments ▸ **Amphientomidae p. 75**
– Paraproct with strong posterior spine; antennae with numerous (usually more than 20) short segments ▸ **Lepidopsocidae p. 61**

31 Fore wing dark, glossy, in strong contrast to body ▸ **Troctopsocidae p. 79**
– Fore wing generally membranous, not thickened and glossy ▸ 32

32 Maxillary palpi with conical sensillum on second segment (paraproct with strong spine; antennae with more than 15 segments) ▸ 33
– Maxillary palpi without conical sensillum on second segment ▸ 34

33 Pulvillus moderately broad, expanded at tip (wings often strongly patterned) ▸ **Psoquillidae p. 67**

– Pulvillus relatively broad (wings usually hyaline or slightly shaded) ▸ **Trogiidae p. 68** (see family diagnoses)

34 Pterostigma not thickened, pulvillus absent (antennae with more than 20 segments) ▸ **Psyllipsocidae p. 72**

– Pterostigma not thickened, pulvillus absent (antennae 15-segmented) ▸ **Pachytroctidae p. 85**

– Pterostigma thickened, pulvillus present (antennae 13-segmented) ▸ 35

35 Wing veins with numerous setae; hind wing margin setose ▸ 36

– Wing veins, at most, sparsely setose; hind wing margin bare except, sometimes, for a few setae on radial fork ▸ 37

36 Fore wing vein M+Cu usually with two rows of setae. Abdomen with ventral eversible vesicles ▸ **Pseudocaeciliidae p. 128**

– Fore wing vein M+Cu with one row of setae. Abdomen without eversible vesicles ▸ **Philotarsidae p. 125**

37 Wing veins and margin glabrous ▸ **Mesopsocidae p. 122**

– Wing veins sparsely setose, hind wing margin sometimes with few setae in radial fork ▸ ***Elipsocidae p. 120**

38 Tarsi 2-segmented ▸ 39

– Tarsi 3-segmented ▸ 55

39 Areola postica in fore wing absent or basally open ▸ 40

– Areola postica in fore wing present, basally closed ▸ 44

40 Pterostigma rectangular ▸ **Ectopsocidae p. 118**

– Pterostigma not rectangular, narrowed basally ▸ 41

41 Fore wing veins and margin glabrous ▸ **Peripsocidae p. 123**

– Fore wing veins and/or margin with at least some setae ▸ 42

42 Fore wing veins occasionally sparsely setose; hind wing margin sometimes with few setae in radial fork ▸ ***Elipsocidae p. 120**

– Fore wing veins more extensively setose; margins setose ▸ 43

43 Fore wing: areola postica absent ▸ **Caeciliusidae** (*Pericaecilius*) **p. 102**
– Fore wing: areola postica open basally (wings coriaceous)
▸ **Amphipsocidae** (*Calocaecilius*) **p. 110**

44 Areola postica in fore wing fused with M for a length ▸ **Psocidae p. 145**
(N.B. some Amphipsocidae (*Taeniostigma*, *Matsumuraiella*) also key here: differ from Psocidae (in which veins almost always glabrous, except *Setopsocus*) by having wing veins strongly setose)
– Areola postica in fore wing linked to M by crossvein ▸ 45
– Areola postica in fore wing free ▸ 47

45 Pterostigma linked to vein Rs by crossvein ▸ **Stenopsocidae p. 110**
– Pterostigma free ▸ 46

46 Pterostigma strongly broadened over distal half; fore wing veins with two rows of setae in basal half (fore wing veins Rs and M linked by short crossvein)
▸ **Amphipsocidae** (*Kodamaius*) **p. 109**
– Pterostigma shallow thoughout length; fore wing veins with one row of setae (fore wing veins Rs and M fused for a short length) ▸ **Hemipsocidae p. 138**

47 Fore wing veins glabrous ▸ **Lachesillidae p. 121**
– Fore wing veins setose ▸ 48

48 Fore wing membrane with dense hairs ▸ 49
– Fore wing membrane usually not setose; if so (rare), venation not reduced ▸ 50

49 Wings 'strap-like', with no fore wing costal break. Vertex rounded (female: gonapophyses reduced) ▸ **Archipsocidae p. 112**
– Wings often broad, sometimes with costal break. Vertex sharp (female: gonapophyses complete) ▸ **Calopsocidae p. 114**

50 Head elongate, labrum elongate with two longitudinal sclerotised ridges ▸ 51
– Head normal, labrum without longitudinal sclerotised ridges ▸ 52

51 Fore wing with 1 anal vein ▸ **Epipsocidae p. 90**
– Fore wing with 2 anal veins ▸ ***Dolabellopsocidae p. 93**

52 Pretarsal claws with a distinct subapical tooth; abdomen without ventral eversible vesicles ▸ **Philotarsidae p. 125**

(N.B. *Lachesilla hirsuta* (Lachesillidae) also keys here, as an unusual member of the family with setose fore wing veins. It is distinct on genitalic features from all Philotarsidae: see couplet 27)

– Pretarsal claws usually without subapical tooth, occasionally with a small subapical tooth; abdomen with ventral eversible vesicles ▸ 53

53 Female gonapophyses complete, external valve setose; male phallosome usually with internal sclerites; pretarsal claws with or without subapical tooth ▸ **Pseudocaeciliidae p. 128**

– Female gonapophyses reduced, external valve bare or with 1-2 setae; male phallosome simple, at most with rugose endophallus; pretarsal claws without subapical tooth ▸ 54

54 Setae on fore wing veins in one rank ▸ **Caeciliusidae p. 93**

– Setae on fore wing veins in more than one rank ▸ **Amphipsocidae p. 105**

55 Body and wings with flattened scales ▸ 56

– Body and wings without flattened scales ▸ 57

56 Paraproct without strong posterior spine; antennae with 15 segments ▸ **Amphientomidae p. 75**

– Paraproct with strong posterior spine; antennae with numerous short segments ▸ **Lepidopsocidae p. 61**

57 Areola postica present in fore wing, venation more or less complete ▸ 58

– Areola postica absent, venation strongly reduced to (usually) two longitudinal veins ▸ **Liposcelididae p. 81**

58 Pterostigma open basally ▸ **Troctopsocidae p. 79**

– Pterostigma not open basally ▸ 59

59 Areola postica fused with M ▸ **Myopsocidae p. 139**

– Areola postica free ▸ 60

60 Areola postica in fore wing long and low; pterostigma not thickened; nodulus absent; hind wing vein M single-branched ▸ **Pachytroctidae p. 85**

– Areola postica in fore wing not long and low (if, rarely in some Psyllipsocidae, this state present, then nodulus present and hind wing vein M 2-branched) ▸ 61

61 Fore wing veins glabrous or very sparsely setose, in which case margin largely glabrous ▸ 62

– Fore wing veins and margin setose ▸ 65

62 Fore wing vein Sc strongly developed, forming closed loop by fusing distally with R1; crossvein from near pterostigma to vein Rs; fore wing vein 2A present ▸ **Prionoglarididae p. 71**

– Fore wing vein Sc not strongly developed, usually short and apparent only near base of wing, ending free in costal cell; pterostigma free; fore wing without vein 2A ▸ 63

63 Areola postica with Cu1a and Cu1b separated well before wing margin; pterostigma with spur vein ▸ **Psilopsocidae p. 144**

– Areola postica normal; pterostigma without spur vein ▸ 64

64 Wing veins and margin glabrous (body length ca 3 mm or more) ▸ **Mesopsocidae p. 122**

– Wing veins sparsely setose; hind wing margin sometimes with few setae in radial fork (smaller than 3 mm) ▸ ***Elipsocidae p. 120**

65 Paraproct with strong spine ▸ 66

– Paraproct without spine ▸ 67

66 Nodulus absent; hind wing vein M simple (wings often strongly patterned) ▸ **Psoquillidae p. 67**

– Nodulus present; hind wing vein M two-branched ▸ **Psyllipsocidae p. 72**

67 Fore wing vein M+Cu usually with two rows of setae; abdomen with ventral eversible vesicles ▸ **Pseudocaeciliidae p. 128**

– Fore wing vein M+Cu usually with one row of setae; abdomen without eversible vesicles ▸ **Philotarsidae p. 125**

Chapter 5

Systematic treatment

by T.R. New

In the following, the scope of each suborder is summarised and further divided into 'groups' ('infraorders') (see previous chapter), which consist of families. Keys are provided for subfamilies and other categories, when appropriate. Within each family or lower segregate, keys are given for all regional genera. For each genus the type species, synonymy and distribution are indicated and typical characters are illustrated. Technical terms can be found in the glossary and the recorded formally described species of each group are listed in chapter 7. Families and other groupings above genus level found only outside South-east Asia are listed in square parentheses.

Suborder Trogiomorpha

The main groups in this suborder are the scaly-winged Lepidopsocidae (many of which resemble small moths in general appearance) and several families containing taxa associated with stored products or cave environments. Many of the species are brachypterous, micropterous or apterous. The five families present in the region include representatives of the two main groups, as below.

Key to the groups of suborder Trogiomorpha

1 Antennae never with secondary annulations on flagellar segments. Head short and broad. Second segment of maxillary palp with conical sensillum on inner side. Fore wing without nodulus, but brachyptery or aptery common. Body and wings sometimes covered with flattened scales. Combined length of hind tibia and tarsus shorter than abdomen. Female: gonapophyses reduced
▸ group ATROPETAE (Lepidopsocidae, Psoquillidae, Trogiidae)

– Antennae sometimes with secondary annulations on flagellar segments. Head long, vertical. Second segment of maxillary palp without conical sensillum on inner side. Fore wing with nodulus; brachypterous individuals with some venation discernible. Scales not present on body or wings. Combined length of hind tibia and tarsus greater than abdomen. Female: gonapophyses reduced, but external valve present as a broad setose lobe
▸ group PSOCATHROPETAE (Prionoglarididae, Psyllipsocidae)

Group ATROPETAE

1 Lepidopsocidae (figs. 75-95)

Diagnosis

Most species of Lepidopsocidae have body lengths of around 2 mm and fore wing length of 2-2.5 mm. Lacinial apex divided, with few teeth. Maxillary palpi with conical sensillum on inner side of second segment (fig. 74); labial palpi 2-segmented. Antennae with numerous segments (more than 20, occasionally up to 50), not annulated. Tarsi 3-segmented, claws curved, usually with one subapical tooth, but sometimes more or none. Fore wing often acuminate, usually (together with body and legs) with covering of flattened scales; pterostigma not thickened; vein Sc often well developed; R-Rs crossvein usually present; Rs and M commonly fused for a length; areola postica long; veins Cu2 and 1A end separately on wing margin. Hind wing vein R1 long; vein M with two branches, these arising separately. Hind wing sometimes absent. Paraproct with strong spine, trichobothrial field not defined. Female: gonapophyses reduced, ventral valve absent; dorsal valve small and often weakly sclerotised, sometimes absent; external valve large, setose. Male: phallosome with parameres divergent anteriorly; aedeagus often complex.

Remarks

Associated primarily with leaf litter and dead foliage, these scaly-winged psocids often run around actively when disturbed. Some species are brightly coloured and, although not closely related to them, may be confused with Amphientomidae (p. 75), which also have scaly wings. The family is taxonomically intricate and four subfamilies have been usually recognised, separated mainly on wing features. Wing scales, broadly resembling those of Lepidoptera, have a variety of forms (fig. 82).

Key to the subfamilies of Lepidopsocidae (based on Smithers, 1990a and Mockford, 2005)

1 Body and wings lacking scales (antennae with about 40 short segments; hind wing with small basal cell; vein R1 arising basal to M2) ▸ **[Thylacellinae]**
– Body and wings with flattened scales ▸ 2

2 Hind wing usually with small basal cell. Antennae with no more than 30 segments; segments about four times as long as wide ▸ 3
– Hind wing without basal cell. Antennae with 30-50 segments; segments about twice as long as wide (wings sometimes reduced) ▸ 4

3 Wing venation relatively complete; hind wing with small basal cell (wings broad, apex usually tapered) ▸ **Perientominae**
– Wing venation reduced to about two main longitudinal veins; hind wing without small basal cell (wings narrow, 'strap-like', apex rounded) ▸ **Parasoinae**

4 Fore wing always with discernible venation. Usually macropterous, but fore wing may be greatly reduced ▸ **Lepidopsocinae***
– Fore wing without traces of veins, reduced to small lobes ▸ **Lepolepidinae**

* Mockford (1993a) resurrected the subfamily Echinopsocinae Roesler to contain *Echinopsocus* Enderlein, *Pteroxanium* Enderlein and *Scolopama* Enderlein, in all of which the wings are usually reduced, and the hind wing absent, in contrast to other genera placed conventionally in Lepidopsocinae – which are usually fully winged. He also raised the genus *Neolepolepis* within this subfamily to contain the North American species earlier placed in *Lepolepis* and which differ from the type species in being macropterous. Mockford (1993a) also supported Roesler's separation of *Lepolepis* s.str. into a distinct subfamily, as here, and as followed by Lienhard & Smithers (2002). In reconsidering the composition of Perientominae, Mockford (2005) suggested that *Parasoa*, formerly included in that subfamily, was sufficiently distinctive to warrant erection of a further subfamily (Parasoinae) to contain it.

Key to the genera of Lepidopsocidae

The following key to all regional genera of Lepidopsocidae indicates subfamily divisions as above, but is of more practical use for these fragile insects in which the antennae may frequently be damaged and scales washed off in liquid preservative. The key characters should be applied cautiously to extralimital taxa. *Thylacella* Enderlein (the only extant genus of Thylacellinae) is omitted, as it is known predominantly from Africa.

1 Antennae with no more than 30 segments, these about four times as long as wide; hind wing with closed basal cell (Perientominae) ▸ 2
– Antennae as above; hind wing without closed basal cell (wings narrow, 'strap-like', venation reduced to about two main longitudinal veins; wings may be reduced to small lobes) (Parasoinae) ▸ *Parasoa*
– Antennae with 30-50 segments, these about twice as long as wide. Wings sometimes reduced; hind wing (if present) without a closed basal cell ▸ *6*

2 Fore wing rounded at apex; broad (vein Sc not broken; hind wing vein R1 arising distal to M1); lateral ocelli close together – closer than distance from ocellus to nearest compound eye margin ▸ *Soa*
– Fore wing pointed at apex; lateral ocelli well separated by more than the distance between each ocellus and nearest compound eye ▸ 3

3 Fore wing terminating as a distinct narrowed and tapered 'tail' ▸ *Papuapsocus*
– Fore wing usually tapered, but never with distinct 'tail' as above ▸ 4

4 Hind wing vein R1 arising distal to M1 (fore wing with vein Sc broken; hind wing veins M1 and M2 arising independently) ▸ *Nepticulomima*
– Hind wing vein R1 arising basal to M1 ▸ 5

5 Fore wing veins R1 and Rs fused for a short length or meeting at a point ▸ *Perientomum*
– Fore wing veins R1 and Rs joined by a crossvein ▸ *Lepium**

6 Fore wing without veins, strongly reduced (antennae with about 50 segments) (Lepolepidinae) ▸ *Lepolepis**
– Fore wing with discernible veins, usually developed normally but if reduced veins still visible ▸ 7

7 Fore wing vein Rs branched. Hind wing usually developed (Lepidopsocinae) ▸ 8
– Fore wing vein Rs simple. Hind wing absent or greatly reduced (Echinopsocinae) ▸ *Echinopsocus*

8 Fore wing vein M 3-branched. Usually macropterous. Ocelli present ▸ 9
– Fore wing vein M 2-branched. Fore wing reduced. Ocelli absent ▸ 10

9 Fore wing with veins Rs and R1 fused for a length (hind wing veins M1 and M2 separate or arising from a common stem) ▸ *Lepidopsocus*
– Fore wing with veins Rs and R1 linked by crossvein; this sometimes missing, so Rs free ▸ *Echmepteryx* s.l.*

10 Fore wing reduced; veins clearly discernible ▸ *Cyptophania*
– Fore wing reduced, veins hardly visible ▸ *Echmepteryx* s.l.*

* *Lepium* and *Lepolepis* are included because of incidence in the eastern Palaearctic, but have not been reported from tropical South-east Asia. *Echmepteryx* (below) is a complex of taxa which some workers treat as distinct subgenera.

Generic information

Cyptophania Banks (figs. 75-76). Type species *Cyptophania hirsuta* Banks.
Generic synonymy: *Pteroxaniella* Karny, type species *Pteroxaniella bifurcata* Karny; *Ptenocorium* Enderlein, type species *Ptenocorium alutaceum* Enderlein. Roesler (1944) treated *Pteroxaniella* as a synonym of *Cyptophania* and *Ptenocorium* as a subgenus; the latter was synonymised by Thornton (1981f).

Generic distribution: Mexico, Hawaiian Islands, western Pacific islands, Seychelles, Christmas Island (Australia).
The genus is included here on the basis of records of the widely distributed *C. marginata* Thornton, Lee & Chui from Christmas Island (Smithers, 1995a). It may well occur also in parts of Indonesia and elsewhere in South-east Asia. In some species (including *C. marginata*), the fore wing venation is hardly visible (the hind wing is reduced to a small scale), and careful checking is necessary to discern it. Other species may have more distinct venation.

Echinopsocus Enderlein (fig. 79, from Enderlein, 1903a). Type species *Echinopsocus erinaceus* Enderlein.
Generic distribution: New Guinea, Australia.
Smithers (1972c) noted that this taxon was described (Enderlein, 1903a) from a single specimen and the type species has not been recorded since then. Information is thus based on Enderlein's fore wing illustration, which appears to represent a very distinctive taxon. A second species, *E. grayi* Smithers, is known from Australia (Smithers, 1984b). It differs from *E. erinaceus* in that the fore wing lacks a long apical extension and in having ocelli, so that the original generic definition is expanded to include this taxon.

Echmepteryx Aaron (figs. 77-78, 84). Type species *Amphientomum hageni* Packard.
Generic distribution: North and South America, Europe, Africa, Seychelles, Mauritius, Sri Lanka, Indonesia, Singapore, Australia, New Zealand.
Echmepteryx s.l. is often divided into several subgenera on basis of wing venation features. Some of these subgenera appear to be very localised and the regional species have been referred to only two subgenera: *Echmepteryx* s.str. and *Thylacopsis*. The key below includes all five described subgenera, to facilitate recognition of any other anomalous taxa that may be discovered in South-east Asia.

– **Key to the subgenera of *Echmepteryx***

1 Hind wing veins M1 and M2 arise from common stem
▸ ***Thylacomorpha* Enderlein**
type species *Thylacomorpha symmetrolepis* Enderlein
Seychelles
– Hind wing veins M1 and M2 arise separately ▸ 2

2 Hind wing Rs stem as long as or longer than R2+3 ▸ 3
– Hind wing Rs stem invariably less than half length of R2+3 ▸ 4

3 Fore wing Rs stem at most as long as R4+5 ▸ **Echmepteryx s.str.**

– Fore wing Rs stem longer than R4+5 ▸ ***Thylacopsis* Enderlein**
type species *Thylax mihira* Enderlein
Sri Lanka

4 Fore wing costal cell normal ▸ ***Loxopholia* Enderlein**
type species *Loxopholia pinnula* Enderlein
Seychelles

– Fore wing costal cell strongly sclerotised ▸ ***Oxypsocus* Tillyard**
type species *Oxypsocus hamiltoni* Tillyard
New Zealand

Whereas some workers have routinely allocated species to one or other of these subgenera, others have not done so, and many species have been described as '*Echmepteryx*' without further characterisation. Smithers (1995a) expressed the more widespread opinion that the subgenera are '... unsatisfactorily defined, on the basis of features of venation which are sometimes difficult to see' and noted that some species, particularly those with reduced hind wings, can not be allocated at present. Several species of *Echmepteryx* (*Thylacopsis*) are very widespread: *Echmepteryx* (*T.*) *lunulata* Thornton, Lee & Chui and *E.* (*T.*) *madagascariensis* Kolbe are among the most widespread of all Lepidopsocidae in the tropics, and *Echmepteryx* s.l. is probably the most diverse lepidopsocid genus in the region.

Lepidopsocus Enderlein (figs. 80). Type species *Lepidopsocus nepticulides* Enderlein.
Generic distribution: Seychelles, Singapore, Indonesia, Christmas Island (Australia), Pacific region (Micronesia, Fiji, Tonga, Hawaiian Islands).
After *Echmepteryx* and *Nepticulomima*, *Lepidopsocus* is the most speciose lepidopsocid genus in South-east Asia. Some species have been allocated to the first-named genus in the past. Thus, at least two species common in the region (*L. marmoratus* (Banks), *L. pretiosus* (Banks)) were described in *Echmepteryx*. As in *Echmepteryx*, many species are structurally very similar and have been diagnosed mainly on body markings. Thus, Thornton's (1981a) treatment of the complex Fijian fauna, of some 20 species, indicates well the kinds of variation in head and other markings that occur. Some Indonesian species are very similar to other Pacific region forms.

Nepticulomima Enderlein (figs. 85-86, 93). Type species *Nepticulomima sakuntala* Enderlein.
Generic distribution: North and South America, United Kingdom (introduced), Africa, Seychelles, Sri Lanka, Japan, western Pacific, Malaysia, Singapore, Indonesia, Australia, New Guinea.
N. biroiana (Enderlein) is relatively widespread, with records from Samoa, New Guinea and Indonesia, and *N. hoesemanni* (Enderlein) has been reported from

Africa and Christmas Island, but other species from the region are known from very few specimens and, mostly, from single localities. The genus is very distinctive on wing venation.

Papuapsocus Mockford (fig. 81). Type species *Papuapsocus caudatus* Mockford. Generic distribution: Papua New Guinea. This monobasic genus is distinguished on the shape of the fore wing, which is unique in the family and more resembles that of some Amphientomidae. It appears to be related most closely to *Nepticulomima* on the position of hind wing vein R1.

Parasoa Thornton (figs. 87-88, 95). Type species *Parasoa haploneura* Thornton. Generic distribution: Malaysia, Singapore, Indonesia, Christmas Island (Australia).
This very distinctive genus is known only by the type species, described from the Batu Caves, Malaysia but later found on vegetation on Panjang, Krakatau Islands (Vaughan et al., 1989). The wing shape and reduced venation are very characteristic and some specimens are brachypterous. The male is known only from a damaged individual from Christmas Island (Smithers, 1995d). Until recently, this genus was included in Perientominae, but Mockford (2005) elevated it to a separate subfamily on a combination of the narrow 'strap-like' wings with reduced venation, the lacinia having three elongate slender tines, and the row of small denticles before the main subapical tooth on the pretarsal claws.

Perientomum Hagen (figs. 89-90). Type species *Amphientomum trichopteryx* Hagen. Generic distribution: Sri Lanka, Indian Ocean islands, Thailand, Indonesia, Australia.
The species are associated with leaf litter or tree bark but are mostly uncommon. As in other genera, head pattern has specific value in *Perientomum*.

Soa Enderlein (figs. 91-92, 94). Type species *Soa dahliana* Enderlein.
Generic distribution: North, Central and South America, Europe (introduced), Africa, Sri Lanka, Malaysia, Indonesia, Pacific region, Australia, New Guinea.
The species-level taxonomy of *Soa* is complex, with several records being merely of '*Soa* sp.' to reflect individuals that differ only very slightly from described species. *Soa flaviterminata* Enderlein has been distributed widely in commerce and records of the genus from North America and Europe are solely of this species, with habitat records such as 'in ship's hold'.
The genus is very distinctive, with the species dark-winged (dark purple) and the fore wing very broad. In natural habitats, they are captured most frequently by beating dead foliage.

2 Psoquillidae (figs. 101-102, 107-108, 114-117, 119-120)

Diagnosis

Body length usually 1.1-2.0 mm, fore wing length up to ca 1.5 mm. Lacinia with apex bifid; second segment of maxillary palp with conical sensillum; labial palpi 2-segmented. Antennae with many segments, not secondarily annulated. Psoquillidae are frequently brachypterous, as in Trogiidae. Wings and body without scales. Fore wing with apex broadly rounded. Pterostigma not thickened; fore wing vein M3 arises in basal half of wing, so M1 and M2 have a long common stem; areola postica long and triangular; veins Cu2 and 1A reach wing margin separately. Hind wing vein M simple. Tarsi 3-segmented; claws without subapical tooth; pulvillus moderately broad, expanded at apex. Paraprocts with strong spine; trichobothrial field not defined but usually some setae with basal rosettes present. Female: subgenital plate simple, rounded; gonapophyses strongly reduced; ventral valve absent, dorsal valve strongly reduced or absent, external valve setose; spermatheca with accessory bodies. Male: phallosome with parameres diverging anteriorly; aedeagus complex.

Remarks

Psoquillidae, a small family, are associated with domestic environments, as well as occurring in leaf litter and on dead foliage in many parts of the world.
Three genera are represented in the region, two being widespread and the third (*Eosilla*) relatively poorly known. All contain few species.

Key to the genera of Psoquillidae

1 Fore wing short and elytriform; veins scarcely visible. Hind wing without vein R1 ▸ ***Eosilla***

– Fore wing not elytriform, although sometimes shortened; veins clearly visible. Hind wing, if present, with vein R1 ▸ 2

2 Fore wing stem of vein Cu1 almost as long as, or longer than, vein Cu1b (wings strongly patterned in black and white) ▸ ***Psoquilla***

– Fore wing stem of vein Cu1 about one third length of Cu1b (wings not strongly patterned in black and white) ▸ ***Rhyopsocus***

Generic information

Eosilla Ribaga (fig. 101). Type species *Eosilla jacobsoni* Ribaga. Generic distribution: Seychelles, Madagascar, Indonesia.
The two species, one described from the Seychelles and the other from Java (Indonesia), are sometimes allocated to separate subgenera as E. (*Empheriella*)

denervosa Enderlein (Seychelles) and *E.* (*Eosilla*) *jacobsoni* (Java). *Eosilla* appears to be very distinctive on wing venation and to be uncommon.

Psoquilla Hagen (figs. 114-117, 119-120). Type species *Psoquilla marginepunctata* Hagen.
Generic synonymy: *Heteropsocus* Verrill, type species *Heteropsocus dispar* Verrill, a synonym of *P. marginepunctata*.
Generic distribution: Hawaiian Islands, North, Central and South America, Europe, Africa, China, Taiwan, Hong Kong, Malaysia, Indonesia, Australia. Distributed widely in commerce.
Psoquilla marginepunctata is one of only two species in the genus and is one of the most widely distributed of all Psocoptera. It has a very distinctive 'black and white' appearance. Often associated with domestic environments, this species occurs also on and under tree bark. Its natural range is unclear, but it has been captured in remote areas of Brazil. Many records, especially from Europe and North America, almost certainly represent localised introductions, as Smithers (1996a) suggested might also be the case for Australia.
The biology of this polymorphic species was described by Broadhead (1961) and Lee (1965, 1987). Brachypterous individuals have the fore wing only about half the length of the body. The other species, *P. infuscata* Badonnel, is known only from Africa.

Rhyopsocus Hagen (figs. 102, 107-108). Type species *Rhyopsocus eclipticus* Hagen.
Generic synonymy: *Deipnopsocus* Enderlein, type species *Deipnopsocus spheciophilus* Enderlein; *Rhyopsocopsis* Pearman, type species *Rhyopsocopsis peregrina* Pearman. These genera were considered subgenera of *Rhyopsocus* by Roesler (1944) but later synonymised by Badonnel (1949a, *Deipnopsocus*) and Mockford & Gurney (1956, *Rhyopsocopsis*), respectively.
Generic distribution: widespread, North and South America, Africa, Micronesia, Christmas Island (Australia).
Rhyopsocus is somewhat elusive and rarely reported from natural environments. It is included here on the record of one individual of the widespread Pacific species *R. pandanicola* Thornton, Lee & Chui from Christmas Island, and so to suggest its incidence in Indonesia and, possibly, elsewhere in the region. Brachypterous individuals of *Rhyopsocus* have been reported.

3 Trogiidae (figs. 96-100)

Diagnosis

Lacinia with apex divided into few large teeth. Antennae with many segments, without secondary annulations. Maxillary palpi with conical sensillum on second segment; labial palpi 2-segmented. Wings absent or strongly reduced. Body and wings without scales. Tarsi 3-segmented; tarsal claws smooth, pulvillus moderately broad. Female: gonapophyses strongly reduced; ventral valve absent; external valve elongate; dorsal valve reduced to a small remnant or absent; spermatheca with complex accessory inclusions. Male: parameres divergent anteriorly; aedeagus complex.

Remarks

Trogiidae includes a variety of small (body length in range of 1.6-2.5 mm) flightless psocids, with shortened fore wings (figs. 96-97) and no hind wings, or with both wings absent. The major genera are widespread, even cosmopolitan, reflecting that they have been transported widely in commerce and their common association with stored products and human habitats. They otherwise occur on tree bark, in leaf litter and in ground habitats such as on and under stones. Because of their economic importance, some taxa have been studied in considerable detail. Generic synonymy is complex, indicating the multiple descriptions which some taxa have received. However, the family is not particularly diverse, with most of the five extant genera having only few species. The family is well-defined and apparently not closely related to Psoquillidae (p. 67). Trogiidae is divided into two subfamilies. Empheriinae is entirely fossil, so that all extant genera belong to Trogiinae. Three of the five genera have been reported from South-east Asia.

Key to the genera of Trogiidae (partly after Lienhard, 1998a)

1 Apical segment of maxillary palp elongate, more than twice as long as wide ▸ *Lepinotus*

– Apical segment of maxillary palp broad, less than twice as long as wide and often strongly broadened to appear 'hatchet-shaped' ▸ 2

2 Hind tibia with two apical spines, no preapical spurs. Fore wing a reduced scale-like structure. (Male without distinct hypandrial brush) ▸ *Trogium*

– Hind tibia with 2-4 apical spines and 1-6 preapical spurs. Fore wing absent or represented by a reduced scale-like structure or small setose lobe. (Male with hypandrial brush) ▸ *Cerobasis*

Generic information

Cerobasis Kolbe. Type species *Cerobasis muraria* Kolbe.
Generic synonymy: *Hyperetes* Kolbe, type species *Hyperetes guestfalicus* Kolbe; *Tichobia* Kolbe, type species *Tichobia alternans* Kolbe; *Albardia* Jacobson & Bianchi, type species *Tichobia alternans* Kolbe; *Myopsocnema* Enderlein, type species *Clothilla annulata* Hagen; *Zlinia* Obr, type species *Zlinia multispinosa* Obr. The above type species of *Hyperetes* and *Tichobia* are synonyms of *Cerobasis guestfalica* (Kolbe).

Generic distribution: North and South America, Europe, Africa, Indonesia, Australia.
Cerobasis guestfalica is charactised by an 'anchor mark' on the frons. It occurs on trees and in leaf litter. It is the only species of this genus reported from South-east Asia (Indonesia: Vaughan et al., 1991).

Lepinotus Heyden (figs. 96-98). Type species *Lepinotus inquilinus* Heyden.
Generic synonymy: *Paradoxides* Motschulsky, type species *Paradoxides psocoides* Motschulsky; *Paradoxenus* Motschulsky, type species *Paradoxides psocoides* Motschulsky; *Cuixa* Navás, type species *Cuixa canaria* Navás; *Heterolepinotus* Obr, type species *Heterolepinotus quadrispinosus* Obr. *Paradoxenus psocoides* and *Cuixa canaria* are both synonyms of *L. inquilinus*. *Heterolepinotus* was treated as a subgenus of *Lepinotus* by Badonnel (1955) but was synonymised with *Lepinotus* by Günther (1974a) and the type species synonymised with *L. patruelis* Pearman by Lienhard (1998a).
Generic distribution: cosmopolitan (domestic species), other species in Europe and Asia. Specific records from tropical South-east Asia are absent, but it is almost certain that the genus will be found in domestic environments in this large area.
Lepinotus are sometimes referred to as 'booklice' when found in domestic environments, although this name is more commonly applied to *Liposcelis*. *Lepinotus* resembles *Cerobasis* in many features, but is separable easily on the form of the maxillary palp. The complex spermatheca of *Lepinotus* spp. is of value for differentiating species.

Trogium Illiger (figs. 99-100). Type species *Termes pulsatorium* Linnaeus.
Generic synonymy: *Atropos* Leach, type species *Termes lignarium* de Geer; *Clothilla* Westwood, type species *Clothilla studiosa* Westwood. Both type species are synonyms of *Trogium pulsatorium*.
Generic distribution: widespread, perhaps cosmopolitan.
Trogium is a small genus and the predominant species (*T. pulsatorium*) is 'domestic'. It has been distributed very widely by human activity and has been reared from Hymenoptera nests and found on trees. Lienhard (1998a) noted it from dead palm foliage and in the 'herb layer', on dead grasses and similar habitats, in Europe. Its area of origin is unknown. This psocid gained early notoriety as the 'death watch' in Europe, by production of audible noises by tapping its abdomen on the substrate. Despite the presumption that it is virtually cosmopolitan, *T. pulsatorium* appears not to have been specifically reported from tropical South-east Asia. As with *Lepinotus*, above, it is almost certain to occur there.

Group PSOCATHROPETAE (as 'PSOCATROPETAE' in many early accounts)

1 Prionoglarididae (as Prionoglaridae in many accounts) (figs. 121-123)

Diagnosis

Head elongate with flat or bulbous postclypeus. Mandibles sometimes greatly extended in incisor region. Lacinia in such taxa reduced to small inconspicuous remnant. Tarsi 3-segmented; pretarsal claws sometimes asymmetrical with anterior claw of each pair having a membranous extension. Macropterous. Fore wing vein Sc strongly developed and fused distally with R1 to form a closed loop; pterostigma scarcely thickened, a crossvein from near its base to Rs; areola postica large; vein 2A present or absent. Paraproct without spine. Female: gonapophyses reduced to broad setose external valve. Male: phallosome unique in order as a large sac with short, posterolaterally directed processes (? outer parameres) and more distal sclerites.

Remarks

Some members of this small family have long attracted attention because of the unique metamorphosis of mouthparts they undergo at the final moult, as shown by *Prionoglaris* Enderlein. At that stage the nymphal mouthparts of normal lacinia and mandibles are transformed to the adult complement of the lacinia being represented by only a tiny remnant, and large sickle-shaped mandibles (fig. 123). The wing venation is characterised by a strongly arched basal length of vein Sc in the fore wing, joining R1 near the base of the pterostigma (fig. 121). Male genitalia, where known, are also unique, in the phallosome consisting of a plate with a distinct spiculate process and lateral 'palpiform' lobes, whose function and homology are unclear.

The family has a highly disjunct distribution, with the four genera found on separate continents. All are variously associated with caves, live under stones or on dead palm foliage. The non-Oriental genera are *Prionoglaris* Enderlein (type species *Prionoglaris stygia* Enderlein, western Palaearctic, three species), *Speleketor* Gurney (type species *Speleketor flocki* Gurney, North America, three species) and *Sensitibilla* Lienhard (type species *Sensitibilla strinatii* Lienhard, Namibia). The systematics of the family was discussed by Lienhard (2000b, 2004a), who emphasised that all species are known from few populations and are regarded as rare, although probably not confined to the sites from which they are currently known. The highly disjunct range may indicate that Prionoglarididae is a relatively old group of Psocoptera. The striking Asian genus was described from Thailand, as a notable addition to the regional psocid fauna.

Generic information

Siamoglaris Lienhard (figs. 121-123). Type species *Siamoglaris zebrina* Lienhard.
Generic distribution: reported only from the type locality, a dry stream bed near Wang Badan cave, Central Thailand.
This genus is known only from the holotype male and appears to be related to *Prionoglaris* (with which it was allocated to the subfamily Prionoglaridinae by Lienhard (2004a)). The two genera are united by the highly enlarged adult mandibles, strongly reduced lacinia and asymmetrical pretarsal claws, all features differing from the conditions in the other subfamily, Speleketorinae. *Siamoglaris zebrina* is a relatively large psocid (fore wing length 3.4 mm) and the specific name refers to the conspicuous stripes across the compound eyes.

2 Psyllipsocidae (figs. 103-106, 109-113, 118)

Diagnosis

Body length ca 1.3-2.0 mm, fore wing length (macropters) to about 2.5 mm. Head elongate. Lacinia narrow, apex divided into few large teeth; maxillary palpi without sensillum on second segment, but a strong bristle sometimes present. Labial palpi 2-segmented. Antennae long, many segments, without secondary annulations. Tarsi 3-segmented; claws with subapical tooth, pulvillus absent. Fore wing: pterostigma not thickened; R1-Rs crossvein sometimes present; areola postica long; veins Cu2 and 1A meet as nodulus on wing margin. Hind wing vein M 2-branched. Brachyptery and microptery common. Paraproct with strong spine; trichobothrial field not defined, but some setae with basal rosettes. Female: gonapophyses reduced; ventral valve rudimentary; dorsal valve small, often membranous; external valve broad, setose. Male: phallosome with parameres divergent apically; aedeagus complex.

Remarks

Until recently, this was the only family of Psocathropetae reported from South-east Asia. The other, Prionoglarididae, was then known by a few peculiar cave-frequenting psocids, one being from Afghanistan and others from North America, Namibia and Europe. Psyllipsocidae are associated mainly with domestic environments and, as with Trogiidae and Psoquillidae, appear to have been distributed widely by human agency. The two genera of Psyllipsocidae from tropical South-east Asia are both widespread. Both are associated with human habitations, have apparently been dispersed widely by commerce, and contain few species. The few other genera of the family have been reported from domestic habitats or similar environments (such as caves) and also contain few species.

Key to the genera of Psyllipsocidae

1 Micropterous, with small wing lobes, lacking any trace of veins ▸ *Psyllipsocus*
– Brachypterous or macropterous, veins present in fore wing ▸ 2

2 Fore wing margin glabrous, but sometimes a few setae on veins. If brachypterous, a few setae present on reduced hind wing ▸ *Psyllipsocus*
– Fore wing margin setose; fore wing reduced, hind wing absent or reduced to small tapered lobe lacking venation ▸ 3

3 Fore wing lanceolate, around 4-7 × as long as wide, with no more than five veins reaching wing margin; vein M simple ▸ *Dorypteryx*
– Fore wing not lanceolate (not more than 3 × as long as wide) but may be slightly acuminate; at least seven veins reaching wing margin; vein M with 2-3 branches ▸ *Psocathropos*

Generic information

Dorypteryx Aaron (fig. 109). Type species *Dorypterx pallida* Aaron.
Generic synonymy: *Dolopteryx* Smithers, type species *Dolopteryx domestica* Smithers.
Generic distribution: widespread, in domestic situations.
Dorypteryx has only three described species. Although it has not been recorded from South-east Asia, it is included here because of the high likelihood that the very widespread domestic *D. domestica* may be found there.

Psocathropos Ribaga (figs. 103-106). Type species *Psocathropos lachlani* Ribaga.
Generic synonymy: *Psocinella* Banks, type species *Psocinella slossonae* Banks; *Axinopsocus* Enderlein, type species *Axinopsocus microps* Enderlein; *Vulturops* Townsend, type species *Vulturops termitorum* Townsend; *Gambrella* Enderlein, type species *Gambrella pilipennis* Enderlein; *Granthakita* Behura & Dash, type species *Granthakita cuttackae* Behura & Dash. '*Psocatropos*' is an unjustified emendation of the original spelling.
Generic distribution: North, Central and South America, Europe, Africa, India, Hong Kong, Thailand, Indonesia, Papua New Guinea, Australia.
This genus of small short-winged psocids is widely distributed, but species level taxonomy is somewhat confused. Lienhard (1998a) listed *P. microps* (Enderlein) as a synonym of *P. lachlani*. These were treated as distinct species by Smithers (1996a) but the synonymy was recognised by Lienhard & Smithers (2002). Both names occur in regional literature for this small saltatorial psocid of very distinctive appearance.

Psyllipsocus Selys-Longchamps (figs. 110-113, 118). Type species *Psyllipsocus ramburii* Selys-Longchamps.
Generic synonymy: *Nymphopsocus* Enderlein, type species *Nymphopsocus destructor* Enderlein; *Parempheria* Enderlein, type species *Parempheria sauteri* Enderlein; *Ocelloria* Weber, type species *Ocelloria gravinymphia* Weber; *Ocellatoria* Weber, type species *Ocellatoria gravinympha* Weber; *Fita* Navás, type species *Fita vestigator* Navás; *Fabrella* Lacroix, type species *Fabrella convexa* Lacroix. All of the above type species except *P. sauteri* are synonyms of *P. ramburii*; further details are listed by Smithers (1996a) and Lienhard & Smithers (2002).
Generic distribution: North, Central and South America, Europe, Africa, Afghanistan, India, China, Taiwan, Japan, Thailand, Malaysia, Australia, New Zealand.
The extensive generic synonymy listed above reflects the extensive polymorphism of *P. ramburii*, ranging from macropterous through brachypterous to strongly micropterous individuals found in this very widespread psocid. *Psyllipsocus ramburii*, however, has not been reported from tropical South-east Asia, although several other species of *Psyllipsocus* have recorded from the region. One North American population was described as *Dorypteryx hageni* Banks, but was synonymised with *P. ramburii* by Mockford (1993); it is noted here to indicate possible confusion over *Dorypteryx* Aaron, a short-winged genus not recorded from South-east Asia (see above). *Psyllipsocus* occurs in domestic environments, caves, leaf litter and similar places. *Parempheria* was raised for species from Japan and Taiwan, and those species are still regarded as distinct, despite generic synonymy. Lienhard & Smithers (2002) noted the possible subgeneric status of *Parempheria*.

Suborder Troctomorpha

The major groups of this suborder in the region are the often brightly coloured scaly-winged Amphientomidae, the so-called 'electrentomoid psocids' (Mockford, 1967b), which are more diverse in the New World, and the true booklice and their close allies.

Key to the groups of suborder Troctomorpha

1 Antennae 11-15-segmented. Head long; ocelli widely spaced; maxillary palpi sometimes with conical sensillum on second segment. Body and wings bare or covered with scales, macropterous or brachypterous. Wings held in tent-shape over the abdomen; nodulus present. Fore legs with femora sometimes with 'comb' of short teeth; pretarsal claws often with more than one subapical tooth
▸ **Group AMPHIENTOMETAE**
(Amphientomidae, Troctopsocidae, [Musapsocidae, Electrentomidae, Compsocidae, Protroctopsocidae])

– Antennae usually 15-segmented, sometimes 14-segmented, very rarely with only 9 or 10 segments. Labial palpi with one segment; maxillary palpi without conical sensillum on second segment. Scales never present on body or wings. Commonly apterous, more rarely macropterous; wings without nodulus, carried horizontally over abdomen. Pretarsal claws with one subapical tooth or (*Belaphopsocus*) lacking tooth. Paraprocts without apical spine and lacking trichobothria. Female: gonapophyses small, glabrous
▸ **Group NANOPSOCETAE**
(Liposcelididae, Pachytroctidae, [Sphaeropsocidae])

Group AMPHIENTOMETAE

1 **Amphientomidae** (figs. 124-146)

Diagnosis

Many are relatively large psocids, with fore wing lengths ranging from ca 2.3 – more than 5 mm. Head usually long, with ocelli distinct. Antennae 15-segmented, with secondary annulations. Lacinial apex divided into small inner and larger outer tooth; maxillary palpi sometimes with conical sensillum on second segment; labial palpi 1- or 2-segmented. Body and wings with flattened scales, sometimes brightly coloured and patterned. Fore wing with pterostigma not thickened; vein Rs connected to M by crossvein; areola postica usually long;

veins Cu2 and 1A meet on wing margin to form nodulus; two anal veins present. Hind wing sometimes lacking basal section of vein Rs. Tarsi 3-segmented; claws long, with one or more subapical teeth. Fore legs often with femoral comb. Female: subgenital plate broad, rounded, usually with internal T-sclerite; gonapophyses complete: ventral and dorsal valves pointed; external valve broad, frequently lobed and without setae; spermathecal opening with sclerifications. Male: phallosome with parameres diverging posteriorly from small anterior median plate; aedeagus open distally, with inner membranous structure.

Remarks

Amphientomidae is the largest and most complex family of this group. In having the body and wings covered with flattened scales, many of the smaller species superficially resemble Lepidopsocidae in general appearance. Like Lepidopsocidae, the adults appear like small moths and hold their wings in a relatively shallow 'tent' over the body. Other taxa are larger and some have the wings falcate or otherwise elaborated. Some are brightly coloured, but the covering scales are detached easily in alcohol or other liquid preservative and diagnostic characters may thus be lost. Enderlein's (1906e) colour pictures demonstrate the spectacular appearance and intricate patterning of these insects.

Before Mockford's (1967b) careful revision of the classification of Amphientometae, this group comprised members of two families, Amphientomidae and Plaumanniidae (= Troctopsocidae, below), with the first comprising four subfamilies. With Mockford's reassignment of the Compsocinae to constitute the family Compsocidae, two of the remaining three subfamilies contain only extant scaly-winged species and the third (Electrentominae, now treated as the family Electrentomidae (Lienhard & Smithers, 2002)) is represented by two fossil genera and several genera from the New World or Africa. Both subfamilies of Amphientomidae are well represented in tropical South-east Asia.

Key to the subfamilies and genera of Amphientomidae (after Smithers, 1990a)

1 Hind wing vein M 2-branched ▸ **Tineomorphinae, 2**

– Hind wing vein M simple or hind wings strongly reduced ▸ **Amphientominae, 3**

2 Ocelli sited very close to compound eyes. Fore wing margin not curved or excavated between vein endings; stem of fore wing vein Rs about as long as R2+3 ▸ *Tineomorpha*

– Ocelli separated from compound eyes by more than the ocellar diameter. Fore wing margin curved between veins: stem of fore wing vein Rs much shorter than length of R2+3 (usually about 1/3 this length) ▸ *Cymatopsocus*

3	Ocelli absent ▸	4
–	Three ocelli present ▸	5
4	Fore wing veins R1 and distal portion of Sc close together ▸	*Stigmatopathus*
–	Fore wing veins R1 and distal portion of Sc divergent ▸	*Pseudoseopsis*
5	Tarsal claws with one subapical tooth ▸	6
–	Tarsal claws with two subapical teeth ▸	7
6	Fore wing with distal section of vein Sc present ▸	*Seopsis**
–	Fore wing with distal section of vein Sc absent ▸	*Stimulopalpus**
7	Fore wing apex strongly acuminate ▸	*Syllysis*
–	Fore wing apex not strongly acuminate ▸	*Paramphientomum*

* See above: placed in Compressioninae by Li Fasheng (2002).

Generic information

AMPHIENTOMINAE

Paramphientomum Enderlein (figs. 124-125, 131-133). Type species *Paramphientomum nietneri* Enderlein.
Generic distribution: Sri Lanka, China, Taiwan, Japan, Singapore, Indonesia.
Roesler (1944) treated *Paramphientomum* and *Hormocoria* Enderlein (type species *Hormocoria tristrigata* Enderlein) as subgenera of *Paramphientomum* and this division is recognised by Lienhard & Smithers (2002), although the precise status of these taxa has not been clarified convincingly. Both putative subgenera have been recorded from Indonesia (Java), as *P. (P.) nietneri* Enderlein and *P. (H.) tristrigatum* (Enderlein). *Paramphientomum* resembles *Amphientomum* Pictet closely, but differs from it in (1) hind wing vein R1 not reaching the wing margin and (2) the distal section of vein Sc absent.

Pseudoseopsis Badonnel (figs. 134-135). Type species *Pseudoseopsis vilhenai* Badonnel.
Generic distribution: Venezuela, Angola, Bhutan, Thailand.
This small genus is included here on the basis of *P. elegans* New from Thailand, but is not otherwise known from the region.

Seopsis Enderlein (figs. 128-129). Type species *Seopsis vasantasena* Enderlein.
Generic distribution: Africa, Nepal, Sri Lanka, Hong Kong, Philippines, Indonesia, Australia.
The relatively small character of the form of the pretarsal claw (with one rather than two subapical teeth) remains the most reliable feature by which to

separate *Seopsis* and *Paramphientomum*, as the species are otherwise very similar in general appearance.
Some species raised in *Seopsis* from Indonesia were transferred to *Stimulopalpus* Enderlein (below) by Smithers (1995d).

Stigmatopathus Enderlein (figs. 126-127). Type species *Stigmatopathus horvathi* Enderlein.
Generic distribution: Malaysia, Indonesia.
This genus is still known only by the type species, which appears to be rare, but possibly widely distributed. Lack of ocelli and of fore wing vein 2A is a distinctive diagnostic character combination for this genus.

Stimulopalpus Enderlein (fig. 130). Type species *Stimulopalpus japonicus* Enderlein.
Generic distribution: Africa, Japan, Singapore, Christmas Island (Australia), Indonesia.
Species of *Stimulopalpus* have been recorded from rock surfaces and beaten from vegetation. Some are very similar to *Seopsis* and Smithers (1995a) discussed the differences between these genera in transferring an Indonesian species (*S. brunneus* (New)) from *Seopsis*. *Stimulopalpus* (1) lacks a comb of spines on the fore femur, (2) has two rather than three ocelli, and (3) lacks the distal section of fore wing vein Sc.

Syllysis Hagen (figs. 136-137). Type species *Amphientomum caudatum* Hagen.
Generic distribution: South America (Paraguay), Sri Lanka, Indonesia.
The anomalous species from Paraguay was described as *Colposeopsis sinipennis* by Enderlein (1910a), but *Colposeopsis* was treated as a subgenus of *Syllysis* by Roesler (1944), with the four other species placed in subgenus *Syllysis*. The species are not well known but the extended apex of the fore wing is characteristic, together with other venational features: distal section of vein Sc absent, vein 2A present, the hind wing vein Sc ends in the wing membrane, and R1 is absent.

TINEOMORPHINAE

Cymatopsocus Enderlein (figs. 138-139, 142, 144, 146). Type species *Cymatopsocus opalinus* Enderlein.
Generic distribution: Malaysia, Singapore.
This genus is apparently rare and Smithers (1972c) noted that it had not then been recorded since it was described (Enderlein, 1903a). Additional species were described by Lee (1974) and New (1975d). The elaborately shaped fore wing and venation are very distinctive.

Tineomorpha Enderlein (figs. 140-141, 143, 145). Type species *Tineomorpha greeniana* Enderlein.
Generic distribution: Angola, Sri Lanka, Singapore, Indonesia.
Tineomorpha is also distinctive but apparently more widely distributed than *Cymatopsocus*. Two species are known from Singapore and only one each from the other regions noted above. Species of *Tineomorpha* are generally rather rare and have been collected in only small numbers.

2 Troctopsocidae (figs. 147-163)

Diagnosis

Fore wing length of macropterous species ca 1.4-4.1 mm. Lacinial apex bicuspid, with median cusp divided and lateral cusps with two or more prominent teeth. Frontal suture absent or weakly defined. Antennae with 11 or 13 segments. Tarsi 3-segmented; claws with subapical tooth and other ventral ornamentation. Usually macropterous, rarely brachypterous with fore wing elytriform and hind wing reduced to small lobe. If macropterous, fore wing pterostigma open or closed basally; areola postica free or linked to M by cross-vein. Hind wing with first section of vein Rs present or absent. Female: subgenital plate rounded, with or without internal 'T-sclerite' near apex; gonapophyses with external valve broad, bilobed, lacking setae; spermatheca sometimes complex. Male: phallosome without external parameres, arms separated posteriorly and connected (sometimes tenuously) by membrane.

Remarks

This family is predominantly Neotropical, but has been reported also from Europe (*Chelyopsocus* Lienhard, *Philedaphia* Lienhard). However, the European taxa are now placed in Protroctopsocidae (see below), leaving true Troctopsocidae only from the neotropics and South-East Asia. It is included here on records of three regional genera, including the remarkable beetle-like genus *Coleotroctellus* Lienhard from South-east Asia. Because of its cryptic habits of living in leaf litter and its very small size (body length only around 1.5 mm), *Coleotroctellus* may have been overlooked elsewhere in the region – as, indeed, may other members of the family. One of the group of 'electrentomoid psocids' discussed by Mockford (1967b), together with his Musapsocidae, Manicapsocidae (now placed in Electrentomidae) and Compsocidae, troctopsocids (s.l.) are the most widely distributed family in this series, with the other families being restricted to the New World (other than for *Manicapsocus* Smithers, Zimbabwe), predominantly in the tropics.
The diagnosis (above) is based on Mockford (1967b), Smithers (1972c) and Lienhard (1988b), with the last of these accounts being particularly important for the regional fauna. Smithers (1972c) divided the family into two subfamilies, Protroctopsocinae and Troctopsocinae, which were discussed further by

Lienhard (1988b). The subfamily Protroctopsocinae was elevated to family status (Protroctopsocidae) by Lienhard & Smithers (2002) and does not occur in the region. All regional genera have fewer than fifteen antennal segments.

The first known species of *Thaipsocus* was previously included within *Troctopsoculus* Mockford. Initially, Lienhard (1988b) also differentiated genera on the basis of having (*Coleotroctellus*) or lacking (*Troctopsoculus*) strong marginal setae on the subgenital plate, but the later discovery of a purported *Troctopsoculus* with such setae (Lienhard, 1990b) invalidated this distinction based on neotropical members of the genus. The three regional genera are closely related, sharing the apomorphy of reduction of the first section of hind wing vein Rs (Lienhard, 1990b). Lienhard & Mockford (1997a) recognised that the Oriental taxa are generically distinct from all neotropical taxa and raised the genus *Thaipsocus* to contain *T. orientalis* (Lienhard) and related species. *Selenopsocus* is very distinctive on the form of the wing and vertex.

Key to the genera of Troctopsocidae

1 Pretarsal claw without lateral pecten. Antenna with 13 segments. Fore wing elytriform in female, normal in male, areola postica free ▸ *Coleotroctellus*

– Pretarsal claw usually with lateral pecten. Antenna with 11 segments. Fore wing normal, areola postica free or linked to M by crossvein ▸ 2

2 Vertex broadly rounded. Fore wing vein Sc ending on radial stem, areola postica free or joined to M3 by crossvein; hind wing vein R1 not fused with R 2+3 ▸ *Thaipsocus*

– Vertex tapered, so head antero-posteriorly flattened to sharp vertical border. Fore wing vein Sc ending on R1; basal part of R1 resembles crossvein, forming angle with radial stem and distal part of R1; areola postica anteriorly concave, due to curvature of first section of vein Cu1a, joined to M by short crossvein; hind wing vein R1 fused with R2+3 (basal part of R2+3 resembling crossvein) ▸ *Selenopsocus*

Generic information

Coleotroctellus Lienhard (figs. 151-154). Type species *Coleotroctellus burckhardti* Lienhard.

Generic synonym: *Sinitroctopsocus* Li Fasheng, type species *Sinitroctopsocus sui* Li Fasheng.

Generic distribution: China, Taiwan (last unpublished), Thailand.

Lienhard & Mockford (1997, following Lienhard, 1988b) showed that this remarkable coleopteriform genus was related to species earlier known as *Troctopsoculus* (now *Thaipsocus*), but differs in the strong sexual dimorphism and the characters noted in the above key. Females of the described species, collected

from plant litter and debris in forests in Thailand and from China, are very similar in general appearance, and the macropterous male (fig. 146) has the base of the pterostigma open and the areola postica free.
Coleotroctellus is much smaller than other known coleopteriform psocids and is not likely to be confused with genera such as *Calocaecilius* (p. 110) and *Calopsocus* (p. 115). However, it may parallel these in participating in some as yet undefined mimicry complex involving small beetles.

Selenopsocus Lienhard & Mockford (figs. 149-150, 156, 158, 160, 162). Type species *Selenopsocus schwendingeri* Lienhard & Mockford.
Generic distribution: Malaysia.
This genus is very distinctive and is known only from the female holotype of the above species. The tapered vertex and unusual wing venation should facilitate recognition. The generic name is derived from the clear 'half moon' pattern on the posterior wing margins.

Thaipsocus Lienhard & Mockford (figs. 147-148, 155, 157, 159, 161, 163). Type species *Thaipsocus siamensis* Lienhard & Mockford.
Generic distribution: Thailand, Sabah and Sarawak (Malaysia), Java.
Only recently distinguished from *Troctopsoculus*, *Thaipsocus* is known from seven described species, all with strongly patterned wings. It appears to be a distinctive regional endemic genus.

Group NANOPSOCETAE

1 **Liposcelididae** (formerly Liposcelidae) (figs. 164-177)

Diagnosis

Body length ca 1-1.5 mm, often around 1 mm; fore wing length (*Embidopsocus*) ca 1.5 mm, males often smaller than females.
Head short, often prognathous; epicranial suture absent or indicated very weakly. Compound eyes reduced (associated with aptery); ocelli often absent. Body may be strongly dorsoventrally flattened. Pronotum divided into three lobes, with median lobe also with median line. In apterous morphs mesonotum and metanotum fused (that is in all males and many females; other females are macropterous); thoracic sternites broad. Lacinia usually bifid, with one or two secondary teeth. Antennae 15-segmented (rarely 9, 10 or 14 segments), with secondary annulations. Tarsi 3-segmented (rarely 2-segmented); hind femur dilated. Wings, if present, elongate and rounded apically, with venation strongly reduced; fore wing with veins R and M unbranched and not reaching wing margin; hind wing with vein R long, not reaching margin. Female: subgenital plate sometimes with small internal T-sclerite; gonapophyses

complete; external valve broad, lobed and without setae. Male: phallosome with inwardly curved parameres, aedeagus complex, extended anteriorly in a narrow process.

Remarks

This family includes the most familiar of all psocids to many people, the domestic 'booklice' belonging to the genus *Liposcelis*, together with a variety of litter-dwelling and subcortical relatives. All are elongate and many are strongly dorsoventrally flattened and apterous. The family is cosmopolitan.

Two of the genera (*Liposcelis*, *Embidopsocus*) reported from the region are both very widespread. However, very few records have been published from South-east Asia and it is virtually certain that many more taxa will be found there with more detailed taxonomic investigation. Species level taxonomy in the family is complex and confused. Other regional genera are much less speciose and are more elusive, also with few records.

Key to the subfamilies and genera of Liposcelididae

1 Hind femur with anterior 'boss' on margin near base; female: subgenital plate with T-sclerite; both sexes apterous, with 2-8 ommatidia in place of compound eye ▸ **Liposcelidinae**, ***Liposcelis***

– Hind femur without anterior 'boss' on margin ▸ **Embidopsocinae 2**

2 Tarsi 2-segmented ▸ ***Belaphopsocus***

– Tarsi 3-segmented ▸ 3

3 Apical segment of maxillary palp normal, not conspicuously expanded ▸ ***Embidopsocus***

– Apical segment of maxillary palp strongly expanded, but not as wide as long ▸ ***Belaphotroctes***

Generic information

Belaphotroctes Roesler. Type species *Eutroctes traegardhi* Ribaga.

Generic distribution: North, Central and South America, Africa, Thailand.

Lienhard (1991) noted that *Belaphotroctes* is known from 'unpublished material' from the Oriental region. According to Lienhard (in litt.) this material was published as '*Belapha* sp.' based on specimens from Thailand (Chiang Dao cave) (Deharveng & Leclerc, 1989; Deharveng & Bedos, 2000). Dr Lienhard has determined this material as *Belaphotroctes ghesquieri* Badonnel, which is widespread in southern Africa and America and sometimes occurs as a pest in food products (Mockford, 1991a). *Belaphotroctes* has 3-segmented tarsi and the apical segment of the maxillary palp is expanded to be considerably wider than the more basal segments, but not as wide as long.

Belaphopsocus Badonnel (figs. 166–167). Type species *Belaphopsocus vilhenai* Badonnel.
Generic distribution: Colombia, Brazil, Paraguay, Africa, Singapore.
'*Belaphapsocus*' is an incorrect subsequent spelling (New, 1971e).
Belaphopsocus murphyi Lienhard (1991) was described from both sexes collected on Sentosa Island, Singapore, and may well occur elsewhere in the region. The generic distribution is apparently highly disjunct. The genus is easily recognisable by the 2-segmented tarsi (as with the reduced number of antennal segments, a likely neotenic feature) and the greatly expanded apical segment of the maxillary palp.

Embidopsocus Hagen (figs. 164–165). Type species *Embidopsocus luteus* Hagen.
Generic synonymy: *Tropusia* Hagen, type species *Atropos oleagina* Hagen; *Stenotroctes* Enderlein, type species *Troctes needhami* Enderlein; *Embidotroctes* Enderlein, type species *Embidotroctes paradoxus* Enderlein; *Trigonosceliscus* Enderlein, type species *Trigonosceliscus leucomelas* Enderlein.
Generic distribution: North, Central and South America, Europe, Africa, India, Sri Lanka, Philippines, Australia.
Despite the very widespread nature of *Embidopsocus*, as noted above, it has not been recorded from most of South-east Asia and also appears to be scarce in Australia. It is not clear whether this simply reflects lack of collecting. Because of its subcortical habit, *Embidopsocus* is not usually captured by beating vegetation and thus may have been overlooked by non-specialist collectors. *Embidopsocus* and its close relatives appear to be particularly diverse in South America (Badonnel, 1973a).

Liposcelis Motschulsky (figs. 168–177). Type species *Liposcelis brunnea* Motschulsky.
Less precise terminology in the past has led to species of this genus being named in *Trogium* auctt., *Atropos* auctt. and *Troctes* auctt., with the last of these sometimes favoured as the correct generic name. See notes by Smithers (1967a) and Lienhard & Smithers (2002).
Generic distribution: cosmopolitan.
The taxonomy of *Liposcelis* is very complex, but one nomenclatural aspect merits particular comment. In the past the name *L. divinatorius* Mueller, also spelled *L. divinatoria* or *L. divinatorium*, has been used widely for species associated with stored products in many parts of the world. Several workers (including Broadhead, 1950a; Lienhard, 1990a; Smithers, 1996a) have pointed out that it is impossible to recognise that species, because the types have not been traced. The name has been applied uncritically to a variety of different species and much 'general' stored products literature has tended to include reference to any 'typical booklouse' as '*L. divinatorius*'. The name is no longer used by psocid

taxonomists and most of the species can be referred more accurately to one or other alternative described species rather than to this *nomen dubium*.

Many species of *Liposcelis* are separated on small details of colour, and pattern of setae. Very careful examination is needed to identify specimens to specific level. Various attempts have been made to subdivide the genus and four main groups of species (designated A, B, C, D) are commonly acknowledged. Groups A and B constitute 'Section I' of the genus and groups C and D comprise 'Section II' (see Badonnel, 1962a, 1963a, 1967b and Lienhard, 1990a, 1998a, for discussion of these divisions). The following key to these groups is taken from Lienhard (1998a), designed for the western Palaearctic fauna but of much wider general application.

If critical identification of *Liposcelis* to species level is needed, it is recommended that specialist help be sought. Because of the considerable economic importance of this genus as pests in stored products systems, such identification may be critical in effecting optimal control measures. The review of taxonomic characters by Lienhard (1998a) is a very pertinent introduction to the identification of many widely distributed species.

– **Key to species groups of *Liposcelis***

1 Posterior border of abdominal tergites 3-4 without intersegmental membrane ('compact' abdomen, fig. 173); tergites fused, approximately uniform in colour, although the segmentation is visible by interruption in the sculpture pattern; posterior margin of each of tergites 5-7 with a membranous pale band, with individual sculpturing different from that of the anterior region of each tergite. Posterior half of prosternum lacking bristles (Section I) ▸ 2

– Posterior border of each of abdominal tergites 3-7 with a pale membranous band, ornamented by indistinct sculpturing different from that of anterior of each respective tergite ('ringed' or 'annulated' abdomen, fig. 172). (In species with abdominal sculpture consisting of large tubercles, the intersegmental membranes at the posterior of tergites 3-4 are often very straight, sometimes represented only by a clear line, but in all these cases the posterior half of the prosternum has a pair of bristles) (Section II) ▸ 3

2 Lateral lobe of pronotum with one or more strong pronotal bristles, with blunt tips and normally scarcely smaller than the humeral bristles, and forming with the humeral bristle a row across the anterior border of the pronotum; behind these there may be several short hairs (fig. 168) ▸ **Group A**

– Lateral lobe of pronotum with only a few small hairs behind the humeral bristle (fig. 169) ▸ **Group B**

3 All prosternal bristles inserted on the anterior half of the prosternum (fig. 170) ▸ **Group C**

– Prosternum with a pair of lateral bristles on the posterior half, more posteriorly than several anterior bristles (fig. 171) ▸ **Group D**

In the absence of specialist regional treatment, it is currently difficult to estimate which species of *Liposcelis* are present in tropical South-east Asia, not least because the few specific records from the region have not been verified. Lienhard's (1998a) keys for western Palaearctic taxa and Mockford's (1993a) treatment of North American species both include many widespread domestic species, in addition to more localised taxa found in leaf litter and under bark, and are the most valuable initial sources of information relevant to the likely regional fauna.

Males are often much rarer than females and, in many species, are unknown, so that published keys are most reliable for female *Liposcelis*. However, it is likely that representatives of all the above species groups occur in the region.

2 Pachytroctidae (figs. 178-188)

Diagnosis

Body length ca 1.4-1.8 mm, fore wing length to ca 1.6 mm. Body not dorsoventrally flattened. Lacinial apex with few teeth. Compound eyes relatively large, even in apterous morphs. Antennae with first 4-5 flagellar segments without secondary annulations. Thoracic sterna narrow; prosternum not lobed; mesonotum and metanotum clearly separate. Hind femur not flattened or otherwise expanded; tarsi 3-segmented. Females macropterous or apterous, males apterous. Wings broad, with apex rounded and veins distinct; pterostigma not thickened; fore wing vein M 2-branched; areola postica long and flat; veins Cu2 and 1A reach wing margin separately. Female: subgenital plate with or without internal T-sclerite; gonapophyses complete, external valve large, not lobed and without setae. Male: phallosome closed anteriorly; parameres flank complex aedeagal structures.

Remarks

Pachytroctidae are small, often apterous psocids commonly associated with leaf litter and domestic environments. They thus occur in similar situations to many Liposcelididae, but are not dorsoventrally flattened and the hind femur is not conspicuously enlarged. Legs and antennae are often relatively long. The general appearance is characteristic, with macropterous individuals holding the broad wings flat over the abdomen. Most species of this family have been referred to either *Pachytroctes* or *Tapinella*, as the two largest genera of the family, but a number of other genera are each represented by one or few species. The most distinctive is the South American *Antilopsocus* Gurney, in which the vertex is ornamented with 'antler-like' projections; in some other genera the vertex is slightly ornamented or medially concave.

The two major genera, *Pachytroctes* and *Tapinella*, are sometimes placed in the subfamilies Pachytroctinae and Tapinellinae respectively, following Enderlein (1905c, 1908d). The integrity of several putative genera is still not wholly clear and *Pachytroctes* has been divided into several subgenera which are regarded by different workers as subgenera or synonyms. These are discussed below. Likewise, *Psylloneura* is sometimes treated as distinct, but Badonnel (1955) noted the possibility that it was a synonym of *Pachytroctes*. *Nanopsocus* is now regarded more generally as valid, but has historically been treated as a subgenus of *Tapinella*.

The genera keyed above include all the Pachytroctidae reported from tropical South-east Asia, although some others may well occur also.

Key to the genera of Pachytroctidae

1 Tarsal claws of each pair differ: one normal, the other 'foliacious' (fig. 188) ▸ *Nanopsocus*
– Tarsal claws of each pair similar in form, normal ▸ 2

2 Macropterous ▸ 3
– Apterous ▸ 4

3 Hind wing with stem of vein Rs much longer than confluent length of R+M ▸ *Tapinella*
– Hind wing with stem of vein Rs about the same length as confluent length of R+M ▸ *Pachytroctes, Psylloneura*

4 Female: subgenital plate with T-sclerite (fig. 184). Male: parameres finely pointed ▸ *Tapinella*
– Female: subgenital plate without T-sclerite. Male: parameres bluntly acuminate ▸ *Pachytroctes*

Generic information

Nanopsocus Pearman (figs. 182-183, 187-188). Type species *Nanopsocus oceanicus* Pearman.

Generic synonymy: *Onychotroctes* Badonnel, type species *Onychotroctes trifasciatus* Badonnel.

Generic distribution: southern United States, tropical America and Africa, Europe, Japan, Malaysia, New Hebrides.

The genus is included here because the type locality (Vanuatu, New Hebrides) is on the regional fringe and because *N. oceanicus* may have been distributed widely in commerce. It appears to be by far the most widely distributed member of this

small genus, which contains five described species. There are no specific recent records, but Mockford (1993a) noted the presence of an undescribed species in Malaysia.
Nanopsocus resembles *Tapinella* in that the female has a T-sclerite on the subgenital plate, and the male clunium has two basal styliform processes. It is regarded as distinct mainly on the very unusual form of the asymmetrical pretarsal claws, with the anterior claw of each pair being highly modified. One African species (*N. trifasciatus* Badonnel) has brachypterous females.

Pachytroctes Enderlein (figs. 180-181, 186). Type species *Pachytroctes aegyptius* Enderlein.
Generic distribution: widespread.
The following have been treated as subgenera, following Roesler (1944): *Psacadium* Enderlein, type species *Psacadium bilimbatum* Enderlein (Taiwan); *Peritroctes* Ribaga, type species *Peritroctes natalensis* Ribaga (South Africa); *Nymphotroctes* Badonnel, type species *Nymphotroctes denisi* Badonnel (Europe); *Neotroctes* Roesler, type species *Pachytroctes brasilianus* Roesler (Brazil). Other workers, including Smithers (1990a) and Mockford (1993a), have treated these as distinct genera. Smithers (1996a) followed others in treating *Psyllotroctes* Roesler (type species *Psyllotroctes plaumanni* Roesler (South America)) as a synonym, although this is doubted by Mockford (1993a) (see also Mockford, 1996a). The regional taxa are treated here as *Pachytroctes* s.l., as the differences between subgenera have not been defined completely and the range of variation in each is largely unknown. *Pachytroctes* s.l. is widespread.

Psylloneura Enderlein Type species *Psylloneura simbangana* Enderlein.
Generic distribution: Uganda, India, Guam, New Guinea.
This genus has been very poorly characterised and Badonnel (1955) suggested that it is also a synonym of *Pachytroctes*, in which case it would take priority as the earlier formalised name. This synonymy has not been undertaken formally, hence listing the genus here as distinct, with regional representation based solely on the type species from New Guinea.

Tapinella Enderlein (figs. 178-179, 184-185). Type species *Tapinella formosana* Enderlein.
Generic distribution: North, Central and South America, Europe, Africa, India, China, Taiwan, Micronesia, Philippines, Indonesia, Australia.
This widespread genus contains many very similar species, some of which have been diagnosed simply on small differences in head and body pattern or markings. Many are also known only from few individuals. The most usual habitats are dead foliage and leaf litter.

Suborder Psocomorpha

Psocomorpha is by far the most diverse of the three suborders and is also the most 'advanced' of them. It includes the great majority of taxa found on vegetation and regarded as typical barklice. Its members are divided amongst four major groups defined as indicated below. These groups differ from more formal superfamilies but have long been used to suggest some degree of alliance between the families within each. They are sometimes regarded as 'infraorders', a convention followed by Lienhard & Smithers (2002).

Major features of the groups of suborder Psocomorpha

1 Head long and vertical, genae long. Labrum with two longitudinal sclerotised ridges on inner side, these often converging anteriorly and fused to anterior margin. Labial palpi short; apex of lacinia often broadened and with several teeth. Tarsi 2- or 3-segmented, claws with subapical tooth. Mostly macropterous, more rarely brachypterous or apterous; fore wing with areola postica present, free; fore wing veins Rs and M usually joined by long crossvein; veins usually setose. Female, gonapophyses reduced ▸ group EPIPSOCETAE (**Epipsocidae [Dolabellopsocidae, Cladiopsocidae, Neurostigmatidae, Ptiloneuridae]**)

– Head normal to short. Labrum without inner sclerotised ridges. Labial palpi broad, often more or less triangular and diverging laterally; lacinia usually narrow apically. Tarsi 2-segmented, claws without subapical tooth, pulvillus broad. Mostly macropterous, rarely brachypterous or apterous. Wing margins and at least some veins usually setose; fore wing areola postica free or connected to M; fore wing veins Rs and M usually fused for a length or meeting at a point. Female: subgenital plate simple, gonapophyses reduced: external valve absent or represented by a small lobe, sometimes with a seta ▸ group CAECILIUSETAE (**Caeciliusidae, Stenopsocidae, Amphipsocidae, [Asiopsocidae]**)

– Head usually short. Labrum without inner sclerotised ridges; most other characters varied and not shared between all families. Unlike the other listed groups of Psocomorpha, which appear to be largely natural groupings of related families, Homilopsocidea was erected (Pearman, 1936) for a series of families that could not be associated easily with members of other groups. As Smithers (1972c) noted 'The result is a heterogeneous assembly of families within the Psocomorpha'. The constitution of Homilopsocidea can thus be defined only at

the family level rather than in more embracing terms
▸ **group HOMILOPSOCIDEA**
(Archipsocidae, Calopsocidae, Ectopsocidae, [Elipsocidae], Lachesillidae, Mesopsocidae, Peripsocidae, Philotarsidae, Pseudocaeciliidae, [Bryopsocidae, Lesneiidae, Sabulopsocidae, Trichopsocidae])

– Head normal to short. Labrum without internal sclerotised ridges. Labial palpi short; lacinia not broadened apically, usually with few teeth. Tarsi 2- or 3-segmented; claws usually with subapical tooth. Most taxa macropterous, venation and wing setation varied; areola postica usually connected to M. Female: gonapophyses usually complete, external valve setose ▸ **group PSOCETAE**
(Hemipsocidae, Myopsocidae, Psilopsocidae, Psocidae)

Group EPIPSOCETAE

1 **Epipsocidae** (figs. 189-192, 194-197)

Diagnosis

Fore wing length mainly in range of 2.5-5.7 mm. Head elongate. Labrum with two sclerotised rods/ridges running entire length of labrum and reaching sides of labrum anteriorly and joined, if at all, by only a weakly sclerotised anterior connection. Antennal scape membranous over much of ventral surface. Tarsi 2-segmented, claws with or without subapical tooth. Anterior and hind femora sometimes with row of cones at bases of setae. Macropterous or brachypterous, females rarely apterous. Fore wing pterostigma without crossveins; areola postica free; macropterous individuals with one anal vein in fore wing. Female: ventral gonapophysis valve present or absent; external valve reduced, represented as small swelling or setal field on dorsal valve.

Remarks

Epipsocidae was formerly considered to include a substantial variety of Psocomorpha with the labrum elongated and bearing a pair of longitudinal thickened ridges (fig. 72). In this sense, as used by many early authors, this broad concept of the family referred most species to *Epipsocus* which, consequently, was accorded a very wide distribution. Major rearrangement of the group by Eertmoed (1973) resulted in recognition of five families within Epipsocetae, three of which do not occur in South-east Asia.

However, even within the subsequent narrowed concept of Epipsocidae, *Epipsocus* has tended to remain as a holding genus with a number of species still *incertae sedis*, in additon to others that are more readily recognisable. In a more restricted sense, true *Epipsocus* appears to be restricted to the American tropics (Mockford, 1998b), with all the South-east Asian and other species formerly placed in it now transferred to other genera. As redefined by Mockford (1998a), the predominant genus in South-east Asia is *Epipsocopsis*.

Some earlier workers recognised subfamilies within Epipsocidae, with the South American *Goja* Navás placed in Gojinae and other genera in Epipsocinae (with *Neurostigma*, before its separation as Neurostigmatidae (Eertmoed, 1973), placed in Neurostigmatinae). No such groups were noted by Mockford (1998a).

Apart from the genera keyed out below, four other genera are known from China or Hong Kong and may occur further south: *Cubitiglabra* Li Fasheng (type species *Cubitiglabra quadripunctata* Li Fasheng) is similar to *Heteroepipsocus* and differs mainly in having supernumerary branches to veins Rs and M in the fore wing, a difference regarded by Mockford (1998a) as of only doubtful generic value (see below). *Hinduipsocus* Badonnel (type species *Hinduipsocus annulipes* Badonnel) is known from India, Nepal and Hong Kong. Males are unknown, females are micropterous. The subgenital plate is extended as a median posterior lobe with two slender apical processes. *Dichoepipsocus* Li Fasheng & Mockford (type species *Dichoepipsocus micropterus* Li Fasheng & Mockford) is known also from a species from Bhutan. A species of the widespread *Bertkauia* Kolbe (type species *Bertkauia prisca* Kolbe) has also been recorded from China.

Key to the genera of Epipsocidae*

1 Fore wing veins Rs and M multi-branched, with at least one branch of each rebranching. (Female: ventral gonapophysis valve absent) ▸ *Dicropsocus*

– Fore wing veins Rs and M normal, any additional veins extremely rare ▸ 2

2 Outer cusp of lacinial apex denticulate, broadened. (Female: ventral gonapophysis valve present) ▸ *Epipsocus*

– Outer cusp of lacinial apex with few or no denticles, may have long acuminate tip ▸ 3

3 Tarsal claws with subapical tooth; hind wing veins Rs and M usually fused for a length, rarely linked by crossvein. Female: ventral gonapophysis valve absent ▸ *Epipsocopsis*

– Tarsal claws without subapical tooth; hind wing veins Rs and M joined at a point or by a short crossvein. Female: ventral gonapophysis valve present ▸ *Heteroepipsocus*

* '*Epipsocus*' is included here to facilitate allocation of anomalous taxa to a 'holding genus' and is not used *sensu stricto*.

Generic information

Dicropsocus Smithers & Thornton (fig. 192). Type species *Dicropsocus montanus* Smithers & Thornton.

Generic distribution: New Guinea, including New Britain.

This genus is apparently closely related to *Epipsocopsis* but is separable from it by lacking the cones on the anterior and hind femora, and by having much more complex wing venation, with multiple branching of fore wing veins Rs and M. The lacinial form occurring in the two genera is similar, and Smithers & Thornton (1977) suggested that *Dicropsocus* 'appears to be an offshoot (of *Epipsocopsis*) which has developed in New Guinea and New Britain'. The three described species are all from that region.

Epipsocopsis Badonnel (figs. 191, 194, 196-197).Type species *Epipsocopsis machadoi* Badonnel. Generic synonymy: *Epipsocidus* Badonnel, type species *Epipsocus (Epipsocidus) angolensis* Badonnel, was regarded as a subgenus of *Epipsocus*, but synonymised with *Epipsocopsis* by Badonnel (1967a).
Generic distribution: Africa, Madagascar, India, Sri Lanka, Thailand, Malaysia, Singapore, Philippines, Indonesia, New Guinea, Solomon Islands, Australia.
Epipsocopsis is very widely distributed in the Old World tropics and is by far the most diverse genus of Epipsocidae found in South-east Asia, from where about 25 species have been described. Some were placed formerly in *Epipsocus*. Many species have the fore wing strongly banded or otherwise marked with brown. The genus was raised originally to contain Epipsocidae, lacking the ventral gonapophysis valve and in which the lacinial tip is not expanded and lacks a row of denticles. These characters are shared with *Dicropsocus* (see above).

Epipsocus Hagen (figs. 189, 190, 195). Type species *Psocus ciliatus* Pictet-Baraban & Hagen.
Generic distribution: Central and South America.
Noted here as a 'holding genus' for unallocated South-east Asian species, notably '*Epipsocus*' *marginatus* Enderlein from New Guinea and '*E. viiv*' (Enderlein) from Indonesia. Any epipsocid with a broad denticulate lacinia, a feature that differentiates it from any other genus included here, should be examined carefully for generic placement, using Mockford (1998a) as a guide for character states. Likewise, micropterous species occur in other parts of the world and any such individuals are highly unlikely to belong to the other genera noted here.

Heteroepipsocus Li Fasheng. Type species *Heteroepipsocus longicellus* Li Fasheng.
Generic distribution: China, ? Philippines, ? Indonesia.
This genus was raised for two species from China and is included here because of the tentative transfer from *Epipsocus* of '*H.*' *inornatus* (Banks) from the Philippines and Indonesia (Mockford, 1998b), following the record from Krakatau, Indonesia, identified as that species by Vaughan et al. (1989). The presence of the ventral gonapophysis valve, a slender non-denticulate lacinia, and veins Rs and M in the hind wing joined by a crossvein, is a sound character combination. Mockford (1998b) commented that *Heteroepipsocus* appears to be closely related to *Cubitiglabra* Li Fasheng, known only from China (females unknown), which

shares the same features of lacinia and hind wing venation. However, *Cubitiglabra* has additional branches to fore wing veins Rs (four branches) and M (six branches).

2 [Dolabellopsocidae] (fig. 193)

Diagnosis

Fore wing length 3.6-4.6 mm. Labrum with sclerotised rods not reaching distal margin of labrum; apex of lacinia narrow, with 1-3 teeth. Tarsi 2-segmented: claws with subapical tooth; pulvillus hooked and bent at an angle. Fore femora with ventral row of short spines on anterior surface. Fore wing with two anal veins; pterostigma long and shallow. Female: ventral gonaphysis valve absent.

Remarks

This family is one of the segregates of Epipsocetae raised by Eertmoed (1973) and is known mainly from Central and South America. It is included here because one genus has been reported from the Oriental region and is sufficiently distinctive to render it an unusual faunal element in the region, and one that may occur elsewhere in South-east Asia. A second genus (*Dimidistriata* Li Fasheng & Mockford, type species *Dimidistriata longicapita* Li Fasheng & Mockford) was described from China, originally in Epipsocidae but later inferred to be in Dolabellopsocidae (Mockford, 1998b).

Generic information

Auroropsocus Eertmoed (fig. 193). Type species *Auroropsocus orientis* Eertmoed.
Generic distribution: India, China.
As noted above, *Auroropsocus*, until recently a monotypic genus but now augmented by several Chinese species (Li Fasheng, 2002), is geographically isolated from most other members of the family. It is also very distinctive, with the fore wing heavily patterned in dark brown and a long setose gonapophysis valve. Fore wing vein Rs is strongly sinuous and the areola postica relatively high. Both these features distinguish *Auroropsocus* from other genera of Dolabellopsocidae. The phallosome is elongate, with distinct outer parameres which do not extend to the apex of the frame.

Group CAECILIUSETAE

1 **Caeciliusidae** (Caeciliidae auctt.) (figs. 198-248)

Diagnosis

Fore wing length within family in range of ca 2.5-4.5 mm, with most species in the lower half of this range. Mandibles usually elongate; galeae large and bulging; lacinial apex varied. Tarsi 2-segmented; claws without subapical tooth, pulvillus broad. Fore wing veins mostly setose, setae short (usually in a single rank) and slanting toward wing apex; veins in distal half of wing usually with setae only on dorsal surface; hind wing margin setose except for basal half of costal edge. Abdomen usually with ventral vesicles (range within family 0-3, usually 2). Female: gonapophyses incomplete, ventral valve often short, slender; external valve absent or reduced to a small lobe (sometimes with long seta) from base of dorsal valve, these structures sometimes fused to form single broad composite external plus dorsal valve.

Remarks

As far as is known, all species of Caeciliusidae are foliage-frequenters and many are associated primarily with living foliage. Most are thus associated with trees, but Mockford (1965b, 1966a, 1969a) showed that some taxa occur particularly on low-growing vegetation such as grasses. Thus, within North American species of the '*Caecilus flavidus*-group', Mockford (1965b) was able to key some species not only on wing length but also on 'occurring primarily on palm foliage' or 'occurring primarily on grasses, sedges and Typhaceae'. Habitat relationships of the South-east Asian taxa are almost wholly unknown other than by sporadic collecting data; most have been captured by beating trees and shrubs and are known from few specimens and localities, so that inferences on habitat 'preferences' are very tentative.

This family until recently was universally known as 'Caeciliidae', with the more recent name being established (Anonymous, 1996) to overcome the junior homonymy of Caeciliidae with a group of amphibians. It has proved one of the most problematical of all psocopteran families to understand, largely because the type genus, *Caecilius* Curtis, has progressively become a depository for a wide range of structurally rather similar taxa from most parts of the world. As a result, *Caecilius* has historically become a 'holding genus' for several hundred species.

Mockford (1989a) presented an outline of a proposed new classification within the family, and this was refined and established in a detailed appraisal (Mockford, 2000, 2001) in which ten new genera were raised and more than 360 new generic combinations established, using a range of characters far more diverse than used by any previous workers. The regional literature therefore includes numerous species described originally in *Caecilius* s.l., but most of these have now been transferred to other genera. Caeciliusidae is cosmopolitan, as was the former concept of *Caecilius*, but more constructive appraisal of generic distributions is now feasible.

Mockford (2000) showed that there has been extensive convergence within the family. For example, distinctive colour patterns have been used to help diagnose certain genera and species groups in the past, but these may have arisen independently in several distinct lineages. Mockford (2000) also noted that some of the characters that indicate relationships of species reliably are cryptic and, as they have been investigated only recently, are absent from most descriptions of new taxa. The great majority of taxa have been described within a broad concept of '*Caecilius*', in which many of the characters now used to diagnose genera are not specifically appraised or noted.

It is also important to appreciate that the family as recognised at present is much more restricted than some historical applications of 'Caeciliidae'. The group was recognised by Kolbe (1880c) as a tribe (Caeciliini), raised to family status by most subsequent workers, with Pearman's (1936a) arrangement much like more modern scope. Roesler's (1944) rearrangement is decidedly less natural. Many of the genera as at present defined have geographical ranges far different from those implied in previous arrangements.

The array of structural characters used in distinguishing taxa within Caeciliusidae is complex and is discussed below (after Mockford, 2000), to indicate the care needed in identifying its members even to subfamily level. Character combinations, rather than single features, are almost always needed and the characters include those of all regions of the body.

Head characters

- The shape of the head capsule can differ in the extent of vertex extension and flattening at the apex and females of genera such as *Dypsocus* and *Protodypsocus* have the vertex more compressed than do the corresponding males;
- ocelli are absent in all Fuelleborniellinae, an African subfamily not yet known from South-east Asia (but noted here because some regional species were described in or attributed to *Fuelleborniella* and transferred later to *Asiocaecilius*), but present in most other taxa;
- the basal two flagellar segments are sometimes thickened and incrassate (fig. 220), in the Asian fauna notably in females of *Dypsocus*;
- the extent of development of the inner clypeal shelf was examined initially by Mockford (2000) and occurs in four states within the family, namely absent, narrow, intermediate and wide (figs. 201-203). It is absent in *Pericaecilius*, narrow in some other Austrocaeciliini and Caeciliusini and in all Epicaeciliini and Maoripsocini, intermediate in *Dypsocus* and wide in most other taxa;
- the labral stylets (figs. 204-205) may be present or absent. They are sometimes very difficult to see, because they can be adpressed against the labrum or be very small;
- the distal inner labral sensilla (figs. 198-200) are always present near the distal margin on the inner surface of the labrum. Mockford (2000) recognised three states, which he designated 'types 1, 2, 3' and which are found respectively in Caeciliusini, some genera of Paracaeciliinae and most other taxa;
- the shape and relative length of the distal segment of the maxillary palp to other segments can also help to diagnose some taxa;

- the tip of the lacinia (figs. 206-213) can be used, with appropriate caution to allow for wear or breakage, to separate some broad taxa of Caeciliusidae, but can be difficult to employ consistently.

Thoracic characters

- Wing venation, particularly of the fore wing, has considerable taxonomic importance. The major taxonomic features noted by Mockford (2000) include the shape of the pterostigma and presence/absence of a spur vein; the shapes of cells r3 and r5 (the latter reflecting the extent of curvature of veins Rs and M), the number of branches of M – usually three, occasionally reduced to two; the presence/absence of vein Cu1a and its relationships to M – usually, a normal free areola postica is present but this may be very small or absent, or linked to M by fusion or a crossvein;
- ciliation of the fore wing, particularly the presence or absence of setae on vein Cu2, and whether one or two ranks of setae are present on veins R and M+Cu1 (rarely, also on Cu2);
- leg characters include the occasional departures from normal cylindrical tibiae of legs 1 and 2, these often being swollen in males of *Valenzuela*. The basal tarsal segment of these legs may be incrassate in *Dypsocus*;
- the presence or absence of the meso-precoxal suture. This is absent in most caeciliusids but present in a few taxa, including some species of *Valenzuela* and *Pericaecilius*.

Abdominal characters

- The number and distribution of ventral abdominal vesicles (p. 18); two are usually present (between segments v-vi, vi-vii), but the number otherwise may be 0, 1 or 3 in different taxa;
- males – three character states are of particular taxonomic relevance:
    - the distribution of setae on the margin of the hypandrium: evenly distributed, as lateral groups only, or lateral plus median groups,
    - the phallosome being closed or open basally, the form of the aedeagal arch (constricted or broad), and the nature of the median endophallus bulb (undivided, partially divided, fully divided),
    - the presence or absence of a papillar field on the epiproct;
- females – the following three character states have taxonomic importance:
    - the form of the subgenital plate, particularly the presence or absence of the lateral apophyses on the margin,
    - the form of the spermatheca, with variations occurring in the width (relative to the duct itself) of the 'neck' region between the bulb and duct, the presence/absence of projections (villi) in the neck region (found only in *Paracaecilius*), and the presence/absence and shape of the sheath surrounding the duct,

- the form of the gonapophyses and presence or absence of a seta on the external valve.

Key to the subfamilies of Caeciliusidae (after Mockford, 2000, modified for regional taxa)

1 Ocelli absent; a single ventral abdominal vesicle; male aedeagal arch with slender apex ▸ **[Fuelleborniellinae]**
(not found in region, but see comment above under head characters)
– Ocelli present (except *Maoripsocopsis*, q.v.); 0, 2 or 3 abdominal vesicles; male aedeagal arch varied ▸ 2

2 Character combination of: three ventral abdominal vesicles; first antennal flagellar segment greatly thickened in both sexes; five distal inner labral sensilla; wings long and slender, with fore wing cell r5 constricted in central region ▸ **[Aphyopsocinae]**
(not found in region, contains only *Aphyopsocus* Smithers, Australia)
– Character combination not as above, features as follows: <u>if</u> three abdominal vesicles, then seven distal labral sensilla present and wings of normal shape with fore wing cell r5 not constricted in central region; <u>if</u> five distal labral sensilla present, then two or no abdominal vesicles, wings of various shapes ▸ 3

3 Two or no ventral abdominal vesicles and five distal inner labral sensilla; lacinial apex sometimes rounded, with small intermediate denticles between main tines ▸ **Caeciliusinae**
– Two or three ventral abdominal vesicles, and seven distal inner labral sensilla; lacinial apex bicuspid, no intermediate denticles between tines ▸ **Paracaeciliinae**

Caeciliusinae

This subfamily is by far the largest of the four and is characterised more fully by the following character set: ocelli present in non-neotenic forms (a very few exceptions); clypeal shelf variable; five distal inner labral sensilla; usually two ventral abdominal vesicles; external gonapophysis valve a lobe from base of dorsal valve, with one or two setae; endophallus 3-lobed, with median lobe of various forms.

Mockford (2000) recognised five tribes in this subfamily, separated as indicated below. The characters used to separate these are sometimes intricate and a later key facilitates identification directly to genus level. The following key thus indicates some of the broader categories in this complex group. The tribal name 'Kodamaiini' used by Mockford was replaced by 'Austrocaeciliini' by Yoshizawa (2001b), following his transfer of *Kodamaius* to Amphipsocidae (see p. 106).

Key to the tribes of Caeciliusinae

1 Distal inner labral sensilla of type 1; clypeal shelf narrow, intermediate or absent; labral stylets absent; fore wing vein Cu1 branched, forming areola postica or joining vein M; spermathecal duct with sheath; median endophallus lobe undivided ▸ **Caeciliusini**

– Distal inner labral sensilla of type 2; clypeal shelf usually broad: if narrow, either (a) labral stylets present, sometimes very small, or (b) fore wing vein Cu1 unbranched or (c) (male) median lobe of endophallus divided and (female) spermathecal duct without a sheath ▸ 2

2 Clypeal shelf broad; apex of lacinia narrow, either bicuspid, bilobed, rounded or flat; either labral stylets absent and maxillary palp segment 4 longer than segment 2, or labral stylets present and maxillary palp segment 4 shorter than segment 2; (male) median lobe of endophallus undivided and (female) spermathecal duct with a sheath ▸ **Coryphacini**

– Clypeal shelf absent or narrow; apex of lacinia moderately to very wide, with many rounded denticles; either labral stylets present (sometimes very small) and fore wing vein Cu1 branched or labral stylets absent and (male) median lobe of endophallus wholly divided and either (a) fore wing vein Cu1 unbranched and (female) spermathecal duct with sheath or (b) fore wing vein Cu1 branched and (female) spermathecal duct without sheath, or (c) both sexes brachypterous (without venation) or apterous, and (female) spermathecal duct with sheath ▸ 3

3 Ventral abdominal vesicles minute or absent; apex of lacinia very broad, with shallow rounded denticles ▸ **[Epicaeciliini]**
(not in Asia; the two genera are from Australia, Europe and Mexico; the Australian *Epicaecilius* might possibly be found in the region)

– Two normal ventral abdominal vesicles; apex of lacinia narrow, not markedly broadened, usually with only one or two incipient tines and not denticulate as above ▸ 4

4 Apex of lacinia with two cusps, usually separated by two or three shallow rounded denticles; (male) endophallus 2-lobed or with median lobe divided, epiproct usually (except *Thorntoniella*) with minute or no field of papillae ▸ **Austrocaeciliini**

– Apex of lacinia broad, flat or with lateral cusp, this with two rows of shallow rounded denticles; (male) median lobe of endophallus only partially divided, epiproct with weak to moderately developed field of papillae ▸ **Maoripsocini**

Taxonomic allocation of regional genera of Caeciliusinae to tribes and subtribes

Caeciliusini: *Caecilius, Dypsocus*

Coryphacini: *Coryphosmila, Protodypsocus, Stenocaecilius, Valenzuela*

Austrocaeciliini:

subtribe Pericaeciliina: *Pericaecilius*

subtribe Thorntoniellina: *Thorntoniella*

subtribe Austrocaeciliina: *Asiocaecilius, Isophanopsis*

Maoripsocini: *Maoripsocopsis*

Key to regional genera of Caeciliusinae

1 Distal inner labral sensilla type 1; clypeal shelf narrow or intermediate; labral stylets absent; fore wing vein Cu1 branched; (male) median lobe of endophallus undivided; (female) spermathecal duct with sheath ▸ **(Caeciliusini)** 2

– Distal inner labral sensilla type 2; clypeal shelf broad <u>or</u>, if narrow, either (a) labral stylets present, sometimes very small, <u>or</u> (b) fore wing vein Cu1 unbranched <u>or</u> (c) (male) median lobe of endophallus divided and (female) spermathecal duct without sheath ▸ 3

2 First two flagellar segments little, if any, broader than succeeding segments; fore wing cell r5 scarcely widened distally; hind margin of vertex usually not flattened and not extended behind eyes (male, no sclerites in membrane above endophallus) ▸ *Caecilius*

– First two flagellar segments thickened and densely setose; fore wing cell r5 widened distally; hind margin of vertex flattened and extended behind eyes ▸ *Dypsocus*

3 Clypeal shelf broad; apex of lacinia slender, of varied form; <u>either</u> (a) labral stylets absent and fourth segment of maxillary palp longer than second segment <u>or</u> (b) labral stylets present and fourth segment of maxillary palp shorter than second segment ; (male) median lobe of endophallus undivided, (female) spermathecal duct with sheath ▸ **(Coryphacini)** 4

– Clypeal shelf narrow or absent; apex of lacinia moderately or very wide, with many denticles; <u>if</u> labral stylets present (even if very small), fore wing vein Cu1 branched; <u>if</u> labral stylets absent, male endophallus with median lobe completely divided and <u>either</u> (a) fore wing vein Cu1 unbranched, female spermathecal duct with sheath, <u>or</u> (b) fore wing vein Cu1 branched, female spermathecal duct without sheath, <u>or</u> (c) both sexes brachypterous (without venation) or apterous, female spermathecal duct with sheath ▸ 7

4 Labral stylets present (fig. 204); maxillary palp segment four shorter than segment two (male basal flagellar segment thickened) ▸ *Stenocaecilius*

– Labral stylets absent (fig. 205); maxillary palp segment four longer than segment two (male basal flagellar segment not, or only very slightly, thickened) ▸ 5

5 Lacinial apex with lateral cusp not thin and flat or drawn into tapered point; wings very variable in shape and pattern ▸ *Valenzuela*

– Lacinial apex with lateral cusp <u>either</u> thin and flat <u>or</u> drawn out into a tapered point; wings rather narrow, with longitudinal banding pattern ▸ 6

6 First flagellar segment of antenna very long, about equal to second and third segments together; wings long and slender, ratio of length : greatest width of fore wing ca 3.5 ▸ *Coryphosmila*

– First flagellar segment of antenna relatively shorter, about 1.4 times length of second segment; fore wing relatively wider (vein M1 reaching margin behind wing apex) ▸ *Protodypsocus*

7 Lacinial apex relatively broad, with lateral cusp and series of shallow rounded denticles; ocelli absent (male, median lobe of endophallus partially divided, apical region continuous, epiproct with weak to moderately developed papillar field) ▸ (**Maoripsocini**) *Maoripsocopsis*
(N.B. males of this genus are unknown; the above characters were used by Mockford (2000) in his tribal division based on *Maoripsocus*, in which ocelli are present)

– Lacinial apex bicuspid, usually with 2-3 shallow rounded denticles between cusps; (male) endophallus 2-lobed or with median lobe divided, epiproct usually weakly papillate or lacking papillar field ▸ (**Austrocaeciliini**) 8

8 Labral stylets absent; <u>either</u> (a) fore wing vein Cu1 branched and (female) spermathecal duct without sheath <u>or</u> (b) fore wing vein Cu1 unbranched and (female) spermathecal duct with short sheath ▸ 9

– Labral stylets present, sometimes very small; fore wing vein Cu1 branched, (female) spermathecal duct with long sheath ▸ **subtribe Austrocaeciliina** 10

9 Fore wing vein Cu1 unbranched; (male) paraproct with papillar field; (female) spermathecal duct with short sheath ▸ **subtribe Pericaeciliina** *Pericaecilius*

– Fore wing vein Cu1 branched, areola postica normal; (male) paraproct without papillar field; (female) spermathecal duct without sheath ▸ **subtribe Thorntoniellina** *Thorntoniella*

10 Wings with transverse banding pattern; pterostigma without a spur vein from hind angle; (female) ventral gonapophysis valve much shorter than dorsal valve, external valve with seta ▸ ***Isophanopsis***

– Wings without a banding pattern; pterostigma with or without a spur vein from hind angle; (female) ventral gonapophysis valve nearly as long as dorsal valve, external valve usually without a seta ▸ *Asiocaecilius*

Generic information

Asiocaecilius Mockford (fig. 218). Type species *Asiocaecilius assamensis* Mockford.
Generic distribution: India, Seychelles, Comoro Islands, Bhutan, China, Singapore, Bali, Lombok, Christmas Island (Australia).
This is a relatively small genus, with seven described species, mostly with obscure dark shading on the fore wing. Most of the species are known from very few individuals and the extent of variation is unknown. A pterostigmal spur vein is sometimes present.

Caecilius Curtis (figs. 206, 216-217, 221, 223, 225, 228). Type species *Caecilius fenestratus* Curtis. Generic synonymy: *Mepachycera* Enderlein, type species *Dypsocus parvulus* Banks.
Generic distribution: following Mockford's (2000) transfer of numerous species from *Caecilius* to other genera, the scope of the genus has been contracted considerably from its former very broad concept. *Caecilius* is now considered an Old World genus, distributed widely through Africa, much of the Palaearctic, and Asia, with an Australian species tentatively referred to it. It was formerly (s.l.) regarded as almost cosmopolitan.
Most of the species since transferred to other genera were described initially as '*Caecilius*', and 'true' regional *Caecilius* species are known from Singapore, Indonesia and the Philippines. Several of these have the fore wing with a longitudinal dark band, often with a transverse connection from this to the pterostigma, and the wing membrane flanking vein R2+3 also dark.

Coryphosmila Enderlein (fig. 219). Type species *Psocus dolabratus* Hagen. Generic synonymy *Coryphocopis* Enderlein, type species *Coryphocopis jacobsoni* Enderlein.
Generic distribution: India, Sri Lanka, China, Taiwan, Singapore, Indonesia, Australia.
Six species have been described, five of them transferred from *Caecilius* (Mockford, 2000). The type species appears to be distributed widely in Asia and is responsible for all the above distributional records other than Australia, China and Indonesia. Three of the species are Indonesian. The genus is a probable endemic in the Asian-Australian region.

Dypsocus Hagen (figs. 207, 215, 220, 222, 224, 226). Type species *Psocus coleoptratus* Hagen.
Generic distribution: India, Sri Lanka, Taiwan, Philippines, Indonesia.
Following Mockford's (2000) rearrangement, only five described species are included currently in *Dypsocus* and all of them are poorly known. The densely setose and thickened basal antennal flagellomeres and the extended flattened dorsal margin of the vertex are a distinctive character combination for this genus within Caeciliusini.

Isophanopsis Badonnel. Type species *Isophanopsis albofasciata* Badonnel.
Generic distribution: Madagascar, Reunion, Comoro Islands, Indonesia, Fiji.
This genus contains three named species, two of them transferred from *Caecilius* by Mockford (2000), who noted also an 'undescribed species from Sumatra', on which basis the genus is included here.The described species are known only from outside the region, as noted above.

Maoripsocopsis Mockford (figs. 209, 230, 235, 237). Type species *Maoripsocopsis sumatrensis* Mockford.
Generic distribution: Sumatra.
Mockford (2000) raised this distinctive genus for a single female collected in Sumatra. It is particularly unusual in lacking ocelli, a feature found otherwise only in neotenic members of the family. The pterostigma has a well defined spur vein and – should it be collected in the future – the male should be recognisable on fore wing features.

Pericaecilius Mockford (fig. 231). Type species *Pericaecilius werneri* Mockford.
Generic distribution: China, Taiwan, Philippines, Indonesia.
This genus is known from three named species (one based on the Indonesian 'Caeciliidae gen. & sp. indet.' of Vaughan et al. (1991), which Mockford (2000) attributed to this genus, as *P. vaughani* Mockford), one from each of the above-listed countries. Later, Mockford (2001) clarified the somewhat confused original description of this species.

Protodypsocus Enderlein (figs. 213, 232, 236, 238-239). Type species *Protodypsocus fissiceps* Enderlein.
Generic distribution: New Guinea, Australia.
Only the type species of *Protodypsocus* is known. It appears to be closely related to *Coryphosmila* and the female genitalia of the two genera are similar. Differences in antennal form and fore wing features appear sufficient to maintain these genera as distinct.

Stenocaecilius Mockford. Type species *Caecilius casarum* Badonnel.
Generic distribution: widespread in most tropical regions – Pacific islands, Central and South America, Atlantic islands, Africa, Seychelles, India, Malaysia, Indonesia, New Guinea, Australia.
Mockford (2000) formally transferred 36 species from *Caecilius* to *Stenocaecilius* and a further nine species were transferred tentatively. *Stenocaecilus* is apparently most diverse in Africa. It is recorded here from records of the widespread *S. analis* (Banks) from Malaysia and of *S. casarum* from Indonesia and (tentatively) New Guinea.

Thorntoniella Mockford (figs. 208, 233-234). Type species *Caecilius laleus* Thornton.
Generic distribution: Assam (India), Indonesia.
Only the type species of *Thorntoniella*, the sole inclusion in Mockford's subtribe Thorntoniellina, is known. It appears to be quite widely distributed in central Indonesia, where it is known from the Krakatau Islands, Java, Bali and Lombok. The elongate fore wing has a long Rs-M fusion and the pterostigma is entirely hyaline (in contrast to most of the posterior half of the wing), unlike most members of the family recorded from Indonesia (Thornton, 1984a).

Valenzuela Navás (figs. 210, 240, 243). Type species *Valenzuela marianus* Navás.
Generic distribution: cosmopolitan.
This genus contains a high proportion of the species placed formerly in *Caecilius* s.l. It was treated as 'Psocida Agnota' by Smithers (1967a) and recognised for correct placement by Meinander (1979a). In his earlier revisionary studies on '*Caecilius*', Mockford (1965b, 1969a) recognised a number of 'species groups', and the members of many of these all belong to *Valenzuela*. South-east Asian species are referable to the following groups, following listing by Mockford (2000): *fasciatus*-group, *mueggenburgi*-group and *sinuofasciatus*-group, with eight other species groups not including any species from the region. These three species groups are separable formally as indicated below.

– Key to regional species-groups of *Valenzuela*

1 Head of female slightly extended and flattened behind compound eye; lacinial apex flat, broadened and rounded, lacking distinct cusps; fore wing long and slender, with longitudinal dark band; pterostigma shallow ▸ ***mueggenburgi*-group**

– Head of female not extended and/or flattened behind compound eye; lacinial apex not as above, usually distinctly but shallowly bicuspid, or peg-like; fore wing of normal shape, with transverse or longitudinal banding pattern or clear; pterostigma angulate, always with some dark part on distal edge or central region ▸ 2

2 Areola postica small and low; (female) spermathecal duct with sheath long, 'carrot-shaped', tapering toward spermapore ▸ *sinuofasciatus*-**group**

– Areola postica of normal size; (female) spermathecal duct with sheath not as above, short to moderate in length ▸ *fasciatus*-**group**

The *mueggenburgi*-group is based on *V. mueggenburgi* (Enderlein), one of the most widely distributed and characteristic Caeciliusidae of the region, and which has been recorded from India, Taiwan, Singapore, Philippines and Indonesia. The only other regional species, tentatively allocated to this group by Mockford (2000), is *V. deceptor* (Banks) (Philippines). Other group members are from China (two species) and Japan (one species).

The *fasciatus*-group is one of the largest in the genus, with very wide distribution. Confirmed regional species are in Indonesia (two) with seven other Indonesian species tentatively attributed by Mockford (2000). In all of them, the fore wing has a transverse banding pattern and angulate pterostigma.

The *sinuofasciatus*-group is based on the Madagascan *V. sinuofasciatus* (Badonnel) and three of the other five species are also from Madagascar. One is from China, and regional representation is solely by *V. albipes* (Thornton) from Bali.

With only one exception (*V. serpentinus* Enderlein, transferred from *Fuelleborniella*), all the regional species in *Valenzuela* were transferred from *Caecilius* by Mockford (2000). In addition to the above groups, Mockford (2000) considered a substantial number of other '*Caecilius*' to belong to *Valenzuela* without attributing them to any species group. Some 20 South-east Asian species, predominantly from Indonesia, were placed in this category. Re-examination of these species (marked with * in chapter 7 under Caeciliusidae) is needed in order to clarify their affinities more precisely. In general, then, and despite its considerable global diversity, *Valenzuela* is not particularly diverse in South-east Asia.

Paracaeciliinae

This subfamily contains five distinctive genera, only two of which occur in South-east Asia, with others from South America, the Mascarene Islands, and the New World. The two regional genera share the following characters within the subfamily: no banding pattern on the fore wing; male endophallus bilobed; female dorsal and external gonapophysis valves fused to produce broad composite valve with one long seta; labral stylets absent.

Key to the genera of Paracaeciliinae

1 Lacinial apex bicuspid, with lateral cusp acuminate at tip; (female) spermatheca with long narrow neck, without internal villi ▸ *Enderleinella*

– Lacinial apex with lateral cusp not acuminate; (female) spermatheca with short broad neck, with internal villi in region of junction with duct ▸ *Paracaecilius*

Generic information

Enderleinella Badonnel (figs. 212, 241, 244, 247). Type species *Caecilius perlatus* Kolbe.

Generic distribution: Brazil, Europe, India, Sri Lanka, China, Singapore, Indonesia, Australia.

This genus is apparently most diverse in China, from where 17 of the 24 species listed by Lienhard & Smithers (2002) were described, with Li Fasheng (2002) subsequently describing a further sixteen Chinese species. Regional representation is by one species each described from Indonesia (*E. calta* (Thornton)) and Singapore (*E. unicolor* (Enderlein)).

Paracaecilius Badonnel (figs. 242, 245-246, 248). Type species *Paracaecilius berlandi* Badonnel. Generic synonymy: *Caecilioidus* Badonnel, type species *Caecilius oxystigma* Badonnel; *Eocaecilius* Badonnel, type species *Eocaecilus wittei* Badonnel; *Badonnelipsocus* Li Fasheng, type species *Badonnelipsocus chebalinganus* Li Fasheng.

Generic distribution: Africa, Madagascar, India, China Japan, Singapore, Indonesia, New Guinea, Australia, New Zealand, Fiji.

Mockford (2000) assigned 22 species firmly and an additional five species tentatively to this genus. One of the Australian species is known from Christmas Island (*P. cardaleae* Smithers). The central region of Indonesia (Java, Krakatau Islands) harbours the three described Indonesian species; one of these (*P. macrops* (Enderlein)) occurs also in Singapore. Two poorly known species have been described from New Guinea.

2 Amphipsocidae (figs. 249-261, 264, 267)

Diagnosis

Fore wing length ca 2.8-5.5 mm.

Mandibles usually elongate; lacinial apex narrow, usually bicuspid. Vertex with pair of shallow concavities, these deeper in males. Tarsi 2-segmented; claws without subapical tooth, pulvillus broad. Fore wing veins strongly setose, setae relatively long and in more than one rank; margin of hind wing usually setose, often including a distinct break on basal costal area. Ventral abdominal vesicles present. Female: gonapophyses reduced, with external valve absent or reduced to small lobe, sometimes with single seta.

Remarks

Amphipsocidae was first designated as a family by Pearman (1936a), who allied it with 'Caeciliidae', Stenopsocidae, Calopsocidae and Polypsocidae (the last-named from Central and South America and not found in Asia, and now treated as a tribe of Amphipsocinae, below), but defined the family only by allocation of representative included genera (*Amphipsocus, Harpezoneura*). Amphipsocidae was redefined by Badonnel (1955), who characterised it within Caeciliusetae as containing taxa having long straight upright setae in more than one rank on fore wing veins, and with the costal margin from the base of the pterostigma to the wing apex densely setose and thickened, together with the pterostigma often having a posterior spur vein. More modern definition and arrangement by Mockford (1978a) has formed the basis for more recent understanding, with the major subsequent change to its constitution being the removal of the subfamily Dasydemellinae to constitute the family Dasydemellidae (Mockford, 1993a). This family was returned to Amphipsocidae as subfamily Dasydemellinae by Yoshizawa (2001b) within a revised diagnosis of the family.

Although *Kodamaius* was included in Caeciliusidae by Mockford (2000) (and by Li Fasheng (2002) in his new subfamily Isophaninae), Yoshizawa (2001b) presented evidence to place it more properly in Amphipsocidae, where it constitutes the subfamily Kodamaiinae.The remaining members of Mockford's 'Kodamaiini' were all retained in Caeciliusidae and the replacement name Austrocaeciliini proposed for these by Yoshizawa.

In the most familiar traditional arrangement (Mockford, 1978a), two main subfamilies are included in Amphipsocidae, one (Calocaeciliinae) represented by only a single genus and the other more complex with several tribes and subtribes erected to reflect its diversity. However, Dasydemellinae is now treated as a subfamily of Amphipsocidae rather than as a distinct family, and the above-noted Kodamaiinae is a recent addition to the family.

Key to the subfamilies of Amphipsocidae

1 Fore wing somewhat elytriform and coriaceous (fig. 257); areola postica open basally, with vein Cu1b absent ▸ **Calocaeciliinae**

– Fore wing usually elongate, sometimes broad; areola postica closed basally ▸ 2

2 Areola postica free; veins Rs and M in fore wing usually fused for a short length ▸ **Amphipsocinae**

– Areola postica linked to M; Rs-M junction in fore wing variable ▸ 3

3 Areola postica linked to M by crossvein; veins Rs and M in fore wing linked by a short crossvein ▸ **Kodamaiinae**

– Areola postica fused with M for a short length; veins Rs and M in fore wing fused for short length ▸ **[Dasydemellinae]***

* No confirmed records from South-east Asia, but one genus, *Matsumuraiella* Okamoto, is found in the eastern Palaearctic (China, Taiwan, Japan) and could possibly be found there.

Amphipsocinae

This is by far the most diverse subfamily and Mockford (1978a) divided it into a series of tribes and subtribes, many with few representatives. Two tribes are represented in South-east Asia, together with four genera.

Regional species have been referred to *Kolbia* Bertkau, the sole genus in the tribe Kolbiini and separable from all the keyed genera by having the lacinial apex flat, rather than bicuspid, and females micropterous and with a bulb in the spermathecal duct extending beyond the sheath. Mockford (1978a) reported '*Kolbea*' (a spelling emended by Enderlein (1901) from the original '*Kolbia*', but later considered invalid; Lienhard (1998a)) as occurring from Europe to Central Asia and Madagascar, and noted that other purported inclusions 'probably belong to other genera'. *Kolbia bakeri* Banks (Philippines) and *K. maculipennis* Karny (Sarawak) remain anomalous. Several African genera otherwise referable to *Amphipsocus* differ from it in having at least three branches to fore wing vein Rs and more than this to fore wing vein M. Any such individuals from tropical South-east Asia should be checked carefully against keys by Mockford (1978a) and Smithers (1990a).

Key to the tribes and genera of Amphipsocinae

1 Femur of fore leg without row of cones; no pterostigmal spur vein ▸ **Dasypsocini, *Dasypsocus***

– Femur of fore leg usually with row of cones; these occasionally lacking in taxa with a pterostigmal spur vein ▸ **Amphipsocini 2**

2 Fore wing vein Cu1a joined to M to produce closed discoidal cell; no basal brush of setae on hind wing costa; (female) spermathecal sheath very broad ▸ **subtribe Taeniostigmatina, *Taeniostigma***

– Fore wing vein Cu1a not joined to M, so areola postica normal; basal brush of setae on hind wing costa; (female) spermathecal sheath not very broad ▸ 3

3 Pterostigma shallow ▸ **subtribe Tagalopsocina, *Tagalopsocus***
– Pterostigma angular, usually with distinct spur vein
▸ **subtribe Amphipsocina, *Amphipsocus***

Generic information

Amphipsocus McLachlan (figs. 249, 256). Type species *Amphipsocus pilosus* McLachlan. Generic synonymy *Rosega* Navás, type species *Rosega picta* Navás.
Generic distribution: Africa, Madagascar, India, Nepal, China, Taiwan, Japan, Malaysia, Singapore, Philippines, Indonesia.
Species of *Amphipsocus* are generally large 'hairy' psocids, with the wings held at rest in a shallow 'tent' over the body. In life the wings can be marked with pale red or pink, especially behind the pterostigma, but this pigmentation does not always persist after preservation, particularly in alcohol. Most species have a distinct spur vein to the pterostigma and fore wing veins Rs and M fused for a short length.
Amphipsocus is by far the most speciose genus of Amphipsocidae in the Old World and many of the purported species appear to be very similar. *Amphipsocus heterothrix* Thornton & Wong, described from Bengal, occurs widely in South-east Asia. The branches of fore wing veins Rs and M, and Cu1a in this species are often somewhat sinuous. *Amphipsocus* species are found most commonly associated with dead foliage.

Dasypsocus Enderlein (figs. 251, 254). Type species *Kolbea solox* Enderlein.
Generic distribution: tropical South America north to southern Mexico, parts of Africa, India, Nepal, Bhutan, Japan, Singapore.
The type species of this unusually widely distributed genus was described from Singapore. Mockford (1978a) believed that the Japanese *D. japonicus* Enderlein, the only other species from the eastern Palaearctic/Oriental region, may be a species of *Amphipsocus* and this has been confirmed recently (Yoshizawa, 2001b). The absence of a row of cones on the anterior femur is perhaps the easiest 'spot' character for differentiation from *Amphipsocus.* The pterostigma of *Dasypsocus* is moderately deep but lacks a spur vein. Branches of veins Rs and M are generally somewhat curved and the Rs-M fusion is usually very short. In both genera the external gonapophysis valve is minute.

Tagalopsocus Banks (figs. 252-253, 255). Type species *Tagalopsocus luzonensis* Banks.
Generic distribution: Assam (India), China, Philippines.
Species of *Tagalopsocus* are large psocids with fore wing length generally exceeding 4 mm. The pterostigma is shallow, without a spur vein, and the hind tibia of some species has a row of spines/bristles on the inner surface. Fore wing vein M3 usually separates from M basal to the radial fork, unlike many *Amphipsocus,* in which the two forks are at almost the same distance from the wing base.

Taeniostigma Enderlein (figs. 261, 264, 267). Type species *Psocus elongatus* Hagen. Generic distribution: Sri Lanka, Bhutan, China, Taiwan, Japan, Vietnam, Philippines, Malaysia, Singapore, Indonesia, New Guinea, Australia.
Taeniostigma species are large psocids, distinguished in the Amphipsocidae by the fusion of fore wing vein Cu1a with M to produce a closed discoidal cell. The pterostigma is long and shallow and veins Rs and M are fused for a short length. Several species are bright green, yellowish brown or yellow in life and all are associated with foliage. The subfamily Taeniostigmatinae was raised by Li Fasheng (2002) to include this genus.
Historically, following Roesler (1944), *Taeniostigma* has been considered allied with *Stenopsocus* Hagen and *Graphopsocus* Kolbe in the family Stenopsocidae (below) on fore wing features. This relationship was dispelled convincingly by Mockford (1978a), although Thornton (1984b) again referred the genus to Stenopsocidae. The type species is very widely distributed, with the name having been applied to specimens from Sri Lanka, through Malaysia to Indonesia and the Bismarck Archipelago. Several species have the antennae strongly banded with black or dark brown and some species (see Smithers (1974c) on Australian taxa) are sexually dimorphic in this feature. Some also have a broad dark pterostigmal stripe along the posterior border. Fore wing vein Rs stem is usually much longer than the branches R2+3 and R4+5.

Kodamaiinae

Generic information

Kodamaius Okamoto (fig. 250). Type species *Kodamaius brevicornis* Okamoto. Generic synonymy *Stenepipsocus* Badonnel, type species *Stenepipsocus collarti* Badonnel; *Epikodamaius* Kuwayama, type species *Epikodamaius ikomai* Kuwayama.
Generic distribution: Africa, India, Bangladesh, Bhutan, China, Taiwan, Japan, Thailand.
Lienhard & Smithers (2002) recognised 22 described species in this genus, most of which have not been found in tropical South-east Asia, and Li Fasheng (2002) decribed a further 22 species from China. *Kodamaius* is included here on the incidence of one species (*K. ikomai* (Kuwayama)) from Thailand. See note on p. 106.

Calocaeciliinae

This subfamily contains only the distinctive genus *Calocaecilius* and the name expresses the superficial resemblance of these insects to Calopsocidae (p. 114) in the form of the vertex (flattened behind the eyes, although this is not as pronounced as in many calopsocids) and fore wing (coriaceous, somewhat elytriform, with the short pterostigma having a well defined 'break' on the costal margin). However, the transverse 'bend' in the fore wings, conferring the appearance of the curved elytra of a small beetle, occurs at the base of the pterostigma rather than at its distal end. Mockford (1974d) speculated that calopsocids and *Calocaecilius* are both members of the same mimicry complex with coccinellid beetles, with which they share bright orange and black colouring. The placement of this genus in a distinct subfamily is due to Mockford (1978a).

Generic information

Calocaecilius Mockford (figs. 257-260). Type species *Calocaecilius decipiens* Mockford.
Generic distribution: Malaysia, Philippines, Indonesia.
The genus was described from the female only, with the male subsequently described by Mockford (1998b). A second species (*C. apertus* Thornton) is known from Bali (Thornton, 1984b) and the genus is at present considered to be endemic to the region. The two sexes are alike in fore wing features, and absence of the basal closure to the areola postica separates *Calocaecilius* from all other Amphipsocidae known.

3 **Stenopsocidae** (figs. 262-263, 265-266)

Diagnosis

Fore wing length ca 3.5-4.5 mm. Mandibles usually elongate; lacinial apex narrow, incipiently bicuspid or with small rounded denticles. Tarsi 2-segmented; claws without subapical tooth, pulvillus broad. Fore wing veins with short setae in one rank; hind wing margin bare or with few setae on radial fork region; fore wing with crossvein from pterostigma to Rs; crossvein from Cu1a to M. Two or three ventral abdominal vesicles. Female: external gonapophysis valve strongly reduced, lacking seta; ventral and dorsal valves slender, elongate; apex of subgenital plate rounded or slightly bilobed; spermatheca with small basal pouch.

Remarks

This small family of foliage-frequenting Psocoptera was for long considered a subfamily of Caeciliusidae and has also included historically some members of the Amphipsocidae (*Taeniostigma*) (see above). Mockford (1978a) placed both widely recognised genera in Caeciliusidae (as Caeciliidae) but later (Mockford, 1993a) allocated them to Stenopsocidae without further comment. Two additional genera (*Cubipilis* Li Fasheng, type species *Cubipilis hamaocaulis* Li Fasheng and *Malostenopsocus* Li Fasheng, type species *Malostenopsocus yunnanicus* Li Fasheng), mainly from China, appear to be very similar to true *Stenopsocus*. The family appears to be particularly diverse in China (Li Fasheng, 2002), with rather few species described from any other part of the world.
Stenopsocidae and Caeciliusidae appear to be closely related and are separated most clearly by the unusual structural features of crossveins between (a) the hind angle of the pterostigma and Rs and (b) the apex of the areola postica and

M. If, rarely, a Cu1a-M crossvein is found in Caeciliusidae, vein M in the fore wing appears 2-branched rather than 3-branched as in all Stenopsocidae. Typically, Caeciliusidae have the hind wing margin setose around most or all of its length, whereas marginal setation in Stenopsocidae is absent or restricted to the radial fork.

Key to the genera of Stenopsocidae

1 Pterostigma long and shallow, expanded slightly opposite crossvein to Rs; crossvein from Cu1a to M long; fore wing margin setose for whole length around to vein Cu2 (fore wing usually not strongly pigmented); three ventral abdominal vesicles ▸ *Stenopsocus**

– Pterostigma rather short, strongly expanded opposite crossvein to Rs; crossvein from Cu1a to M short; fore wing margin setose only as far as wing apex (fore wing often strongly patterned); two ventral abdominal vesicles ▸ *Graphopsocus*

* *Cubipilis* Li Fasheng and *Malostenopsocus* Li Fasheng are not differentiated from *Stenopsocus* in this account.

Generic information

Graphopsocus Kolbe (figs. 263, 266). Type species *Hemerobius cruciatus* L. Generic synonymy: *Teratopsocus* Reuter, type species *Teratopsocus maculipennis* Reuter.
Generic distribution: North and South America, Europe, Africa, Canary Islands, Sri Lanka, China, Japan, Singapore, Philippines.
The type species, *G. cruciatus*, is one of the most strongly patterned and abundant foliage-frequenting psocids in Europe. Several other described species resemble *G. cruciatus* in general appearance and it is uncertain whether some of them, including taxa from South-east Asia, are properly attributed to *Graphopsocus* or belong to *Stenopsocus*. Two nominal species are possible members of the regional fauna and both are poorly known: *G. infirmus* Banks (Philippines) and *G. subaequalis* Banks (Singapore). *Graphopsocus* (or *Stenopsocus uniformis* (Hagen) is widespread, with records from Sri Lanka, Philippines and Java.

Stenopsocus Hagen (figs. 262, 265). Type species *Psocus immaculatus* Stephens.
Generic distribution: Europe, China, Taiwan, Japan, Vietnam, Malaysia, Philippines, Indonesia, New Guinea, Australia.
Two main forms occur in *Stenopsocus*, namely species with unmarked wings and those with the distal posterior border of the pterostigma strongly darkened. Most species have fore wing vein Cu2 bare, but in a few species this vein is setose, a feature given generic significance by Li Fasheng (2002). One Indonesian species (*S. cubitalis* Thornton) with this feature was transferred to *Malostenopsocus* by Li Fasheng (1992d), but it is here included, without any formal presumption

of generic position, in *Stenopsocus*. However, the general pattern of venation is very similar in all species, and species are separated mainly on differences in marking and body colour. In life, most species of these relatively large foliage-frequenting psocids are greenish or yellowish with dark brown markings. *Stenopsocus uniformis* (Hagen), variously placed in *Graphopsocus* and *Stenopsocus*, is widespread. About half a dozen other species are known from South-east Asia. *Stenopsocus* s.l. appears to be particularly diverse in China, from where nearly 100 species have been described (Li Fasheng, 2002).

Group HOMILOPSOCIDEA

As noted earlier this is not a natural group, but contains a variety of psocid families which have not fitted 'neatly' elsewhere in the suborder. The families are treated here without implication of them being closely related except where relationships have been confirmed or inferred from phylogenetic analysis, as discussed in context for each.

1 Archipsocidae (figs. 268-273)

Diagnosis

Body length usually ca 1.2-1.8 mm, fore wing length (macropters) to ca 1.5 mm. Lacinial apex slender, bicuspid, the outer cusp usually broader than the inner cusp, and may be denticulate. Tarsi 2-segmented; claws without subapical tooth, pulvillus broad. Fore wings usually coriaceous, with venation largely obscure, especially in apical half of wing. Areola postica usually shallow; membrane of wing setose, longer setae around wing margin, setae on medial cell area of margin usually in two crossing series. Hind wing membranous, venation reduced; veins glabrous; veins R, M, Cu each with one branch. Alary polymorphism common (males micropterous or apterous; females macropterous or with wings variously reduced). Female: subgenital plate rounded or transverse; gonapophyses either (a) with ventral valve absent, dorsal valve slender, external valve broad with long setae, or (b) entirely absent. Male: hypandrium simple, transverse or rounded; phallosome with external parameres small or absent, endophallus usually without defined sclerites. Paraprocts simple; no large trichobothrial field, usually only one or two trichobothria present.

Remarks

Archipsocidae is a widely distributed family of small, predominantly tropical Psocoptera. The few species described from South-east Asia probably do not reflect their true diversity in the region, as the general appearance of most species is very similar and they can be confused easily without detailed examination. Archipsocidae are small, brown, 'hairy' psocids with wings held in a shallow 'tent' over the body. The most conspicuous feature is their presence in large colonies on trees, living under extensive silken web sheets. Linked to this habit of bark-dwelling, many species are polymorphic, with macropterous, brachypterous and apterous individuals present in the same colony, and frequent exhibition of neoteny.

The family includes two distinct subfamilies (Badonnel et al., 1984): Pararchipsocinae (p. 44) (hind wing without a closed basal cell, or this very small) are known only from the New World, Archipsocinae (hind wing with a distinct closed basal cell) are widespread. Both described genera of Archipsocinae are also widely distributed and both are present in South-east Asia. They are distinguished on genitalic features.

Key to the genera of Archipsocidae

1 Female: external gonapophysis valve broad and setose; dorsal valve slender. Male: phallosome elongate ovoid, apex a broad arch (fig. 272) ▸ *Archipsocus*

– Female: gonapophyses entirely absent. Male: phallosome with aedeagal arch constricted to form a median arched area (fig. 273) ▸ *Archipsocopsis*

Generic information

Archipsocus Hagen (figs. 268-272). Type species *Archipsocus puber* Hagen.

Generic distribution: tropical and near-tropical regions of North, Central and South America, Africa, Taiwan, Malaysia, Singapore, Indonesia, Australia.

Many species of *Archipsocus* are very hard to distinguish from each other on general appearance. All are small (body length usually 3 mm or less) and antennal flagellar segments VI and X each have a conspicuous discoidal sensillum bearing a long filament. Species have been separated mainly on details of colour pattern and (more reliably) genitalic features, but few have been recorded from South-east Asia. *Archipsocus recens* Enderlein appears to be quite widely distributed, with records from Taiwan, Singapore and Java. *Archipsocus sanurensis* Thornton and *A. alternatus* Vaughan et al. were described from Indonesia.

Archipsocopsis Badonnel (figs. 31, 273). Type species *Archipsocus mendax* Badonnel. Erected originally as a subgenus, later elevated to full generic status by Badonnel (1966a).

Generic distribution: North, Central and South America, Africa, Sri Lanka, Indonesia, Australia.

Loss of gonapophyses in *Archipsocopsis* is associated with viviparity. As in *Archipsocus*, the species are very similar in appearance. An additional character for generic recognition is that antennal flagellar segments VI and X each have a coeloconic sensillum with a short central cone (fig. 31).

The Sri Lankan *A. fernandi* (Pearman) was recorded from Bali by Thornton (1984a). It is the only specific record of the genus from the region.

2 Calopsocidae (figs. 274-294)

Diagnosis

Fore wing length ca 4.3 – more than 7 mm. Head (except in *Callistoptera* and *Torrepsocus*) with vertex tapered to an acute transverse crest dorsally, usually strongly emarginate medially. Lacinial apex divided, with one tine much broader than the other. Fore wing broad, membrane and margin setose, often convex and coriaceous; usually with accessory venation, sometimes as an extensive reticulate network of secondary veinlets and, at least, with supernumerary venation in pterostigmal area. Hind wing broad, venation in distal half with more than one rank of setae, marginal setae in apical region crossing each other. Female: subgenital plate simple or bilobed, the lobes sometimes associated only loosely with main body of plate; gonapophyses complete, external valve with large rounded lobe. Male: hypandrium simple; phallosome elongate with well developed outer parameres; strong internal rodlike sclerotisation on endophallus; clunium and epiproct usually ornamented with comblike row of teeth.

Remarks

The Calopsocidae are one of the most characteristic psocopteran families of tropical South-east Asia and are confined to the Oriental and Melanesian regions, with greatest diversity in New Guinea. They have a very distinctive appearance: many are relatively large and dark psocids.
The family has sometimes been considered to comprise two subfamilies. In that system, Callistopterinae includes only *Callistoptera*, a genus placed earlier in its own family (Callistopteridae), but transferred to Calopsocidae by New (1977i). Calopsocinae includes the other seven genera currently recognised (New, 1978d; Thornton & Smithers, 1984). However, the subfamily division acknowledged by Smithers (1990a) was not used by Thornton & Smithers (1984) who, however, noted that *Callistoptera* lacks the highly modified vertex found in most other genera except *Torrepsocus*. The following key is directly to genus level.
Six of the eight genera listed below are, as far as present information implies, endemic to New Guinea, where five named species of *Neurosema* also occur. Altogether, 26 of the 33 described species are known from New Guinea, with 24 not known from elsewhere. Nothing is known of the biology of Calopsocidae. All known species have been captured by beating vegetation, mostly in forests, or in light traps or Malaise traps. Capture localities range from almost sea level (mangroves at 2 m) to well over 1000 m above sea level. It is possible that Calopsocidae are associated with dead foliage.

Key to the genera of Calopsocidae

1	Vertex rounded dorsally, without median emargination ▸	2
–	Vertex strongly narrowed to form transverse ridge dorsally, with strong median emargination ▸	3
2	Hind wing vein Rs simple ▸	*Callistoptera*
–	Hind wing vein Rs 2-branched ▸	*Torrepsocus*
3	Fore wing with anterior apical lobe, caused by emargination of postapical border, at or near position of vein M reaching margin ▸	*Neurosema*
–	Fore wing without anterior apical lobe as above ▸	4
4	Anterior and posterior margins of fore wing almost parallel; apical half of wing little, if any, wider than basal half ▸	*Nemupsocus*
–	Posterior margin of fore wing curved; wing widest subapically ▸	5
5	Areola postica of fore wing acutely angled at apex ▸	*Calosema*
–	Areola postica of fore wing not acutely angled at apex ▸	6
6	Fore wing vein M 5-branched; fore wing vein Rs 4- or 5-branched (male epiproct as long as, or longer than, broad) ▸	*Dendropsocus*
–	Fore wing vein M 3-branched; fore wing vein Rs 2- or 4-branched (male epiproct broader than long) ▸	7
7	Fore wing narrower in basal half than distally; length : breadth at nodulus more than 3 ▸	*Cyclopsocus*
–	Fore wing ovoid; length : breadth at nodulus less than 2.9 ▸	*Calopsocus*

Generic information

Callistoptera Enderlein (figs. 274-275, 285). Type species *Callistoptera anna* Enderlein.
Generic distribution: New Guinea.
This genus is known only by the type species, which appears to be rare, as very few specimens are known. It is also very distinctive and is among the largest Calopsocidae, with the fore wing length exceeding 5 mm.

Calopsocus Hagen (figs. 276-284, 286). Type species *Psocus infelix* Hagen. Generic synonymy: *Dirla* Navás, type species *Dirla javana* Navás, see Thornton & Smithers (1984); *Mindaus* Navás, type species *Mindaus irretitus* Navás.
Generic distribution: Sri Lanka, Thailand, Malaysia, Philippines, Indonesia, New Guinea, Solomon Islands, Australia (including Christmas Island), New Caledonia.

Calopsocus, with eighteen described species, is by far the most diverse and widely distributed genus in the family, extending from Sri Lanka to Australia and New Caledonia, and has representatives in many parts of South-east Asia. The species are alike on wing shape, but differ considerably in the complexity of secondary fore wing reticulate venation and patterning (several examples are therefore depicted here), with some taxa having the fore wing markedly convex. *Dirla* was distinguished earlier on the fore wing reticulate venation being limited to the region anterior to vein Rs, but this character is not now regarded as having generic value and no clear genitalic differences are found between *Calopsocus* and the species placed formerly in *Dirla*.

The female antennal flagellum of several species is incrassate and this feature was used by Roesler (1944) to give *Mindaus* subgeneric status. The more widespread nature of this feature was noted by Thornton & Smithers (1984), who designated several species groups within *Calopsocus*, as follows:

1. the '*marginatus*-group' (*C. marginatus*, *C. javanus*, *C. furcatus*), from West Malaysia, Java, New Guinea, with a posterior hyaline margin to the fore wing membrane, the male clunial comb with a row of setae immediately anterior, and the female dorsal gonapophysis valve apically rectanguloid with a longitudinal sclerotised vane at right angles to the plane of the main valve;
2. the '*infelix*-group' (*C. infelix*, *C. guttatus*, *C. solomonensis*, *C. koridensis*) lacks the hyaline posterior border to the fore wing membrane and the setae anterior to the male clunial comb. The female dorsal gonapophysis valve is lobed apically, with a broad V-shaped incision on the posterior margin. *Calopsocus infelix* is the most widely distributed member of the genus, being known from Sri Lanka, Bangladesh, West and East Malaysia, Indonesia and Papua New Guinea. Females of the above species are very similar to each other and can be identified only by association with the more distinctive males;
3. the '*rizali*-group' (*C. rizali*, *C. pallistigma*, *C. ovatus*) share a similar restricted pattern of fore wing venation, a multiple comb on the male clunium, and a widened region about half the length of the long pair of endophallic sclerites. All are known only from the Philippines.
4. The '*reticulatus*-group' (*C. reticulatus*, *C. reticuloides*) have a very broad fore wing with relatively tapered apex and a very fine pattern of reticulate venation. The lateral parts of the rugose field of the male ninth tergite are linked posteriorly. The two species are both from New Guinea.

Other species are not yet allocated to species groups in this way, although many were allied with the '*rizali*-group' to comprise collectively the '*rizali*-section', by which the other three species groups designated by Thornton & Smithers (1984) comprised a collective '*infelix*-section'. Testing of several cladograms of *Calopsocus* species (Thornton & Smithers, 1984) suggested that several of the species characters (a hyaline posterior fore wing band, incrassate female antennal flagella and configuration of two pairs of long and one pair of short endophallic sclerites) had each arisen several times in the genus.

Calosema Thornton & Smithers (fig. 288). Type species *Calosema browningae* Thornton & Smithers.
Generic distribution: Papua (Irian Jaya).
This monotypic genus was distinguished from *Calopsocus* by the unusual shape of the areola postica, other features of fore wing venation, and the presence of groups of marginal teeth on the anterior of the male epiproct. The external gonapophysis valve is broadly rounded and the dorsal valve has a deep apical indentation. This genus appears intermediate between *Calopsocus* and *Neurosema*.

Cyclopsocus Thornton & Smithers (fig. 287). Type species *Cyclopsocus hyalinus* Thornton & Smithers.
Generic distribution: New Guinea.
The five species of this genus are separable by differences in wing venation and colour, together with features of the male clunium and phallosome. The wing shape is not as exaggerated as in many other calopsocids and resembles that of many other Psocoptera. Accessory venation is confined to the anterior fore wing region, in front of vein Rs. The membranes of both wings are setose. Thornton & Smithers (1984) noted the similarities to *Dendropsocus* (below), which also has a 'generalised' wing shape. The two genera differ as indicated in the key above, as well as in male clunial ornamentation: in *Cyclopsocus*: this comprises a posterior comb of one or two rows of teeth and an 'omega-shaped' pattern of tubercles, whereas in *Dendropsocus* the surface ornamentation is rugose.

Dendropsocus Thornton & Smithers (fig. 289). Type species *Dendropsocus swartensis* Thornton & Smithers.
Generic distribution: Papua (Irian Jaya).
A monotypic genus, probably closely related to *Cyclopsocus*, as noted above.

Nemupsocus New (fig. 291). Type species *Nemupsocus lobatus* New.
Generic distribution: Papua New Guinea.
Known only from the type species, *Nemupsocus* is distinctive on the very deeply notched vertex and the shape of the fore wing.

Neurosema McLachlan (figs. 292, 294). Type species *Neurosema apicale* McLachlan.
Generic distribution: Philippines (a single individual noted by Thornton & Smithers (1984), not named), New Guinea (all known species, five of them named).
The fore wing shape of *Neurosema* is very distinctive (New 1978b, although the apical lobe is only weakly developed in one species, *N. alobum* Thornton & Smithers) and the species are separable to some extent by differences in fore

wing hyaline area distributions. All are large (fore wing length ca 5-7 mm). The male clunial comb is weakly defined and the hypandrium 3-lobed with the lateral lobes strongly hooked. The female dorsal gonapophysis valve is very deeply divided apically.

Torrepsocus Thornton & Smithers (figs. 290, 293). Type species *Dirla pulleni* Smithers. Generic distribution: New Guinea.
Torrepsocus is known from two species and is unusual in the family in having the vertex unmodified. It is characterised by the long fore wing with a strong indentation at the apex of the pterostigma (as in *Callistoptera*), the very limited secondary venation and strong sclerites of the male penial bulb.

3 Ectopsocidae (figs. 295-304)

Diagnosis

Fore wing length ca 2-2.5 mm. Lacinial apex slender, bicuspid, usually with the outer cusp broader than the inner cusp. Vertex setose. Tarsi 2-segmented; claws without subapical tooth, pulvillus broad. Fore wing without areola postica (vein Cu1 unforked); pterostigma almost rectangular, not expanded in central region; setae on veins usually short and well-spaced. Hind wing generally lacking setae, but some may be present on margin of radial fork. Veins Rs and M usually joined by crossvein. Male: clunium often ornamented considerably, sometimes with strong posterior expansion; hypandrium simple; phallosome complex, endophallus with asymmetrical sclerites. Female: subgenital plate with median lobe or more or less bilobed at margin; gonapophyses with ventral and dorsal valves more or less reduced, sometimes absent; external valve broad and rounded, often short, always setose.

Remarks

Ectopsocidae are small, generally brown psocids which are sometimes very common on foliage, many of them being particularly abundant on dead leaves. They lay eggs in batches on foliage and nymphs sometimes aggregate under loose webbing. Ectopsocidae are mostly less than about 2.5 mm long and hold their wings in a very shallow tent over the body. The wings often appear spotted, sometimes because the apex of the main veins is shaded with brown, but in many other species all veins are unshaded.
The two genera found in South-east Asia are both very widespread, and many of the species are superficially similar in appearance. Their recognition can be confused by alary polymorphism, and genitalic examination is necessary to confirm specific identity. No formal divisions between family and genus level are recognised, but several genera raised by Li Fasheng (2002) for Chinese species need further appraisal.

Ectopsocidae was treated formerly as the subfamily Ectopsocinae of the Peripsocidae (p. 123), with the two groups associated mainly by lacking the areola postica in the fore wing. However, on genitalic and other features, Ectopsocidae are clearly distinct. The venational combination of rectangular pterostigma and no areola postica is distinctive for the family within the Psocomorpha with 2-segmented tarsi.

Key to the genera of Ectopsocidae

1 Male: clunium with row of teeth or other ornamentation. Female: all three pairs of gonapophyses present; subgenital plate usually bilobed and setose ▸ *Ectopsocus*

– Male: clunial ornamentation more complex, often with a row of teeth but in addition may be extensive asymmetrical sclerotisation. Female: gonapophyses represented by only external valves; subgenital plate usually with median posterior lobe ▸ *Ectopsocopsis*

Generic information

Ectopsocus McLachlan (figs. 295-299, 301, 303, 304).Type species *Ectopsocus briggsi* McLachlan. Generic synonymy: *Micropsocus* Enderlein, type species *Micropsocus waterstradti* Enderlein; *Chaetopsocus* Pearman, type species *Chaetopsocus richardsi* Pearman; *Interpsocus* Edwards, type species *Interpsocus brunneus* Edwards (Note: *Interpsocus* was raised on the hind wing venational feature of veins Rs and M being fused for a short length rather than being linked by a crossvein. This feature seems simply to be a relatively unusual condition within *Ectopsocus* and genitalic features are of similar form throughout).

Generic distribution: cosmopolitan.

As implied above, *Ectopsocus* is very widespread and species have been described from many parts of the world. Taxa allied to *E. briggsi* are amongst the most widespread of all 'non-domestic' psocids – even allowing for possible misidentifications with very similar taxa, '*E. briggsi*' has been recorded from North and South America, Europe, much of Africa, India, Australia, New Zealand. Numerous species of *Ectopsocus* occur in South-east Asia and most of them are macropterous.

Ectopsocus venosus Lienhard from Sabah (East Malaysia) is unique in its proliferation of fore wing veins (figs. 297-298): in this species, Rs has five branches and M six or seven branches. Several South-east Asian species, including *E. comitus* Thornton & Wong, *E. maindroni* Badonnel and *E. cinctus* Thornton, were transferred by Li Fasheng (2002) to his new genus *Estipulaceus* Li Fasheng, which appears to be very similar to true *Ectopsocus*.

Ectopsocopsis Badonnel (figs. 300, 302). Type species *Ectopsocus balli* Badonnel. Generic distribution: North America, Africa, Europe, Hong Kong, Taiwan, Japan, Malaysia, Indonesia.

There are few specific records of *Ectopsocopsis* from South-east Asia, but it is included here because a number of taxa still formally placed in *Ectopsocus* have not been examined critically. Females in particular of the two genera are distinctive and can be diagnosed easily on genitalic structure. *Ectopsocopsis cognata* Thornton & Wong has been recorded from Malaysia and *E. decorata* (Thornton & Wong) from Indonesia. *Ectopsocopsis* is much less diverse than *Ectopsocus*.

4 [**Elipsocidae**] (figs. 305-308)

Diagnosis

Fore wing length ca 2.6 mm. Lacinial apex often moderately wide, bicuspid, usually with outer cusp wider than inner cusp, and usually without denticles. Tarsi 2- or 3-segmented; claws with subapical tooth, pulvillus slender or with slight dilation toward apex. Macropterous or with sexual alary dimorphism: females of many genera and males of some are apterous, micropterous or brachypterous. Brachypterous morphs are associated with neotenic features. Macropterous morph: pterostigma and areola postica of normal form (areola postica sometimes absent); short setae in single rank on veins; hind wing margin glabrous except, in some taxa, for few setae on radial fork. Abdomen without ventral eversible vesicles. Male: hypandrium simple or bilobed; phallosome simple, endophallus without distinct sclerites but radular area sometimes generally sclerotised. Female: subgenital plate broad, often with two incipient or more pronounced apical lobes with long setae; gonapophyses complete, ventral and dorsal valves usually with apical processes, external valve large, broad, and often almost as long as dorsal valve.

Remarks

Elipsocidae have not been recorded from the region covered in this review, but the family is included here because of the presence of a species of *Elipsocus* in Hong Kong, suggesting the strong likelihood that their putative absence in other parts of tropical South-east Asia may reflect undercollecting. The family is diverse in Australia, and is generally better represented in temperate regions than in the tropics.

Generic information

Elipsocus Hagen (figs. 305-308). Type species *Elipsocus westwoodii* McLachlan (a synonym of *E. pumilis* (Hagen)). Generic synonymy *Cabarer* Navás, type species *Cabarer fasciatus* Navás.

Generic distribution: North and Central America, Europe, Africa, Sri Lanka, China, Hong Kong.

Elipsocus species are sometimes abundant on trees in the northern temperate region, but there is little information on the genus elsewhere. *Elipsocus oligotrichus* Thornton from Hong Kong differs from many 'typical' *Elipsocus* in that fore wing vein Cu2 is glabrous rather than setose (see fig. 17 in Thornton, 1959b), and the vein setae are short and confined to the basal half of the wing. The female subgenital plate lobes are not well-differentiated, but are incipiently distinct, and the species was regarded as 'rare' in Hong Kong (Thornton, 1959b). The species are typically bark-frequenting psocids.

5 **Lachesillidae** (figs. 309-312)

Diagnosis

Fore wing length ca 1.8-2.2 mm. Lacinial apex bicuspid, with the two cusps more or less equal in size, sometimes denticulate. Tarsi 2-segmented (except in the South American *Eolachesilla* Badonnel: 3-segmented); claws with subapical tooth, pulvillus slender or slightly expanded. Vertex and frons usually with numerous setae. Fore wing glabrous or with setae short and very sparse*; hind wing glabrous except (rarely) for a few small marginal setae on the radial fork; areola postica normal. Hind wing veins Rs and M fused for a length. Male: hypandrium with distinct 'claspers' laterally and converging on midline; phallosome open posteriorly, often elongate and very variable in form; endophallus not evident. Female: subgenital plate broad, transverse or sometimes incipiently lobed; gonapophyses reduced; only external valves present, these usually elongate with rounded apex, and setose.

This diagnosis is largely for Lachesillinae. The only genus recorded in South-east Asia is *Lachesilla* and this seems to be represented only by a few species. *Lachesilla* is particularly diverse in the New World and a number of well defined species groups are recognised (García Aldrete, 1974a), mainly on characteristic wing and genitalic features. García Aldrete (1988a) believed that the poor representation of Lachesillidae in the Oriental Region is genuine, rather than reflecting inadequate collecting and suggested that the '*Lachesilla* niche' may there be occupied by the very diverse Ectopsocidae.

Remarks

Lachesillidae are closely related to Elipsocidae and the two families may be phylogenetic sister groups (Mockford & Sullivan, 1986). The family was established by Pearman (1936a) to include *Lachesilla* Westwood. The higher classification was refined by Mockford & Sullivan (1986), who defined two subfamilies: Lachesillinae (including *Lachesilla*, *Hemicaecilius* Enderlein, *Homoeolachesilla* Li Fasheng and *Nadleria* Badonnel & García Aldrete) and Eolachesillinae (a New World group containing a suite of genera, most of which are confined to Central and South America). Lachesillidae are relatively small psocids. Only *Lachesilla* has been recorded from South-east Asia.

* *L. hirsuta* Lienhard (Thailand) is exceptional in having a row of relatively well-developed long hairs on most fore wing veins, and sometimes with two rows on vein 1A, and Cu2 bare (Lienhard, 2002).

Generic information

Lachesilla Westwood (figs. 309-312). Type species *Termes fatidicus* Linnaeus. Generic synonymy: *Pterodela* Kolbe, type species *Hemerobius pedicularius* Linnaeus; *Leptopsocus* Reuter, type species *Leptopsocus exiguus* Reuter; *Terracaecilius* Chapman, type species *Terracaecilius pallidus* Chapman.
Generic distribution: North, Central and South America, Europe, Africa, Madagascar, Russia, Himalayan region, Pakistan, China, Japan, Thailand, Malaysia, Australia, Polynesia, Micronesia.
Lachesilla species are mostly associated with dead foliage and many are relatively small 'brown' psocids, somewhat resembling ectopsocids in size but with the wings held much more steeply over the body. As in Ectopsocidae, the wings are usually unmarked or with the membrane near the tip of the veins darkened. They can occur in substantial numbers (although Lachesillidae have been reported very infrequently from South-east Asia). *Lachesilla pedicularia* is globally widespread and is the only regional species reported in extensive recent collections. *Lachesilla brinchangensis* New & Lee was described from Malaysia and *L. hirsuta* Lienhard from Thailand.

6 Mesopsocidae (figs. 313-316)

Diagnosis

Fore wing length (males) ca 3.8-4.2 mm, females of most species micropterous. Lacinial apex bicuspid, both cusps rugose, the outer one the larger. Macropterous (or with females micropterous). Tarsi 3-segmented (sometimes 2-segmented in neotenic forms); claws with a subapical tooth, pulvillus slender or with apex slightly dilated. Wings glabrous; sometimes with few minute marginal setae, in hind wing these (if present) confined to radial fork. Pterostigma and areola postica of normal form; fore wing with one anal vein. Abdomen without ventral eversible vesicles. Male: hypandrium simple; phallosome without sclerites in endophallus. Female: subgenital plate usually with distal process in two parts – a short neck with two sclerotised bars, and a rounded or pointed distal lobe. Gonapophyses complete, ventral and dorsal valves each with distal pointed dorsal process; external valve broad, setose.

Remarks

Mesopsocidae do not occur commonly in South-east Asia but are included here because of an isolated record from Singapore (New & Thornton, 1976) and because they 'surround' the area, with species in China and Australia. However, the family is most diverse in Africa and the western Palaearctic region, with representatives also in North America, Chile and Japan. The predominant genus of the family is noted below. Most species are bark-frequenting psocids.

The family is difficult to separate clearly from Elipsocidae. The former family Psoculidae shares many features with some mesopsocid genera and, following Badonnel & Lienhard (1988b), is now included within Mesopsocidae.

Generic information

Mesopsocus Kolbe (figs. 313-316). Type species *Hemerobius unipunctatus* Müller. Generic synonymy *Trocticus* Bertkau, type species *Trocticus gibbulus* Bertkau; *Holoneura* Tetens, type species *Elipsocus laticeps* Kolbe; *Labocoria* Enderlein, type species *Mesopsocus diopsis* Enderlein.

Generic distribution: North America, widely in Palaearctic to Hong Kong, Africa, Singapore (unidentified species).

As Badonnel & Lienhard (1988) commented, '*Mesopsocus*' had become somewhat of a holding genus for a considerable variety of Mesopsocidae and, even after that family revision (in which eleven genera were recognised), 45 of the 66 species then known were still referred to *Mesopsocus*, rendering it by far the largest genus in the family.

Mesopsocus species are generally large. Females of many species are micropterous. In macropterous specimens, veins R and (M+Cu1) are fused for a short length at the base of the hind wing, whereas these veins are separate from their origins in other genera. *Mesopsocus hongkongensis* Thornton, with macropterous females, is known from Hong Kong, Taiwan and Japan. The Singapore species may be closely related. *Mesopsocus* lack a placoid sensillum at the apex of antennal flagellum segment vi, in contrast to several otherwise similar genera.

7 Peripsocidae (figs. 317-327)

Diagnosis

Fore wing length ca 1.8-2.4 mm. Lacinial apex slender, bicuspid, with outer cusp usually broader than inner cusp. Vertex smooth or with few small setae. Tarsi 2-segmented; claws with subapical tooth, pulvillus slender. Fore wing without areola postica; pterostigma of normal form, more or less expanded in central region. Wing setation usually reduced or absent, occasionally more conspicuous; hind wing glabrous, with veins Rs and M fused for a length. Male: hypandrium simple; phallosome a closed frame, with pair of poriferous parameres; aedeagal arch usually with a distinct apical constriction or process; endophallus with symmetrical sclerites. Female: subgenital plate often with distinct, commonly bilobed, apical region; gonapophyses complete; dorsal valve large, usually setose at apex; external valve large, setose.

Remarks

Peripsocidae are common on dead foliage, bark or (more rarely) low vegetation, and can sometimes occur in very large numbers. Eggs are laid in batches, sometimes communally (New, 1985b).

As with Ectopsocidae, with which they were formerly associated by being placed in the same family (as the subfamily Peripsocinae), there are few genera. The predominant one (*Peripsocus*) is very widespread. Most species are within the size range of 2-4 mm long and are variously brown or grey. The wings are held in a steep 'tent' over the body. In many species the fore wings are fumose or marked with diffuse brown bands or spots.

The diagnosis below encompasses the former subfamily Peripsocinae, which is now of family rank (see also Ectopsocidae, above). Two genera occur in the region (see remarks below, concerning additional genera separated by Li Fasheng, 2002).

The two genera are very similar in general appearance and can be distinguished reliably only on genitalic structures, as above. Many regional species currently placed in *Peripsocus* need more detailed examination to ensure their correct generic placement. The third main genus included in the family is *Kaestneriella* Roesler, from the New World. It is most similar to *Peripsocus* (Mockford & Wong, 1969) and some species have conspicuous setae on the fore wing veins, and fore wing cell r5 tends to be relatively wider than in *Peripsocus*. The Chinese *Coniperipsocus* Li Fasheng now includes several species described as *Diplopsocus* by Li Fasheng & Mockford (1993). The genera *Bicuspidatus* Li Fasheng, *Cycloperipsocus* Li Fasheng, *Periterminalis* Li Fasheng, *Properipsocus* Li Fasheng and *Turriperipsocus* Li Fasheng are all separated on rather small details and are difficult to recognise unambiguously; they are not separated from *Peripsocus* in this account.

Key to the genera of Peripsocidae

1 Male: clunium with caudal row or 'comb' of teeth or other expansion/ornamentation. Phallosome without broad basal plate: if basal plate present, this no broader than remainder of phallosome and commonly narrower. Endophallic sclerites often 'anchor-shaped'. Female: subgenital plate usually with distinct median apical lobe, this simple or incipiently divided at apex; gonapophyses with external valve usually at least 0.4 times length of dorsal valve ▸ *Peripsocus*

– Male: clunium straight, not expanded, and lacking any row or comb of teeth. Phallosome with broad basal plate, often expanded anteriorly and always at least as broad as rest of phallosome frame. Endophallic sclerites not 'anchor-shaped'. Female: subgenital plate without distal median lobe, margin with two more or less distinct lobes; gonapophyses with external valve relatively small, about 0.25-0.3 times length of dorsal valve ▸ *Diplopsocus*

Generic information

Diplopsocus Li Fasheng & Mockford (figs. 319-320, 322, 324, 326). Type species *Diplopsocus cupressicolus* Li Fasheng & Mockford.
Generic distribution: India, China, Hong Kong, Malaysia, Singapore, Indonesia, Philippines, Australia.
This genus was raised to accommodate 21 new species from China and some species from elsewhere that were previously placed in *Peripsocus*. Additional rearrangements may prove necessary as additional taxa are examined more critically. The characters used to separate the genera in the above key appear to be robust.

Peripsocus Hagen (figs. 317-318, 321, 323, 325, 327). Type species *Psocus phaeopterus* Stephens. Generic synonymy *Peripsocopsis* Tillyard, type species *Peripsocopsis maorica* Tillyard.
Generic distribution: widespread, perhaps cosmopolitan.
By far the more diverse of the two genera noted here, *Peripsocus* is believed to occur throughout tropical South-east Asia. Each of the genera *Periterminalis* Li Fasheng and *Turriperipsocus* Li Fasheng contain single species transferred from *Peripsocus* by Li Fasheng: *Periterminalis hiatus* (Thornton & Wong), Malaysia, and *Turriperipsocus chiangensis* (New) from Thailand, respectively, in addition to a few non-regional species.

8 **Philotarsidae** (figs. 328-339)

Diagnosis

Fore wing length ca 2.2-3.8 mm. Lacinial apex bicuspid, the outer cusp broad. Tarsi 2- or 3-segmented; claws with subapical tooth; pulvillus slender. Macropterous, or occasionally brachypterous. Fore wing with areola postica free; margin and veins setose; M + Cu usually with one rank of setae; at least some hind wing veins setose; marginal setae of both wings may cross in opposing ranks. Abdomen without ventral eversible vesicles. Male: hypandrium simple, without paired lateral lobes; aedeagus rounded at apex; external parameres not, or scarcely, protruding beyond tip of aedeagal arch. Female: subgenital plate with simple apical lobe; gonapophyses complete, external valve large and setose.

Remarks

The Philotarsidae have been studied extensively during the last few decades, with many new South-east Asian species described and with considerable progress toward a satisfactory higher classification within the family. Thornton (1981g) noted the progress from Smithers' (1967a) enumeration of 27 species in the family to about 150 species, with a very high proportion of these from the Australian and West Pacific regions (see Thornton & New, 1977a; Thornton &

Smithers, 1974, 1977; Thornton et al., 1977). The complex history of the family was discussed by Smithers (1972c) and Thornton (1981g), while Mockford (1984d) noted the confusion between some philotarsid genera and some pseudocaeciliid genera arising from over-reliance on tarsal segment number as a defining family characteristic. Mockford thus arranged the array of genera included in Philotarsidae by Thornton (1981b) among three families: Philotarsidae, Pseudocaeciliidae (p. 128) and Bryopsocidae (p. 45), the last raised to contain the New Zealand *Bryopsocus*. The major genera included in Thornton's subfamily Zelandopsocinae were transferred to Pseudocaeciliidae (together with the later-described *Novopsocus* Thornton), so that a more modern perspective of the scope of the family limits it to Thornton's 'Philotarsinae'. Although older acccounts and discussions of Philotarsidae may incorporate any of the above-mentioned taxa, Mockford's concept of the family (as followed by Smithers, 1990a) is used here.

The genus *Philotarsus* Kolbe does not occur in South-east Asia. It is separable from *Haplophallus* by (female) the dorsal gonapophysis valve pointed rather than rectangular, and with a small apical projection, and (male) phallosome with internal sclerites. *Philotarsus* is found in the northern hemisphere and the Caribbean area, with a disjunct (and unconfirmed) record from South America, so that it is largely (? or entirely) Holarctic. Species from other regions and described in *Philotarsus* have all been transferred to other genera.

Key to the genera of Philotarsidae

1 Apex of antenna normal, with whorl of apical setae. Female: subgenital plate without distinct subapical sclerite(s) ▸ ***Haplophallus***

– Apex of antenna strongly attenuated, with single apical seta (fig. 331). Female: subgenital plate with distinct subapical sclerite(s) ▸ 2

2 Fore wing 'strap-like', with sides subparallel; vein Cu2 without setae. Male: hypandrium simple. Female: dorsal gonapophysis valve smoothly rounded at apex ▸ ***Aaroniella***

– Fore wing of more typical shape, widest at level of pterostigma; vein Cu2 with setae. Male: hypandrium bilobed at apex. Female: dorsal gonapophysis valve rectangular ▸ ***Philotarsopsis***

Generic information

Aaroniella Mockford (figs. 329, 331, 333, 334, 337). Type species *Elipsocus maculosus* Aaron.

Generic distribution: widespread. Major distributional centre is from Madagascar eastward through the Oriental region, New Guinea, Australia and New Zealand, to South and Central America. *Aaroniella* is apparently absent from much of western North America, mainland Africa, India and much of the Palaearctic region. European and North American records are probably recent invasions (Thornton, 1981g; Lienhard, 1998a).

The 'strap-like' fore wing shape of *Aaroniella* is very distinctive, but some individuals are brachypterous. In a few species this represents sexual dimorphism, with brachypterous females and macropterous males. In addition, many of the species have the setae on the wing veins arising from dark spots, at least in the basal half of the wing. Vein Rs is sinuous after separating from M; fore wing vein Cu2 is glabrous; the various fore wing cells often contain patches of brown or grey pigment. The antennal segments typically have white apices. The wing shape and attenuated apical flagellar segment are a reliable diagnostic combination for the genus and apply to all morphs. All the species appear to be bark dwellers.

Thornton (1981g) recognised several species groups in *Aaroniella*. The *pulchra*-group comprises three species from South-east Asia and these are clearly closely related. A large sister-group, the *guttulata*-group, is the most diverse array, with species from the Indian Ocean to western Pacific, including the Philippines, Indonesia and New Guinea. Still another group, unnamed by Thornton, includes species from Central and South America, New Guinea and Lord Howe Island. Members of the *pulchra*-group have discrete patches of pigment in the fore wing apical cells and the male fore wing has setae on dark spots only in the basal half of the wing. The endophallic sclerites are large and often with dentations. In contrast, members of the *guttulata*-group lack endophallic sclerites or have only a single pair of small sclerites, the male fore wing setae are on dark spots along the whole length of the wing and there are no distinct pigment spots in the apical fore wing cells. The third group noted above does not have fore wing setae arising from dark spots.

Haplophallus Thornton (figs. 328, 332, 335, 338). Type species *Haplophallus orientalis* Thornton.

Generic distribution: South America, Africa, Sri Lanka, Hong Kong, Malaysia, Singapore, Indonesia, New Guinea, Australia, New Zealand, western Pacific region.

Haplophallus orientalis is relatively widespread, with records from Africa, Hong Kong and Indonesia, but most other species appear to be more restricted in distribution. Thornton (1981g) designated two species groups within *Haplophallus*, as follows, with most regional species belonging to the *orientalis*-group. The *sinus*-group includes one species from the Solomon Islands and further into the Pacific, and species from the Australian region and South America.

– **Key to the species-groups of *Haplophallus***

Note that these groups are defined solely on characters of the female genitalia and that males can not be allocated to one or other group other than by association with females.

1 Female: subgenital plate with apical lobe continuous with disc; dorsal gonapophysis valve rectangular, with very short apophysis; external gonapophysis valve oval ▸ ***orientalis*-group**

– Female: subgenital plate with apical lobe a distinct sclerite, separated from disc by a distinct suture; dorsal gonapophysis valve rounded apically, with distinct subapical apophysis; external gonapophysis valve usually tapered ▸ ***sinus*-group**

Philotarsopsis Tillyard (figs. 330, 336, 339). Type species *Philotarsopsis delicata* Tillyard. Generic synonym: *Latrobiella* Thornton, type species *Haplophallus bundoorensis* New.

Generic distribution: Chile, Falkland Islands, New Guinea, Solomon Islands, Australia, New Zealand, New Caledonia.

Most species of *Philotarsopsis* were described earlier in *Haplophallus*, with a few (mainly those from New Guinea) in *Aaroniella*, and were transferred formally to *Latrobiella* by Thornton (1981b). The genus is more closely related to *Aaroniella*, with which it shares the attenuated antennal apex, than to *Haplophallus*. Most species are more familiar in the literature as members of *Latrobiella*, with the above synonymy established by Lienhard & Smithers (2002).

9 Pseudocaeciliidae (figs. 340-385)

Diagnosis

Most species are in the size range of fore wing length 1.9-3.2 mm. Lacinial apex bifid, with lateral tine rather slender. Tarsi 2- or 3-segmented; claws usually without subapical tooth (occasionally a small tooth present), pulvillus broad. Fore wing with pterostigma usually rather shallow; areola postica usually rather small and shallow; fore wing veins with more than one row of setae; M-Cu stem usually with two rows of setae; marginal setae long and crossing in two ranks behind wing apex; Cu2 glabrous; vein M with two or three branches. Hind wing with setae on branches of Rs and M; margin setose, with setal ranks crossing. Abdomen with ventral eversible vesicles. Male: hypandrium usually with paired lateral lobes or processes; phallosome pointed apically, with external parameres generally extending beyond apex of aedeagal arch; endophallus often with sclerotisations. Female: subgenital plate simple or with apex bilobed; gonapophyses complete, ventral valve often broad; dorsal valve with distal acuminate process; external valve with strong setae.

Remarks

Pseudocaeciliidae are associated mainly with foliage, with most species occurring on either living or dead leaves on trees. In life, many are brightly coloured, in pale green, yellow or straw-brown to match their foliage substrates and many have the fore wing strongly patterned with brown bands or other markings. Brachyptery has been reported in some taxa, but most are macropterous. The considerable variety of claw and pulvillus features in the family denotes taxa associated variously with foliage or bark-dwelling. Some species are true foliage-frequenters (Turner, 1973a), that lay batched eggs on foliage and have well-formed ventral abdominal vesicles.

The family includes a complex array of genera, some of which have a history of transfer between families. Thus, the entire subfamily Zelandopsocinae were initially included in Philotarsidae (Thornton, 1981g, 1984b) and transferred to Pseudocaeciliidae by Mockford (1984d). The 'traditional' character of tarsal segment number used to separate the two families was shown by Mockford not to be reliable, and either family can have species with two or three tarsal segments, whereas much earlier literature separated the families on this character alone (i.e. 'tarsi 3-segmented': Philotarsidae, or 'tarsi 2-segmented': Pseudocaeciliidae). *Bryopsocus* (p. 45) was regarded by Mockford (1984d) as intermediate between the two families. In addition, Pseudocaeciliidae and Calopsocidae (p. 114) also have many features in common.

The predominant subfamily in this family, Pseudocaeciliinae (formerly considered as the entire family), contains an array of rather similar genera, many of them designated originally by Lee & Thornton (1967). All these genera are known predominantly from South-east Asia and the Pacific region. Several of these are very distinctive, but others have not remained as narrow as the original authors intended, so that – for example – *Heterocaecilius* has become a holding genus for a variety of taxa and its scope has recently received comment by Li Fasheng (2000), New (1993) and Lee & New (1992). It was designated originally as a 'holding group' (Lee & Thornton, 1967). Three subfamilies are recognised, but one (Electropsocinae) is represented solely by an amber fossil species (*Electropsocus* Roesler).

The major traditional diagnostic features for the family have been (1) bifid lacinia, (2) numerous setae on head and thorax, (3) fore wing vein setae in two rows, and (4) crossing of setal ranks on apical wing margins, together with (5) male with paired hypandrial lateral sclerites and (6) female subgenital plate bilobed (Yoshizawa, 1995), but genitalic examination is necessary to separate clearly this family from others noted above.

Key to the extant subfamilies of Pseudocaeciliidae

1 Tarsi 2-segmented ▸ **Pseudocaeciliinae**
– Tarsi 3-segmented ▸ **Zelandopsocinae**

Pseudocaeciliinae

Until recently, this group constituted the entire family Pseudocaeciliidae, designated first by Pearman (1936a) with *Pseudocaecilius* Enderlein as a 'typical genus'. A much broader family concept by Roesler (1940c) included three subfamilies (Pseudocaeciliinae, Archipsocinae, Peripsocinae), with Pseudocaeciliini one of seven tribes in Pseudocaeciliinae. That tribe isolated the then-known genera that constitute 'modern Pseudocaeciliidae', and was raised to subfamily rank by Roesler (1944) within a very broad concept of Pseudocaeciliidae that included many of the families of Homilopsocidea. Badonnel (1951) augmented the family *sensu stricto* by addition of *Pseudoscottiella* Badonnel and that arrangement was followed by Lee & Thornton (1967), who revealed the diversity of genera somewhat resembling *Pseudocaecilius* in general appearance, recognising seven distinct genera in addition to *Heterocaecilius*. Some of the genera are very difficult to diagnose, as discussed below, and the following key can be used only with caution, as noted at its end.

Key to the genera of Pseudocaeciliinae

1 Fore wing vein M unbranched, of unusual form, as Cu1a fused anteriorly with M to constitute a cell occupying much of the posterior wing margin ▸ *Allopsocus*
– Fore wing vein M with two or three branches; areola postica of normal form ▸ 2

2 Fore wing vein M 2-branched ▸ *Mepleres**
– Fore wing vein M 3-branched ▸ 3

3 Fore wing Rs stem short, less than half length of R4+5 ▸ *Lobocaecilius*
– Fore wing vein Rs relatively longer, at least half length of R4+5, commonly as long as or longer than R4+5 ▸ 4

4 Areola postica very tall, triangular ▸ *Mesocaecilius*
– Areola postica shallow, often almost semicircular ▸ 5

5 Fore wing Rs stem sinuous ▸ 6
– Fore wing Rs stem straight or almost straight ▸ 7

6 Fore wing Rs stem distinctly S-shaped ▸ *Ophiodopelma*
– Fore wing Rs arched only at base ▸ *Allocaecilius*

7 Male: anal area of fore wing with strong papillate field; endophallus with numerous thorn-like sclerites (female: gonapophysis valves not lobed) ▸ *Phallocaecilius*

– Male: anal area of fore wing without strong papillate field (if, rarely, papillae present, endophallus with rod-like sclerites); endophallus not as above (female: gonapophysis valves varied) ▸ 8

[females can be identified reliably only if associated with males]

8 Male: endophallus unsclerotised. Female: dorsal and ventral gonapophysis valves not lobed ▸ *Pseudocaecilius*

– Male: endophallus usually with rod-like sclerites. Female: dorsal and ventral gonapophysis valves strongly lobed and with spatulate setae ▸ *Heterocaecilius* s.l.** 9

**See note below for use of the following:

9 Tarsal claws without subapical tooth; areola postica usually more than twice as long as high ▸ 10

– Tarsal claws with subapical tooth; areola postica usually less than twice as long as high ▸ 11

10 (Females not identifiable reliably)
Male: posterior margin of hypandrium divided into two sections, with different processes ▸ *Heterocaecilius* s.str.

– Male: posterior margin of hypandrium complete, not divided ▸ *Neocaecilius*

11 Female ▸ 12

– Male ▸ 13

12 External gonapophysis valve 2-lobed, setae mainly on the (smaller) outer lobe ▸ *Levucaecilius*

– External gonapophysis valve 1-lobed ▸ *Chorocaecilius*, *Orbocaecilius*

13 Posterior margin of hypandrium symmetrical, with small lateral processes and a small median prominence ▸ *Chorocaecilius*

– Posterior margin of hypandrium complete, not divided ▸ *Orbocaecilius*

* *Mepleres* was placed in the new subfamily Meplerinae by Li Fasheng (2002), based on the 2-branched fore wing medial vein, with the subfamily including also *Scottiella* Enderlein.

** This key, up to couplet 9, leaves a residue of genera comprising *Levucaecilius* and the various subdivisions of *Heterocaecilius* as separated by Li Fasheng (2000). These are difficult to separate, simply because of incomplete knowledge. The female of *Levucaecilius* is distinct from the others on the external gonapophysis valve being

2-lobed and with setae mainly on the smaller outer lobe. This genus appears to be clearly distinct on female features, but the male has not been differentiated adequately from that of related taxa. Hypandrial form in described species is of the general *Chorocaecilius* form (couplet 13), but considerable variety occurs. The other genera exhibit a mosaic of characters. Females of the other genera included have the external gonapophysis valve with only one lobe, but many females will not be identifiable reliably to genus unless they are associated unambiguously with their corresponding males, and not all of these are known. These genera are treated with *Heterocaecilius* in the following account. Note also the comments under *Levucaecilius* in relation to Li Fasheng's genera. *Orbocaecilius* was recognised as a synonym of *Heterocaecilius* by Lienhard & Smithers (2002), but *Neocaecilius* and *Chorocaecilius* were listed as distinct.

Generic information

Allocaecilius Lee & Thornton (figs. 340, 344, 348, 353, 355). Type species *Allocaecilius heterothorax* Lee & Thornton.
Generic distribution: India, Hong Kong, Malaysia.
This genus was raised to accommodate the type species from India, together with one from Hong Kong and three from Malaysia, and has since been augmented by further West Malaysian species. However, it has not yet been reported from East Malaysia or Indonesia, although it may well occur there. It is one of several genera which are easily confused with *Pseudocaecilius* on wing features, but the basal arching of fore wing vein Rs is distinctive. The dorsal and ventral gonapophysis valves are lobed and the male endophallus contains only rod-like sclerites. The tarsal claws are smooth.

Allopsocus Banks (figs. 341, 345, 349). Type species *Allopsocus marginalis* Banks.
Generic distribution: Sri Lanka, Philippines, Malaysia.
Allopsocus is very distinctive, indeed unique amongst Psocoptera, on its fore wing venation (fig. 341). It was long enigmatic and its family status regarded as *incertae sedis*, with Banks (1920) allying it with Pachytroctidae and related forms. Allocation to Pseudocaeciliidae (New & Thornton, 1976) revealed that it is distinctive also in having the dorsal gonapophysis valve lobed and ventral valve slender, and tarsal claws smooth. The male is unknown.
Allopsocus appears to be rare and none was found in extensive West Malaysian collections studied by Lee & New (1992).

Heterocaecilius Lee & Thornton (figs. 356-360). Designated as a 'collective group' and no type species designated by Lee & Thornton (1967); type species later designated by Mockford (1993a): *Heterocaecilius minotus* Lee & Thornton (given as '*minutus*' [sic] by Mockford (1993a)). Generic synonym: *Orbocaecilius* Li Fasheng, type species *Orbocaecilius paulicellus* Li Fasheng (see Lienhard & Smithers, 2002).
Generic distribution: widespread throughout South-east Asia and Pacific region: India, Japan, Hong Kong, Malaysia, Singapore, Philippines, Indonesia, New Guinea, Micronesia, New Zealand.

This collective group was raised to accommodate a variety of taxa broadly resembling *Pseudocaecilius* in wing venation but which did not fit into any of the more limited genera raised and appraised by Lee & Thornton (1967). Its members reflect the mosaic nature of some of the genera in this family. Since 1967, *Heterocaecilius* has been treated widely as a 'holding genus', to which an increasing variety of regional taxa have been attributed. These species are unified by wing venation, both the ventral and dorsal gonapophysis valves being lobed, and the subgenital plate bilobed, but are more variable in the form of the male hypandrium. The illustrations given here are of the type species.

Lee & Thornton (1967) noted a number of possible groupings of species that they considered not to warrant separate generic status at that time. Smithers (1995a) raised *Levucaecilius* (below) to contain a relatively well-defined segregate. However, Li Fasheng (1999, 2000), in considering Chinese species, restricted *Heterocaecilius* to a narrower concept (as indicated in the above key) and raised three additional genera as subdivisions within the former broad scope: *Neocaecilius* Li Fasheng (type species *Heterocaecilius campanula* Lee & Thornton, from Caroline, Marshall and Gilbert Islands), *Chorocaecilius* Li Fasheng (type species *Heterocaecilius diogenes* Lee & Thornton, New Zealand), and *Orbocaecilius* Li Fasheng, leaving *Heterocaecilius* s.str., for which Li Fasheng designated the type species as *Heterocaecilius simplex* Lee & Thornton (Fiji). This designation was deemed invalid by Lienhard & Smithers (2002).

Li Fasheng (2000) transferred a number of previously described species to each of his new genera, so that the following distributions were implied: *Neocaecilius* (China, Malaysia, Micronesia, Fiji), *Chorocaecilius* (China, New Zealand), *Orbocaecilius* (Nepal, China, Hong Kong, Malaysia, Australia, Fiji, Caroline and Marshall Islands), with the taxa he designated in *Heterocaecilius* s.str. spanning China, Hong Kong, Fiji, Philippines and Malaysia. These four genera were designated, as in the above key, on hypandrial and tarsal claw characters, but many females can not be separated confidently.

The holding genus nature of *Heterocaecilius* thus remains useful, pending further critical examination of the many taxa of closely similar pseudocaeciliids in South-east Asia. Lee & Thornton (1967) noted that the presence/absence of a subapical tooth on the tarsal claw and modification of the pulvillus might be expected to be a common parallel feature and lack generic value in this group, so that the primary separation feature employed by Li Fasheng (2000) may need further validation before his genera can be appraised fully.

Levucaecilius Smithers (figs. 343, 346, 350). Type species *Pseudocaecilius greenwoodi* Karny.

Generic distribution: Fiji, Christmas Island (Australia).

This genus was erected for a group of species considered as Fijian endemics and included in *Heterocaecilius* by Lee & Thornton (1967), together with a new species from Christmas Island described by Smithers (1995a). The most distinctive

feature is the division of the external gonapophysis valve into two lobes with most (in many cases, all) the setae on one of these lobes. This group is indeed distinctive, but there are problems with harmonising some of Li Fasheng's generic allocations with *Levucaecilius*. Thus, the basis for *Levucaecilius* is the '*greenwoodi*-group' of Lee & Thornton (1967), from which *H. greenwoodi* was placed by Li Fasheng in *Orbocaecilius*, *H. dardanus* Lee & Thornton allocated to *Neocaecilius*, and *H. simplex* became the (invalidly designated) type species of *Heterocaecilius* s.str.
Li Fasheng apparently overlooked Smithers' (1995d) account, but it is clear that Li Fasheng's generic concepts in this group differ from those of other workers who have published on Pseudocaeciliidae. *Levucaecilius* is included here because of the close biogeographical affinity of Christmas Island to Indonesia and the essentially Indonesian character of its biota.

Lobocaecilius Lee & Thornton (figs. 342, 347, 351-352, 354). Type species *Lobocaecilius cynara* Lee & Thornton.
Generic distribution: China, Malaysia, Singapore, Indonesia, parts of Pacific region (Caroline Islands, Mariana Islands, Tahiti, Fiji, Hawaiian Islands).
The short stem of fore wing vein Rs is diagnostic for this genus. The tarsal claws lack a subapical tooth and the male endophallic sclerites comprise only small spines. Both ventral and dorsal gonapophysis valves are lobed. The hypandrium is complex and variable, usually with 2-4 unusual 'finger-like' sclerites posteriorly, and the male epiproct is commonly tuberculate.

Mepleres Enderlein (figs. 361, 365, 369, 371). Type species *Mepleres meandricus* Enderlein. Generic synonymy *Pseudoscottiella* Badonnel, type species *Pseudoscottiella megops* Badonnel; *Meniscopsocus* Li, type species *Meniscopsocus sinicus* Li.
Generic distribution: Jamaica, Africa, India, Bangladesh, China, Taiwan, Japan, Malaysia, Indonesia, Australia, New Zealand, Fiji, Micronesia.
Most of the species now included in *Mepleres* were decribed originally in *Pseudoscottiella* – then considered a widespread genus characterised by the fore wing vein M being 2-branched – and the above generic synonymy was established by Yoshizawa (2000a). *Mepleres* was described originally in the caeciliid series, but assigned to Pseudocaeciliidae by Badonnel (1969) and Vaughan et al. (1991). Li Fasheng's (1993) genus *Meniscopsocus* was allocated to Trichopsocidae, but reappraised by Yoshizawa (2000a), leading to the above synonymy.
Yoshizawa formally transferred all species described under *Pseudoscottiella* to *Mepleres* and the species allocated to *Meniscopsocus* by Li Fasheng (1993a), by transfer from *Pseudoscottiella*, also to *Mepleres*. The genus thereby includes all pseudocaeciliids with a 2-branched fore wing media found in South-east Asia. *Scottiella* Enderlein (Seychelles) shares this feature, but the areola postica is semicircular rather than more flattened and elongate and the fore wing costa is not thickened.

Mepleres is diverse and widely distributed in the region, and some species have the fore wings very strongly patterned.

Mesocaecilius Okamoto. Type species *Mesocaecilius quadrimaculatus* Okamoto.
Generic distributiion: Sri Lanka, China, Taiwan, Malaysia, possibly Singapore, Indonesia.
The type species of *Mesocaecilius*, which is only known from Taiwan, is characterised by the very large areola postica, which is steep and triangular; setae are present on the wing membrane. The genus is included here because of ambiguity over the generic placement of '*Pseudocaecilius*' *ornatus* Enderlein (from Singapore). Lee & Thornton (1967) noted that this species has a very similar wing pattern to *M. quadrimaculatus* (as noted also by Badonnel, 1967a) and that further examination might reveal a relationship between the two species. *Mesocaecilius* was transferred to the new subfamily Mesocaeciliinae by Li Fasheng (2002).
In *Mesocaecilius*, the ventral and dorsal gonapophysis valves are both lobed and tarsal claws have a small subapical tooth. *Pseudocaecilius ornatus* was transferred to *Heterocaecilius* by Thornton (1984a) on the basis of specimens from Bali. Thornton's (1984a) *Heterocaecilius rhadnus* (Bali) is similar, but neither of these species (or *Pseudocaecilius paraornatus* New, from Sri Lanka) have the very high areola postica depicted by Okamoto (1910) for *Mesocaecilius*, and this character remains the major basis for maintenance of the genus.

Ophiodopelma Enderlein (figs. 362, 366). Type species *Ophiodopelma ornatipenne* Enderlein. Generic distribution: China, Taiwan, Japan, Sri Lanka, Philippines, East Malaysia, Indonesia, New Guinea.
The fore wing venation of *Ophiodopelma* is very characteristic, both in the strongly sinuous Rs stem and the acute anterior angle of the areola postica. The species are among the smallest and most 'delicate-looking' pseudocaeciliids, with some having dark markings or stripes in the fore wing cells around the wing apex. Both ventral and dorsal gonapophysis valves are lobed; the hypandrium lacks accessory sclerites of projections and the endophallus has long sclerites. The tarsal claws are smooth.
One species from Sabah, *O. glyptocephalum* Lienhard (1985c), has remarkable head morphology, with the male having apparently unique sensory areas on the frons. The function of these is unknown.

Phallocaecilius Lee & Thornton (figs. 363, 367, 370, 372). Type species *Pseudocaecilius hirsutus* Thornton.
Generic distribution: China, Hong Kong, Japan, Indonesia, Christmas Island (Australia).
This genus was raised for an apparently isolated species, *Phallocaecilius hirsutus*, with remarkable development of endophallic sclerites as a series of thorn-like projections on a median 'ribbon'. This species, originally from Hong Kong, is

known now from Sumatra, Java and the Krakatau Islands (Vaughan *et al.*, 1989) and Christmas Island (Smithers, 1995d). The Indonesian *Heterocaecilius brevifurca* Thornton (Bali) may be closely related, perhaps synonymous.

The phallosome form is unique in Pseudocaeciliidae, but females are 'more typical pseudocaeciliids'. According to Lee & Thornton (1967) and Yoshizawa (1996a), *P. hirsutus* is viviparous.

Pseudocaecilius Enderlein (figs. 364, 369, 373). Type species *Pseudocaecilius elutus* Enderlein. Generic synonymy *Hageniella* Enderlein, type species *Psocus zonatus* Hagen.

Generic distribution: North and Central America, Africa, India, Sri Lanka, Nepal, Bhutan, China, Taiwan, Hong Kong, Japan, Korea, Malaysia, Singapore, Philippines, Indonesia, New Guinea, Australia, Pacific region (including Micronesia, Tonga, Fiji, Hawaiian Islands, Galápagos Islands).

The type species is a synonym of *P. citricola* (Ashmead), which is virtually pantropical, with numerous synonyms, but most species reported from South-east Asia appear to be much more localised. Many differ only in details of wing pattern or genitalia.

The name *Pseudocaecilius* was formerly applied in a much wider sense than at present, encompassing many of the taxa in the above genera with a 3-branched fore wing media. *Pseudocaecilius* and *Ophiodopelma* were treated as subgenera of *Pseudocaecilius* by Roesler (1944). Genitalic features noted in the above key define the current scope of the genus.

Zelandopsocinae

Key to the genera of Zelandopsocinae

The key indicates the likely close relationships between *Austropsocus* and *Zelandopsocus*, with some species transcending character states found in most species of each genus. *Novopsocus* resembles *Austropsocus* in that the female subgenital plate lobes do not overlap and the phallosome lacks the ribbon-sclerite and has only membranous endophallic sacs. As Thornton et al. (1977) noted, parallel apomorphies within the genera of Zelandopsocinae can make accurate assessment of relationships, and definition of species groups, difficult.

1 Head capsule flattened, so vertex with sharp posterior edge; postclypeus scarcely protruding; fore wing long and narrow, nearly parallel-sided ▸ **Novopsocus**

– Head capsule normal; fore wing of normal shape, not exceptionally narrow ▸ 2

2 Female: subgenital plate 2-lobed, with lobes setose and overlapping; epiproct with dorsal setose flap. Male: middle lobe of hypandrium 'mushroom-shaped'; phallosome with spines, endophallic sacs and ribbon-sclerite ▸ *Zelandopsocus*

– Female: subgenital plate posterior lobe with 1-2 setae on each lobe, may be only incipiently divided; epiproct usually without dorsal flap. Male: middle lobe of hypandrium not 'mushroom-shaped'; phallosome lacking ribbon-sclerite, usually lacking endophallic sacs (some exceptions) ▸ *Austropsocus*

Generic information

Austropsocus Smithers (figs. 375, 379, 381, 384). Type species *Austropsocus insularis* Smithers.

Generic distribution: New Guinea, Australia, New Zealand and offshore islands, New Caledonia.

This genus was raised to contain a brachypterous species, but most others appear to be entirely macropterous. About 32 species have been described, mostly from Australia or New Zealand. Two species (*A. novoguinensis* (Thornton & Smithers), *A. perforatus* Thornton & Smithers) are known from New Guinea.

Novopsocus Thornton (figs. 374, 377, 378, 383). Type species *Zelandopsocus stenopterus* Thornton & Smithers.

Generic distribution: New Guinea.

The unusual features of the type species were noted by Thornton & Smithers (1977) and it was later transferred to *Austropsocus* by Thornton (1981g) before becoming the basis of a separate genus (Thornton, 1984c). In addition to the highly unusual form of the wings, the male basal antennal flagellar segment is greatly enlarged and flattened.

Zelandopsocus Tillyard (figs. 376, 380, 382, 385). Type species *Zelandopsocus formosellus* Tillyard.

Generic distribution: Australia, New Zealand, New Caledonia.

Probably not found in South-east Asia, but noted here because two New Guinea species were originally decribed in the genus (Thornton & Smithers, 1977). One is the type species of *Novopsocus*, above, and the other was transferred to *Austropsocus* by Thornton (1981g); this species, *A. novoguinensis* (Thornton & Smithers) is somewhat different from most members of the genus and manifests the 'exceptional' character states given in parentheses in couplet 2 of the above key.

Group PSOCETAE

The four families included in this series comprise two (Hemipsocidae, Psilopsocidae) that are relatively small, each containing a single regional genus, and two (Myopsocidae, Psocidae) that are much more dominant components of the regional psocid fauna.

1 Hemipsocidae (figs. 386-391)

Diagnosis

Fore wing length ca 2.5-2.8 mm. Lacinial apex divided, with inner cusp preapical and smaller than outer cusp. Head with vertex acute, with a strong median emargination. Tarsi 2-segmented, claws with small subapical tooth, pulvillus broad. Fore wing: areola postica joined to M by crossvein; vein M 2-branched; veins Rs and M often meeting at a point; veins (except Cu2) with one row of setae, margin setose. Hind wing glabrous. Male: hypandrium simple; epiproct with rugose area, often more or less triangular; phallosome frame narrow anteriorly, open or weakly closed posteriorly, endophallus with rugose sclerotisation. Female: subgenital plate simple, rounded posteriorly; gonapophyses complete; ventral and dorsal valves relatively broad, pointed; external valve large, triangular, with very few setae.

Remarks

Hemipsocidae is a small family of tropical psocids, usually associated with foliage and leaf litter. Several species appear to be widespread and common. They are very distinctive in appearance. Although placed by some earlier workers (Badonnel, 1951; Smithers, 1972c) in the Homilopsocidea, they are now placed in the Psocetae (Mockford, 1974a, 1993a), in which the family appears to be rather isolated and distinctive.

All regional taxa belong to the genus *Hemipsocus*, one of only three genera in the family. The other clearly diagnosed one is *Anopistoscena* Enderlein (Seychelles), which has the distinctive feature of the distal portion of fore wing vein Cu1a missing, so that the areola postica is fused with cell m3. It was redescribed and figured by Smithers (1994c). The third genus, *Metahemipsocus* Li Fasheng, is known from China. It is very similar to *Hemipsocus* and only doubtfully distinct.

Generic information

Hemipsocus Selys-Longchamps (figs. 386-391). Type species *Psocus chloroticus* Hagen. Generic distribution: North, Central and South America, West Indies, Cuba, Africa, India, Sri Lanka, Seychelles, Japan, Bhutan, Taiwan, Vietnam, Thailand, Malaysia, Singapore, Philippines, Indonesia, New Guinea, Australia (including Christmas Island), Samoa, Vanuatu, Hawaiian Islands.

The general appearance of living *Hemipsocus* is of small, pale brown, fast-running and fluttering moths. Two main groups of species occur in the area. In one, typified by *H. luridus* Enderlein, the fore wing vein setae are sited on dark spots (fig. 386). In the other group, species such as *H. chloroticus* (Hagen) have the wings wholly pale (fig. 387). The species of both groups differ on details of body colouration and genitalic structure, but their taxonomy is still somewhat confused. Both the above-named species are very widely distributed in the region.

In some specimens of *H. luridus* from Christmas Island, the fore wings are somewhat shortened, not reaching the apex of the abdomen (Smithers, 1995d).

2 Myopsocidae (figs. 392-410)

Diagnosis

Fore wing length commonly up to about 4.5 mm, occasionally larger.
Lacinial apex with small outer tine and broader inner cusp with several 'teeth'. Tarsi 3-segmented; claws with strong subapical tooth, pulvillus moderately broad. Fore wing strongly patterned with grey/brown, often with margin having short alternating dark and pale longitudinal bars. Pterostigma normal; vein Cu1a fused with M to constitute closed discoidal cell. Both wings glabrous; sometimes an extended 'flap' on fore wing vein Cu2. Male: hypandrium symmetrical, simple; phallosome a closed frame (often with median longitudinal sclerotised bar) or pair of rods; epiproct often with processes or lobes; paraproct with pointed process(es). Female: subgenital plate with elongate tapered apical lobe from broad discal region; gonapophyses complete; ventral and dorsal valves with long slender apices; external valve rounded or quadrate, with numerous setae.

Remarks

Myopsocidae are a family of moderate-sized to large psocids and are apparently closely related to Psocidae. They are intricately patterned grey or brown and associated with tree bark. The family is widely distributed, although associated primarily with tropical regions, and diverse in South-east Asia. A number of recent studies on the fauna of New Guinea and nearby areas (Smithers & Thornton, 1974b, 1979) and Indonesia (Thornton & Browning, 1992) have added greatly to perspective of the regional fauna.
Myopsocidae has been recognised as distinct and placed in Psocetae in most classical schemes (Pearman, 1936a; Roesler, 1944; Badonnel, 1951) and this arrangement has been followed without any serious doubt by later authors. However, generic status in the family has proved much more complex to evaluate and many species have had a series of different generic attributions in their taxonomic history (see Smithers, 1975c). Some explanation of this dilemma and history of confusion is given here, to help explain the 'current position'.
As Mockford (1982) commented, the four genera raised by Enderlein (*Lophopterygella, Lichenomima, Phlotodes* and *Rhaptoneura*) for long contained all species except *Myopsocus* included in the family. They were raised on wing features (the last three on venational characters) that have not always proved reliable for consistent separation. Considerable confusion occurred when (1) Roesler (1944) synonymised *Phlotodes* and *Rhaptoneura* at the genus level whilst retaining them as subgenera, (2) Badonnel (1955) maintained these on genitalic features but without specifying the features involved, and (3) Smithers (1964d) assigned all

species falling into *Myopsocus*, *Lichenomima*, *Phlotodes* and *Rhaptoneura* to *Myopsocus*, pending closer examination of the types. Badonnel's (1967a) arrangement (following Enderlein, 1910a) of treating all species in which hind wing veins Rs and M are joined by a crossvein to *Myopsocus* (with *Lichenomima* a synonym), and all those in which these veins are fused for a length to *Phlotodes* (with *Rhaptoneura* a synonym) (cf. figs. 393-394), was followed by most workers, until Mockford (1982) re-examined the type material of the type species of *Myopsocus*, *M. unduosus* (Hagen). He showed that the venational character noted above was not correct, following an error by Enderlein (1910a). Mockford concluded that *Myopsocus* has hind wing veins Rs and M fused for a length, so that *Phlotodes* and *Rhaptoneura* are in fact synonyms of *Myopsocus*. In contrast, true *Lichenomima* species have these veins linked by a crossvein. Mockford therefore believed that this genus was probably valid, rather than being a synonym of *Myopsocus*. Li Fasheng (2002) used the form of the hind wing Rs-M linkage as a basis for delimiting subfamilies: Myopsocinae (these veins fused) or Lichenomiminae (veins linked by a crossvein).

A large number of described species of Myopsocidae, from many parts of the world, were thus re-assigned by Mockford (1982) and Smithers (1985c): in total, 74 species were then re-assigned to *Myopsocus* and fifteen to *Lichenomima*. Thornton & Browning (1992) considered that some aspects of generic allocation in the family are still unsatisfactory, but the above notes help to provide perspective for generic allocation within this distinctive family.

Most recently, Lienhard (2004b) investigated more closely the taxa superficially placed in *Lophopterygella* on the basis of the apical posterior margin of the fore wing being 'scalloped' between veins, and described *Thorntonodes*, distinguished on highly unusual hypandrial characters. He also noted single 'scalloped-wing' species of *Myopsocus* (*M. horripilis* Lienhard; Thailand, Sabah) and *Lichenomima* (*L. sinuosa* Lienhard; Indonesia: Sumatra).

Although not recorded from South-east Asia, it is pertinent to note also a genus described by Smithers (2004) to contain three unusual Australian species. This, *Nimbopsocus* Smithers (type species *Nimbopsocus thorntoni* Smithers) resembles *Myopsocus* in having hind wing veins Rs and M fused for a length, but differs from true *Myopsocus* in the form of the phallosome (an elongate rectangular frame with median rod and two, parallel lateral arms, fig. 406) and hypandrium.

Key to the genera of Myopsocidae (based on Thornton & Browning, 1992)

1	Hind wing veins Rs and M fused for a length ▸	2
–	Hind wing veins Rs and M joined by a crossvein ▸	4
2	Ocelli absent. Fore wing rather narrow ▸	*Mouldsia*
–	Ocelli present. Fore wing of normal shape ▸	3

3 Male: phallosome without posterior U-shaped extension. Female: subgenital plate with narrow tapered apical process with apical setae ▸ *Myopsocus*

– Male: phallosome with posterior U-shaped extension. Female: subgenital plate with distal apical lobe tapered to short rounded apex, without long apical process and lacking setae ▸ *Smithersia*

4 Fore wing margin along apex and posterior margin smoothly rounded (exception: *Lichenomima sinuosa* with slightly scalloped margin). Male: hypandrium simple, with no distinct apical lobe. Female: subgenital plate with transverse distal sclerite ▸ *Lichenomima*

– Fore wing margin along apex and posterior margin scalloped between veins ▸ 5

5 Male: hypandrium with short apical lobe. Female: subgenital plate with weakly tapered truncate or slightly bifid apex ▸ *Lophopterygella*

– Male: hypandrium with strongly developed basal lobe, directed anteriorly and ventrally prominent (figs. 409-410). Female: subgenital plate with distinct rounded apical lobe ▸ *Thorntonodes*

Generic information

Lichenomima Enderlein (figs. 397, 400). Type species *Lichenomima conspersa* Enderlein.

Generic distribution: North and South America, Galapagos Islands, Cuba, Africa, India, Nepal, China, Hong Kong, Japan, Thailand, Malaysia, Indonesia, New Guinea, Australia, Fiji.

Following Mockford (1982, 1993a), *Lichenomima* is regarded as a genus distinct from *Myopsocus*. It is diagnosed by having hind wing veins Rs and M joined by a crossvein and the presence of a small distinct truncate apical lobe on the subgenital plate, with the plate never ending in a single tapered process. The phallosome lacks a median rod. It is one of the most diverse genera in the family; Thornton & Browning (1992) noted one species (*L. merapi* Thornton) from Java which is intermediate between the genera in having the '*Lichenomima* venation' and '*Myopsocus* subgenital plate' structure.

Lophopterygella Enderlein (fig. 394, 407-408). Type species *Lophopterygella camelina* Enderlein. Generic synonymy: *Festona* Navás, type species *Festona lunata* Navás.

Generic distribution: Haiti, Africa, China, Taiwan, Vietnam, Thailand, Laos, Singapore, Indonesia, Philippines, New Guinea, Australia, Micronesia.

Lophopterygella is widely distributed, as indicated above, and broadly tropicopolitan with greatest documented diversity in the South-east Asian region. It is distinguished by the hind wing veins Rs and M being joined by a crossvein, but primarily on the pronounced scalloping of the apical and posterior fore wing margin. Correlated with this wing condition are (1) apex and subgenital plate weakly tapered, truncate or slightly bifid and (2) hypandrium with a short setose apical lobe and phallosome lacking any median rod or distinct endophallic sclerites. See comment under *Thorntonodes*, which was distinguished from *Lophopterygella* only recently. As with *Myopsocus*, the consistency of the intricate fore wing markings may be of considerable use in initial recognition of some species: Thornton & Browning (1992) provided a 'field key' to Indonesian species based on this character.

Thorntonodes Lienhard (figs. 409-410). Type species *Lophopterygella lobata* New & Thornton.
Generic distribution: India, China, Thailand, Singapore.
Members of this genus resemble closely *Lophopterygella* in superficial appearance, but are distinct on the highly unusual form of the male hypandrium, in which a strong basal lobe protrudes anteriorly and is prominent ventrally in lateral aspect, in contrast to the much simpler structure of *Lophopterygella* (cf. figs. 407, 409).

Mouldsia Smithers (fig. 395). Type species *Mouldsia barbarae* Smithers.
Generic distribution: New Guinea, Solomon Islands, Australia.
Mouldsia is highly unusual among Psocoptera in that the macropterous adults lack ocelli. It thus differs from all other Myopsocidae known. The wings are relatively long and narrow, with the type species (from Queensland, Australia) having the wings at least three times as long as wide (Smithers, 1978a). The hind wing anal vein becomes obsolete well before reaching the wing margin. Associated genitalic features are:

1. the subgenital plate with a short distal setose process and
2. the hypandrium being well sclerotised and spiculate and the male clunium with lateral 'flaps' and a median lobe. Few species are known (three are described) and the genus is apparently confined to the east of the region, as implied above.

Myopsocus Hagen (figs. 392, 399, 401, 403-405). Type species *Psocus unduosus* Hagen.
Generic synonymy *Phlotodes* Enderlein, type species *Myopsocus kolbei* Enderlein; *Rhaptoneura* Enderlein, type species *Rhaptoneura dispar* Enderlein.
Generic distribution: North America (southern parts, possibly adventive), Cuba, Jamaica, Central and South America, Europe (Mediterranean region), Africa, Seychelles, Aldabra, Madagascar, India, Thailand, Malaysia, Singapore, Philippines, Indonesia, New Guinea, Solomon Islands, Australia, New Zealand, Melanesian region.
In the sense used here (following Mockford, 1982), *Myopsocus* is by far the largest genus of Myopsocidae, with many of the taxa having very intricate fore wing patterning which presumably aids crypsis on tree bark. The principal recognition features are the form of hind wing veins Rs and M (fused for a length),

predominant grey/brown colouration, and the following combination of genitalic features:

1. subgenital plate with a narrow apical extension, almost always with two apical setae (*M. toxeres* (Smithers & Thornton) from New Guinea is exceptional),
2. phallosome usually a closed ring with a median internal rod (other conditions noted by Thornton & Browning (1992) are
   a. frame open but median rod present and
   b. frame open and median rod absent), and
3. hypandrium shallowly lobed and may have strong lateral processes.

Rarely, the hypandrium is more complex. The putative differences from *Lichenomima* are

1. hind wing venation,
2. the frequent presence of a median phallosome rod and
3. the narrow apical process to the subgenital plate.

The systematics of *Myopsocus* are complex, because many of the species are known only by one sex, but several authors have emphasised the consistency of wing pattern features as useful for association of sexes and recognition of some taxa, together with the details of facial markings (Smithers & Thornton, 1974b; Thornton & Browning, 1992). Four possible species groups were designated by Smithers & Thornton (1974b). Thornton & Browning (1992) allocated several Indonesian species to one or other of these. The groups are based as follows and are noted here for their possible value in helping to delimit segregates within this very large and complex genus. See also *Nimbopsocus*, from Australia (p. 140, fig. 406).

- *Group I*. Pterostigma strongly concave; subgenital plate convex; dorsal gonapophysis valve tapered; hypandrium with marginal sclerifications; male epiproct with rugose area and not overlying ninth tergite; phallosome frame closed, with median rod.
- *Group II*. Pterostigma strongly concave; subgenital plate convex; dorsal gonapophysis valve tapered; hypandrium with marginal sclerifications; male epiproct circular and with extension overlying ninth tergite; phallosome frame open, external parameres enlarged apically, with median rod.
- *Group III*. Pterostigma with R1 more or less straight basal to posterior angle; subgenital plate concave; dorsal gonapophysis valve tapered; hypandrium with extensive sclerification, hind margin sometimes complex; male epiproct with anterior extension overlying ninth tergite; phallosome frame ring-like, open or closed, without median rod.
- *Group IV*. Pterostigma concave; subgenital plate concave; dorsal gonapophysis valve broad-based, abruptly narrowed to curved, almost filamentous, apical half; hypandrium extensively sclerotised; male epiproct with

extension overlying ninth tergite; phallosome frame ring-like, open, with median rod.

South-east Asian representation of these groups includes species from Groups I and II; Group III occurs from Africa to New Caledonia, without specific recognition in South-east Asia, and Group IV is from Australia and New Zealand. The last two groups may thus be restricted geographically (as recognised by Smithers & Thornton, 1979), but a number of regional species have not been assigned to any group and clearer allocation is necessary for better biogeographical interpretation. The groups may also be important in reflecting future generic segregates. Thornton & Browning (1992) suggested that 'true' *Myopsocus* should be restricted to its type species, which differs clearly from other taxa placed in the genus. If this course of action were adopted, the 90 or so other species currently placed in the genus would need reassignment. Such a step could be taken only after detailed re-examination of many taxa, but the possible 'holding genus' status of *Myopsocus* must be acknowledged. Lienhard (2004b) reported a species from Thailand (*M. horripilis* Lienhard) with the unusual feature of having expanded and apically indented setae on the mesonotum.

Smithersia Thornton (figs. 393, 396, 398, 402). Type species *Psocus hermosus* Banks. Generic distribution: Singapore, Java, Sumatra.
This small genus (with four described species) is known only from tropical South-east Asia and may thus be regarded as a distinctive regional endemic genus of Myopsocidae. The wings are relatively narrow and very finely 'granulated' with markings, but ocelli are clearly present and the genus is distinctive also on the complex spurs on each side of the hypandrium. The lack of apical setae on the subgenital plate may also be characteristic.

3 **Psilopsocidae** (figs. 411-417)

Diagnosis

Fore wing length ca 3.2-5.4 mm. Lacinial apex broad, with several tines. Tarsi 3-segmented; claws with subapical tooth, pulvillus broad. Fore wing glabrous; areola postica large, with Cu1 forking well before reaching wing margin and basal region of areola postica thereby often long and slender. Pterostigma with basal margin concave, strongly broadened near apex and usually with short spur vein from posterior angle; veins Rs and M meeting at a point or fused for a short length. Male: hypandrium simple; phallosome a closed ring-like frame, without distinct outer parameres. Female: subgenital plate with narrow elongate apical lobe with few setae; gonapophyses complete; ventral valve long; dorsal valve broad, with long narrow apical process; external valve broad and setose.

Remarks

Psilopsocidae is a very small family and contains a single widely-distributed genus with few described species. The taxonomic history of the genus *Psilopsocus* was discussed by Mockford (1961a), who noted Enderlein's initial placement of it in the Mesopsocidae and Roesler's (1944) subsequent erection to independent family status, then placed in Epipsocetae. Mockford noted the considerable resemblance between Psilopsocidae and Myopsocidae within Psocetae. The biology of *Psilopsocus*, primarily known from the Australian *P. mimulus* Smithers but likely to be more general, is highly unusual (Smithers, 1995b: nymphs are phragmotic (Smithers, 1963a)) and are wood borers that tunnel into the ends of pine twigs. *Psilopsocus* is thus the first known wood-boring psocopteran. Adults do not share this habit and are found on tree bark, where oviposition occurs. The distinctive nymphs have the apex of the abdomen very strongly sclerotised, with the paraprocts and epiproct together closing off the anus so that the rear of the insect is well-protected in the tunnel. Similar, but less pronounced, sclerotisation occurs in nymphs of the Philippine *P. nebulosus* Mockford (Smithers, 1995a).

Generic information

Psilopsocus Enderlein (figs. 411-417). Type species *Psilopsocus nigricornis* Enderlein. Generic distribution: South Africa, Malaysia, Philippines, New Guinea, Australia.

All species of *Psilopsocus* appear to be rare and are known from few individuals. All have the fore wing intricately patterned with grey, and wing markings alone can be used to key the species, although venational and genitalic differences also occur. Most species are known from very limited areas, but it is not known if this simply represents serendipitous collecting. The four New Guinea species were reviewed by Smithers & Thornton (1973) and a key to the seven species then known was given by New & Lee (1991a).

4 Psocidae (figs. 418-518)

Diagnosis

Lacinial apex with few teeth. Tarsi 2-segmented; claws with a, usually strong, subapical tooth (rarely absent), pulvillus slender. Wings usually glabrous, very rarely with few setae on main fore wing veins or on margin of hind wing radial fork. Basal length of fore wing vein Sc usually evident; fore wing vein Cu1a usually fused with M for a length, more rarely meeting it at a point or linked by short crossvein, to produce a large discoidal cell in the central region of the wing; hind wing veins Rs and M fused for a length; fore wings sometimes strongly patterned. Male: paraproct with strong curved or lobed process(es); hypandrium sometimes asymmetrical, usually ornamented and often with complex apophyses or spines, may incorporate sternite of abdominal segment viii; phallosome closed or open, sometimes represented only by paired rod-like sclerites. Female: subgenital plate usually with a median lobe, setose at apex; gonapophyses complete; ventral valve slender and pointed; dorsal valve broader and sometimes with apical apophysis; external valve large and strongly setose, often transverse; spermathecal opening often strongly sclerotised.

Remarks

Psocidae is one of the largest families of Psocoptera in the region. It contains many of the largest psocid species, many of them with strongly patterned wings. All the species are associated primarily with tree or shrub bark. The family, and many of the genera, are widespread and, because of their relatively large size, Psocidae are collected more often than many other psocids by non-specialists.

The family is taxonomically complex and its members differ greatly in size. Particular Cerastipsocinae have fore wing length exceeding 6 mm, but others (such as many *Trichadenotecnum*, p. 165) have fore wing length in the range of 2.5-3.5 mm. The recently described *Thyrsopsocopsis* (p. 168), with fore wing length greater than 8 mm, is the largest psocid in South-east Asia.

Smithers (1972c) noted that, in terms of expressing intra-family relationships 'Of all families in the order, the Psocidae is perhaps the most difficult to deal with, although so many species have been described.' This is in part due to many early workers allocating species uncritically to the genus *Psocus*, so that this genus contained a great variety of taxa united simply by having a closed discoidal fore wing cell and 2-segmented tarsi. See further discussion under *Psocidus* (p. 160).

In many taxonomic treatments of this family, four subfamilies are recognised. Thyrsophorinae (formerly regarded as a distinct family, Thyrsophoridae) contains the largest of all Psocoptera and are almost entirely Neotropical. The other three subfamilies are widespread. The conventional representation of these, as Amphigerontiinae, Cerastipsocinae and Psocinae, as used by most workers until recently, was challenged by Mockford (1993a), who combined Cerastipsocinae and Psocinae as a broader 'Psocinae' to reflect that they may be much more closely related to each other than either is to Amphigerontiinae. This arrangement was followed by Lienhard & Smithers (2002), but the more traditional divisions have been retained in some accounts, such as that by Lienhard (1998a). Allocation of genera to the traditional subfamilies has been generally clearcut, but Smithers (1990a) keyed Psocidae directly to genus level, so overcoming occasional difficulties of subfamily or (below) tribal allocation of some genera.

A major advance by Mockford (1993a) was to define tribes within his broad concept of Psocinae (that is, all genera attributed previously to Cerastipsocinae and Psocinae), based on the North American fauna and using genitalic features as a primary basis for divisions. The former Cerastipsocinae thereby became the

Psocinae tribes Cerastipsocini and Metylophorini, and the Old World Cycetini (as recognised by Roesler, 1944) constitutes a third tribe in this series (see also Smithers, 1972c). The former Psocinae includes the tribes Psocini and Ptyctini. All five tribes of Psocinae *sensu* Mockford (1993a) occur in South-east Asia, but a number of regional genera are rather difficult to allocate clearly to one or other of these and could be considered *incertae sedis*. The arrangement here follows that in Lienhard & Smithers (2002). Alternative keys to traditional subfamilies and to Mockford's arrangement are given here. The taxonomic history of division within the family is summarised in table 5.

TABLE 5
Summary of history of taxonomic arrangement of 'Psocidae'.

BROADER USE
'Psocidae': initial use equivalent to whole order 'Psocoptera' as used by recent workers.
'Psocidae' divided into five tribes, collectively encompassing whole of Psocoptera.
FAMILY LEVEL USE
Psocidae as a distinct family within order, sensu modern authors, introduced by Pearman (1932a, 1936a). Thyrsophoridae there retained as separate family.
Psocidae applied in wider sense, to incorporate also Hemipsocidae by Roesler (1944); this usage not generally accepted thereafter, with Hemipsocidae treated more widely as separate family.
Subfamily divisions of Psocidae listed by Smithers (1972a) as Amphigerontiinae, Antipsocinae (poorly characterised and not generally considered valid, based solely on *Antipsocus* Roesler, Costa Rica), Cerastipsocinae (with tribes Cerastipsocini, Metylophorini, Cycetini), Psocinae. Thyrsophoridae and Hemipsocidae treated as distinct families. [Note: *Antipsocus* is a synonym of *Anomopsocus* Roesler, Lachesillidae, unusual in that family in having a closed fore wing discoidal cell.]
Thyrsophoridae reduced to a subfamily (Thyrsophorinae) of Psocidae by New (1978a).
Family categorised into four subfamilies (Amphigerontiinae, Cerastipsocinae, Psocinae, Thyrsophorinae) by Smithers (1990a).
Rearrangement by Mockford (1993a) based on North American taxa: Amphigerontiinae sensu previous authors retained; Cerastipsocinae and Psocinae combined as enlarged 'Psocinae' with tribes Cerastipsocini, Metylophorini, Psocini, Ptyctini (Cycetini and Thyrsophorinae not noted, as absent from North America).
Lienhard & Smithers (2002) followed Mockford's arrangement, incorporating the taxa in parentheses above, in a world appraisal.

Key to the subfamilies of Psocidae

1 Fore wing veins R4+5 and M linked by crossvein or (some Neotropical taxa) fused for a short length. Pterostigma elongate, narrowed to become very shallow in distal half. Basal part of fore wing vein Sc ending free in membrane or meeting R1 (large species, fore wing ca 8 mm, narrow) ▸ **Thyrsophorinae**

– Fore wing veins R4+5 not connected or fused (if, in *Cycetes* only, these veins meet, pterostigma not conspicuously shallow and basal part of fore wing vein Sc reaches costal margin). Pterostigma relatively deep, not abruptly shallowed over apical half ▸ 2

2 Male: hypandrial sclerites encompassing abdominal sternites viii and ix, so apex of male abdomen appears 'encapsulated'; parameres open distally (no closed phallosome frame); apophysis of male paraproct broadly rounded. Female: apex of dorsal gonapophysis valve a slender tapered process. Fore wing vein M comprising distal border of discoidal cell usually convex ▸ **Amphigerontiinae**

– Male: hypandrial sclerites confined, or largely confined, to sternite ix; sternite viii never wholly sclerotised, so male abdomen apex not 'encapsulated'; parameres either forming a closed frame, or frame open posteriorly, never divergent or thickened distally; apophysis of male paraproct with pointed apex. Female: apex of dorsal gonapophysis valve variable. Fore wing M comprising distal border of discoidal cell usually concave or straight, only rarely convex
▸ **(Psocinae, sensu Mockford)**
▸ 3

3 Antennae usually much longer than fore wing (ca. 1.5 times). (Male: hypandrial sclerites usually symmetrical, always confined to sternite ix ; phallosome frame closed. Female: apex of dorsal gonapophysis valve rounded or with short point, without pronounced style.) ▸ **Cerastipsocinae**

– Antennae usually only slightly longer than fore wing or shorter than fore wing. (Male: hypandrial sclerites usually asymmetrical, sternite viii sometimes with small sclerotised area, as above; phallosome frame closed or, more rarely, open. Female: apex of dorsal gonapophysis valve generally with well-defined short style) ▸ **Psocinae**

Key to main subfamilies and tribes of Psocidae (excluding Thyrsophorinae) (based on Mockford, 1981, 1993a; Endang et al., 2002)

N.B. Couplet 1 as couplet 2 above, because Amphigerontiinae is the same entity in both keys.

1 Male: hypandrial sclerites encompassing abdominal sternites viii and ix, so apex of male abdomen appears 'encapsulated'; parameres open distally (no closed phallosome frame); apophysis of male paraproct broadly rounded. Female: apex of dorsal gonapophysis valve a slender tapered process. Fore wing vein M comprising distal border of discoidal cell usually convex ▸ **Amphigerontiinae**

– Male: hypandrial sclerites confined, or largely confined, to sternite ix; sternite viii never wholly sclerotised, so male abdomen apex not 'encapsulated'; parameres either forming a closed frame, or frame open posteriorly, never divergent or thickened distally; apophysis of male paraproct with pointed apex. Female: apex of dorsal gonapophysis valve variable. Fore wing M comprising distal border of discoidal cell usually concave or straight, only rarely convex
▸ **(Psocinae, sensu Mockford)**
▸ 2

2 Fore wing vein R4+5 recurved to touch vein M ▸ **Cycetini**

– Fore wing vein R4+5 free ▸ 3

3 Male: hypandrium usually with a pair of small sclerotised areas on sternite viii; interface of male clunium and epiproct either a straight line or epiproct protruding over clunium. Female: dorsal gonapophysis valve terminating in a slender process, abruptly narrower than basal region of valve ▸ **Ptyctini**

– Male: hypandrium without pair of small sclerotised areas on sternite viii; interface of male clunium and epiproct either a straight line or clunium projecting over epiproct. Female; dorsal gonapophysis valve terminating in either a gradually tapered process or without an apical process ▸ 4

4 Male: hypandrium usually symmetrical. Female: subgenital plate with main pigmented/sclerotised median region meeting lateral sclerotised areas perpendicularly to form T-shaped pattern.* Maxillary palp with fourth segment short and broad, length ca 2.6-2.8 times width (antennae very long, usually ca 1.5 times fore wing length) ▸ **Cerastipsocini**

– Male: hypandrium usually asymmetrical. Female: subgenital plate with main pigmented/sclerotised median region at angle of more than 90° to lateral sclerotised area, to form Y-shaped pattern. Maxillary palp with fourth segment long, length ca 3.5-4.5 times width (antennae either long, as above, or of similar length to fore wing) ▸ 5

5 Antennae at least 1.5 times length of fore wing. Male: paraproct either with no lobe lateral to distal process or lobe bearing trichobothrial field. Female: dorsal gonapophysis valve rounded distally, without apical process; external gonapophysis valve not longer than ca length of dorsal valve ▸ **Metylophorini**

– Antennae at most as long as fore wing, or only slightly longer. Male: paraproct with shoulder or lobe lateral to distal process and distal to trichobothrial field.

Female: dorsal gonapophysis valve with at least a short apical process; external valve a large plate, often extending nearly half length of dorsal valve ▸ **Psocini**

* see comment under *Sundapsocus*, p. 156.

Amphigerontiinae

This subfamily is distinctive because of the apomorphous development of the male abdominal apex and reduced phallosome, these constituting a diagnostic character combination. It appears to be a natural group without need for tribal subdivision. Females are sometimes more difficult to diagnose, although the form of the dorsal gonapophysis valve is characteristic. Many Amphigerontiinae have the wings hyaline or fumose, without strong banding or spotting patterns. The eight genera reported from South-east Asia range from being widely distributed to very circumscribed in distribution. They have been separated mainly on relatively small features of wing venation and genitalic structures. Most Amphigerontiinae are relatively small.

Key to the genera of Amphigerontiinae* (after Endang et al., 2002)

1 Fore wing veins Rs and M connected by a crossvein ▸ 2
– Fore wing veins Rs and M meeting at a point or fused for a length ▸ 4

2 Discoidal cell of fore wing parallel-sided ▸ *Anomaloblaste*
– Discoidal cell of fore wing narrowed posteriorly ▸ 3

3 Discoidal cell externally convex. Ocelli absent. Male hypandrium simply rounded ▸ *Euclismioides*
– Discoidal cell externally concave. Ocelli present. Male hypandrium with hind margin in three lobes ▸ *Amphigerontia*

4 Vertex of pterostigma smoothly rounded ▸ 5
– Vertex of pterostigma not smoothly rounded ▸ 6

5 Fore wing vein M+Cu1a somewhat sinuous. Male hyandrium asymmetrical ▸ *Blastopsocus*
– Fore wing vein M+Cu1a not sinuous. Male hypandrium symmetrical ▸ *Neoblaste*

6 Posterior border of pterostigma convex over apical third, basal two thirds curved ▸ *Indoblaste*
– Posterior border of pterostigma not as above ▸ 7

7 Posterior third of pterostigma border curved, basally concave ▸ **Blaste (Euclismia)**
– Posterior half of pterostigma border convex, basally straight ▸ **Javablaste**

* Note that the genera *Elytropsocus* and *Kaindipsocus*, both from New Guinea, were treated in this subfamily by Lienhard & Smithers (2002). They are listed here as Psocinae: Ptyctini (p. 161).
The above key is based primarily on wing characters that may appear rather trivial, and reinforced by hypandrial characters. Females of some genera are difficult to differentiate unless associated clearly with males. The taxonomic history of this group undoubtedly contains errors of generic placement and many of the problems are discussed by Endang et al. (2002). One considerable source of confusion is that many of the taxa have been treated variously as distinct genera or as subgenera of *Blaste* s.l. (see below), so that a number of described taxa have been allocated uncritically to one or other level. The genus *Epiblaste* Li Fasheng (type species *Epiblaste glandacea* Li Fasheng) is known from a single Chinese species and is not included here.

Generic information

Amphigerontia Kolbe (figs. 418, 428). Type species *Psocus bifasciatus* Latreille.
Generic distribution: North, Central and South America, Europe. Additional taxa placed in the genus are from South America, Africa, China, Taiwan, Japan, Myanmar, Vietnam, Indonesia, but the correct generic placement of these needs confirmation.
In some instances, re-examination has led to generic transfers – for example, the Fijian *Psocus (Amphigerontia) vitiensis* Karny was transferred to *Ptycta* by Thornton (1981a) and subsequently to *Copostigma* by Smithers (1985a).
The genus is included here on the basis of few species, whose generic placement needs confirmation. *Amphigerontia lata* Enderlein was described from Java. The type specimen has not been re-examined critically and it is possible, even likely, that it may belong elsewhere. Prudently, however, it can not be claimed that *Amphigerontia* is either present or absent in Indonesia – although it has not been found in extensive recent collections. *Amphigerontia namiana* Navás was described from Vietnam and two species (*A. feai* Ribaga, *A. incerta* Ribaga) from Myanmar.

Anomaloblaste Endang, Thornton & New (figs. 419, 422, 425). Type species *Anomaloblaste spinivalva* Endang, Thornton & New.
Generic distribution: Java, Bali, Lombok.
The two species on which this genus is based are, perhaps, the closest taxa to 'true *Amphigerontia*' found recently in the region, but differ from it in the very long crossvein between fore wing veins Rs and M and the shorter anterior margin to the areola postica. The genus was recognised as distinct (but not named) by Endang & Thornton (1992), but males are still unknown, so that its affinities can not be appraised fully. Females differ from 'true *Amphigerontia*' in the narrower apical lobe to the subgenital plate and the very long dorsal and ventral gonapophysis valves, together with a relatively short external valve.

Blaste Kolbe (figs. 420, 424, 426, 429). Type species *Blaste juvenilis* Kolbe.
Generic distribution: North, Central and South America, Africa, Europe, Arabia, India, Sri Lanka, Mongolia, China, Japan, Thailand, Malaysia, Indonesia, New Guinea, Australia, New Zealand.
The above extensive distribution is based on that cited by Smithers (1996a). It reflects the complex nature of *Blaste* s.l. and could be modified substantially by inclusion/exclusion of the several taxa regarded as subgenera by Roesler (1943), namely *Lasiopsocus* Enderlein (type species *Lasiopsocus michaelseni* Enderlein from South Africa and Australia, which is treated as a distinct genus by Smithers, 1996a), *Blastopsocus* Roesler (type species *Psocus variabilis* Aaron, from North America, listed as a distinct genus by Smithers & Thornton, 1981a, and Mockford, 1993a), *Blastopsis* Badonnel (type species *Blaste (Blastopsis) triangularum* Badonnel from Angola), *Blastopsocidus* Badonnel (type species *Blaste (Blastopsocidus) maculata* Badonnel), *Euclismia* Enderlein (type species *Psocus quadrimaculatus* Latreille from Europe, retained as a subgenus by Lienhard, 1998a) and *Euclismiopsis* Badonnel (type species *Blaste (Euclismiopsis) machadoi* Badonnel from Africa and China). Subgenera recognised by Lienhard & Smithers (2002) are *Blaste*, *Blastopsis*, *Euclismia* and *Euclismiopsis*, but many included species can be allocated only as '*Blaste* s.l.'. The status of several of these segregates is unclear. Badonnel (1955) regarded his taxa as subgenera. However, Thornton (1960), in noting the need for revision of the complex, suggested the likelihood that the subgenera would be accorded full generic status. His key to entities in the complex did not distinguish between genera and subgenera other than referring to *Neoblaste* as a new genus. Later workers on the Amphigerontiinae in South-east Asia have tended to regard these taxa and others such as *Euclismioides* as full genera, and that use is followed here. Many of the older-described taxa have not been re-examined critically, so that some referred to *Blaste* s.l. may not be *Blaste* s.str.
Blaste is not common in South-east Asia; species have been recorded from Thailand, Malaysia and Singapore, together with a single record (as *Blaste (Euclismia)*) from Indonesia.

Blastopsocus Roesler (figs. 421, 423, 427, 430). Type species *Psocus variabilis* Aaron (then known as *Euclismia variabilis*, see Enderlein, 1925a).
Generic distribution: North, Central and South America, New Guinea.
Originally designated as a subgenus of *Blaste*, generic status for *Blastopsocus* was proposed by Mockford (1953b), followed by Smithers & Thornton (1981a), who added the New Guinea *B. angustus* Smithers & Thornton to the two North American species then comprising the genus. Mockford (1993a) included three North American species in the genus. The true placement of *B. angustus* needs confirmation, as acknowledged by Smithers & Thornton (1990), not least because of the unusual generic distribution of the genus resulting from its inclusion.

Euclismioides Smithers & Thornton (figs. 431, 438, 442). Type species *Euclismioides inocellata* Smithers & Thornton.
Generic distribution: New Guinea.
This unusual monotypic genus lacks ocelli. It may be endemic to New Guinea. In having a well-defined fore wing Rs-M crossvein, *Euclismioides* resembles *Amphigerontia*, but the hypandrium is simple and unornamented rather than lobed posteriorly. The dorsal and ventral gonapophysis valves are both long, the external gonapophysis valve has a strong tapered lobe and the subgenital plate lobe is extended, rounded and reinforced by an internal 'bar'.

Indoblaste Endang & Thornton (figs. 434-435, 439, 443, 457). Type species *Indoblaste lienhardi* Endang & Thornton (subsequent designation by Endang et al., 2002).
Generic distribution: Indonesia.
This genus typically has the apical part of the posterior border of the pterostigma convex, unlike *Blaste* and *Javablaste* (Endang et al., 2002). The male genitalia are unusual and characterised by the hypandrium having an apical pair of 'pegs' and lateral projection with medially directed lobes. The parameres are slender, without hooks, and are fused or separate anteriorly. Dorsal and ventral gonapophysis valves are both long; the external valve has a tapered lobe; the apical lobe of the subgenital plate is elongate and bears setae on the surface rather than just marginally, as in most other regional genera. The spermapore plate has an unusual pattern of sclerotisation, characterised by Endang & Thornton (1992) as 'a posterior scabbard-shaped sclerite'. A somewhat similar structure is found in typical European *Amphigerontia*, although it is there less pronounced.

Javablaste Endang,Thornton & New (figs. 432-433, 441, 445, 458). Type species *Javablaste aidab* Endang, Thornton & New.
Generic distribution: Indonesia.
Javablaste was raised for males of two species and unallocated females which do not correspond with other taxa in the '*Blaste*-group'. It is most similar to the European *Neopsocopsis* Badonnel, but differs in the form of the subgenital plate. The hypandrium has a median lobe and spinous lateral lobes, considerably more complex than those of true *Amphigerontia*, and the wing venation also differs from that genus: the discoidal cell is broad but narrowed posteriorly.

Neoblaste Thornton (figs. 436-437, 440, 444, 456, 510). Type species *Neoblaste papillosa* Thornton.
Generic distribution: China, Hong Kong, Korea, Singapore, Indonesia.
This genus appears to be especially well-represented in Indonesia, from where eight of the thirteen described species are known. Thornton (1960) diagnosed the genus on wing venation and male genitalic structure, with the latter used as key characters to recognise *Neoblaste* by the symmetrical hypandrium and the

broadly rounded lateral lobes with accessory sclerites, and the free parameres with apical hooks. The subgenital plate has a rounded to transverse median lobe, setose along the margin.

Psocinae

This diverse subfamily contains a considerable variety of forms and members of some groups are particularly large and striking in appearance. The five tribes (sensu Mockford and the above key) are treated separately below.

Cerastipsocini

Members of this tribe are large and often strongly patterned. They were united previously by the combination of very long slender antennae and the short broad apical segment of the maxillary palp. Historically, many species have been referred to *Cerastipsocus* Kolbe, but those from South-east Asia were in the putative subgenera *Clematoscenea* and *Sigmatoneura* (p. 157), now acknowledged as generically distinct. Removal of these taxa from *Cerastipsocus* s.l. leaves *Cerastipsocus* s.str. as a purely New World genus with the defining character suite noted by Mockford (1975b, 1981) not known in any Old World cerastipsocine.

There has also been historical confusion with *Eremopsocus* McLachlan and its putative subgenera *Syngonosoma* Kolbe, *Dinopsocus* and *Podopterocus*, as noted below (see also New, 1978d). Mockford (1975b) considered that the Asian *Dinopsocus* and *Podopterocus* should be retained as distinct genera pending further studies, and that *Eremopsocus* (and its possible synonym *Syngonosoma*) is confined to the New World. This arrangement was followed by Lienhard & Smithers (2002), but both *Dinopsocus* and *Podopterocus* are now recognised as synonyms of *Sigmatoneura*, in the tribe Metylophorini (p. 157).

Several anomalous genera from China allocated to this group by Li Fasheng (2002) are not appraised here.

Key to the genera of Cerastipsocini

1	Fore wing veins in basal half of wing setose ▸	*Setopsocus*
–	Fore wing veins all glabrous ▸	2
2	Fore wing veins R2+3 and R4+5 diverging basally at an angle of 90° or more (fore wing cell M3 much narrower than other median cells, with veins M3 and distal portion of Cu1a almost parallel) ▸	*Clematoscenea*
–	Fore wing veins R2+3 and R4+5 diverging basally at an angle of ca 60° ▸	3

3 Fore wing veins Rs and M meeting at a point or linked by short crossvein; first portion of Cu1a longer than second portion ▸ *Psococerastis*

– Fore wing veins Rs and M fused for a length; first portion of Cu1a shorter than second portion ▸ *Sundapsocus*

Generic information

Clematoscenea Enderlein (figs. 446, 449, 451-452, 454). Type species *Psocus lemniscatus* Enderlein.
Generic distribution: Nepal, Malaysia, Indonesia, New Guinea, Solomon Islands.
This genus is very distinctive on its conspicuous wing pattern and fore wing venation, particularly in the shape of cell M3. Its allocation as a subgenus of *Cerastipsocus* by Roesler (1944) is not currently accepted. *Clematoscenea* appears to be particularly diverse in New Guinea, from where ten species are known. Thirteen species of *Clematoscenea* have been described.

Psococerastis Pearman (figs. 459-460, 464, 469, 471, 515). Type species *Phryganea gibbosa* Sulzer.
Generic distribution: widespread, except for North America, Australian region and the Melanesian arcs.
Psococerastis is diverse, with more than 80 species recognised. Roesler (1944) recognised the subgenera *Psococerastis* and *Dactylopsocus* Roesler (South America) in *Psococerastis*, but these entities are more conventionally treated as distinct genera. The definition of *Psococerastis* was appraised by Mockford (1981), who noted the considerable value of fore wing markings for initial recognition of species. This has historically been the major feature used to differentiate species in South-east Asia, where the variety of large, strongly patterned forms known as the '*taprobanes*-group' has proliferated (New, 1978g). Numerous species in this complex have been raised on relatively small differences in wing pattern and small genitalic details, and uncertainty over association of the sexes, which may differ in intensity of pigmentation or manifest sexual dimorphism to the extent that they may not be associated in field collections (Mockford, 1981, on New World *Psococerastis*) has contributed to confusion over species limits.
The *taprobanes*-group is widely distributed in the region (New, 1978g) and many of the constituent species may have only very limited ranges in South-east Asia. Only one species (*P. collessi* Smithers & Thornton) has been recognised from New Guinea.

Setopsocus Smithers & Thornton (figs. 462, 470). Type species *Setopsocus magnus* Smithers & Thornton.
Generic distribution: Singapore, Indonesia, New Guinea.
This unusual genus was raised to accommodate a New Guinea species of Psocidae with setae on the veins of the basal half of the fore wing. The Singapore *Psocidus hirsutus* New was transferred to it by Endang et al. (2002), with this

species then recorded also from Java. The genus is small but relatively widely distributed in South-east Asia.
Three other genera of Psocidae have setae on fore wing veins, but none of *Lasiopsocus* Enderlein (Amphigerontiinae; Australia, South Africa), *Chaetopsocidus* Badonnel (Colombia) or *Diplacanthoda* Enderlein (Metylophorini; Madagascar) has been recorded from the region.

Sundapsocus Smithers (figs. 463, 466-467, 472). Type species *Sundapsocus robertsoni* Smithers.
Generic distribution: Christmas Island.
As noted on p. 134, Christmas Island, although 'politically Australian', has very strong biotic affinities with Indonesia. *Sundapsocus*, known from a single species, has not yet been found in nearby parts of Indonesia. It is rather similar to *Psococerastis*, as Smithers (1995d) noted. Smithers (1995a) also commented on the form of the subgenital plate in *Sundapsocus*. The pigmented area has lateral arms that diverge anteriorly rather than lying transversely as in other cerastipsocines. On this feature, used also by Mockford (1981) to define tribes, *Sundapsocus* falls into Metylophorini. Smithers allocated it to Cerastipsocini on the balance of other features, but the placement of the genus remains somewhat anomalous. The maxillary palp (with the last segment length less than three times the width) also suggests placement in Cerastipsocini.

Cycetini

This tribe comprises a single genus, known only from South-east Asia and northern Australia. It is very distinctive.

Generic information

Cycetes Enderlein (figs. 461, 465, 468). Type species *Cycetes thyrsophoroides* Enderlein. Generic synonymy *Goya* Navás, type species *Goya pictus* Navás.
Generic distribution: Indonesia, Australia.
The wing venation of *Cycetes* is unique among Psocidae (other than Thyrsophorinae). The Indonesian *C. thyrsophoroides* is not common (New, 1976a), but has been collected widely in the archipelago, from Java, Bali, Komodo, Sumbawa and Timor. The antennae are long, as in Cerastipsocini, and the basal part of vein Sc reaches the costa in both wings. Female genitalia are distinctive: the three colourless areas of the subgenital plate and the ventral prolongation of the external gonapophysis valve are both highly unusual features. Smithers (1977a) noted the possible synonymy of *Goya pictus* with *C. thyrsophoroides*. Only two species are known, the second being *C. collessi* Smithers from Australia.
Cycetes has not yet been reported from the Asian mainland and appears to be a regional endemic genus.

Metylophorini

Metylophorini appear to be closely related to Cerastipsocini and, as noted above, some features of *Sundapsocus* transcend the conventional boundaries of these tribes. The shape of the maxillary palp has traditionally been used to separate these groups. At the generic level, Metylophorini are less diverse than Cerastipsocini and only two genera are represented in South-east Asia. Both are widespread and are large, strongly marked Psocidae. Placement of *Sigmatoneura* in Metylophorini was confirmed by Smithers (1980b), in contrast to Roesler's (1944) placement as a subgenus of *Cerastipsocus*. However, Li Fasheng (2002) raised a new subfamily, Sigmatoneurinae, to contain *Sigmatoneura* and divided it into two subgenera: *Sigmatoneura* s.str. (apical lobe of subgenital plate about as wide as long; hypandrium without a posterior lobe) and *Longifolia* Li Fasheng (apical lobe of subgenital plate more than twice as long as wide; hypandrium with long posterior lobe).

Key to the genera of Metylophorini

1 Fore wing vein Sc ending free in wing membrane; veins Rs and M fused for a short length or meeting at a point; areola postica trapezoidal (basal and distal portions of vein Cu1a not converging strongly toward M) ▸ *Metylophorus*

– Fore wing vein Sc meeting costa; veins Rs and M connected by crossvein; areola postica an isosceles triangle (basal and distal portions of vein Cu1a converging strongly toward M) ▸ *Sigmatoneura*

Generic information

Metylophorus Pearman (figs. 476-477, 481, 483-484). Type species *Psocus nebulosus* Stephens.
Generic distribution: North, Central and South America, Europe, Africa, India, Bhutan, China, Japan, Indonesia.
Metylophorus species typically have the fore wing predominantly brown or grey and fumose hind wings. The species differ in details of wing pattern and genitalia, but the sexes are sometimes hard to associate. Most regional species have been recorded from only few individuals. They appear to be relatively rare taxa in the area. *Metylophorus* is not known from New Guinea or Australia and until recently the most easterly records of the genus were from Bhutan (New, 1978f; Smithers & Thornton, 1990).

Sigmatoneura Enderlein (figs. 447-448, 450, 453, 455, 473-475, 480, 482, 485, 509). Type species: *Cerastipsocus subcostalis* Enderlein. Generic synonymy *Scaphopsocus* Smithers, type species *Scaphopsocus phaeotherus* Smithers; *Dinopsocus* Banks, type species *Dinopsocus atratus* Banks; *Podopterocus* Banks, type species *Podopterocus longicornis* Banks.

Generic distribution: Africa, China, Taiwan, Japan, Malaysia, Brunei, Singapore, Indonesia, Philippines, New Guinea, Australia.

The history and systematics of this genus are complex and confusing but have been clarified by Yoshizawa et al. (2005).

The very striking sexual dimorphism of species of *Sigmatoneura*, with males having the fore wing hyaline and females with this wing strongly patterned in dark brown (cf. figs. 473-475), has led to problems in associating the sexes, with most species described only from one or other form. *Scaphopsocus* was raised to contain a male from Tanzania and its synonymy was effected by Smithers (1976b) in a paper in which the earlier history of the genus was discussed. The evaluation of the significance of incrassate flagellar segments in this group has been partially responsible for this confusion, as this feature is exhibited differentially in different species of the complex.

Podopterocus has a similar taxonomic history to that outlined below for *Dinopsocus* and was regarded as distinct primarily on the apparently distinctive apomorphy of a greatly enlarged hind tibia. Considerable difficulty has occurred in interpreting the significance of the tibial expansion and also of the irregularly incrassate first flagellar segment of the antennae. Hind tibial expansion, to varying species-specific extents, occurs only in males, with the hind leg of some taxa (and of all females known) normal. Only the type species has been described in *Podopterocus*. *Dinopsocus* was also treated as a subgenus of *Eremopsocus* by Roesler (1944), together with *Podopterocus*, apparently on the basis of the basal antennal flagellar segment being thickened and incrassate. However, this antennal form occurs also in '*Podopterocus*'. Mockford (1975b) advocated retention of *Dinopsocus* as a distinct genus, pending further genitalic examination. The fore wing of *Dinopsocus* is narrow and 'pointed' toward the apex. Two species were described under that genus name. Both *Podopterocus* and *Dinopsocus* are known only from tropical South-east Asia and are regarded as regional endemics.

Yoshizawa et al. (2005) demonstrated clearly that these psocids constitute a monophyletic group, based on both molecular and autapomorphic morphological characters.These autapomorphies are

1 female fore wing dark brownish black, with veins R2+3, R4+5 and first section of Cu1a white,
2 vein R4+5 strongly sinuate, and
3 sexual dimorphism in wing colour.

These authors concluded that the various forms exhibit a mosaic of characters and can not be convincingly diagnosed as separate taxa.To illustrate this variation, illustrations are provided of individuals earlier diagnosed as *Podopterocus* (figs. 447-448, 453, 455) and *Sigmatoneura* (figs. 473-475, 480, 482, 485, 509).

Psocini/Ptyctini

As noted earlier, some genera within these groups, the former subfamily Psocinae in its entireity, are difficult to allocate to one or other tribe. Placement has been hampered by relatively uncritical addition of species to several genera, particularly within Ptyctini (Smithers, 1985a), so that these genera (although defined initially on features of one or few species) have been artificially expanded to accommodate taxa which might be placed more accurately elsewhere and their limits have thus become blurred.

The major feature separating the two tribes is the form of the male clunium-epiproct interface and use of this character has served, for example, to exclude a number of species allocated formerly and uncritically to *Trichadenotecnum* (p. 165; Ptyctini; Yoshizawa, 2001a). A few are now placed in *Atrichadenotecnum* (p. 160; Psocini). As another example, discussed below (p. 163), *Ptycta* is recognised widely as a polyphyletic holding genus (Thornton, 1984a) and, again, *Psocidus* (intended as a holding genus: Pearman, 1934b) was erected to contain the large number of species placed earlier in *Psocus* Latreille, but excluded from it by a narrower definition of *Psocus* (Pearman, 1932a) and which could not be confidently allocated elsewhere.

Many such species, from many parts of the world, remain *incertae sedis*. *Psocidus* (p. 160) has been redefined subsequently in a more restricted sense to include a group of taxa related to the type species.

A number of species of *Psocus* s.l. described from the Philippines by Banks are still anomalous, as noted below. It is doubtful whether any of them belong truly to *Psocus*, in the sense applied to the European type species (*Psocus bipunctatus* L.) and its close relatives (see Lienhard (1998a) for diagnosis and discussion of *Psocus*). *Psocus* s.str. appears not to occur in tropical South-east Asia.

Psocini

Psocini is by far the less diverse of these two tribes in the region and only two regional genera are placed formally within it.

Key to the genera of Psocini

1 Fore wing usually heavily pigmented, often with numerous spots/patches along or between branches of veins R and M ▸ *Atrichadenotecnum*

– Fore wing mainly hyaline. If shaded, markings entirely or predominantly in basal half of wing ▸ ***Psocidus* s.str.***

* This placement differs from that of Lienhard & Smithers (2002), who listed *Psocidus* in Ptyctini following unpublished advice from E.L. Mockford.

Generic information

Atrichadenotecnum Yoshizawa (figs. 478-479). Type species *Atrichadenotecnum quadripunctatum* Yoshizawa.
Generic distribution: Sri Lanka, Japan, Taiwan, Hong Kong, Malaysia, Indonesia.
Atrichadenotecnum was raised to contain two species earlier allocated to *Trichadenotecnum*, and three new species. Endang et al. (2002) attributed two Indonesian species to this genus, despite some differences in the phallosome form from the state used by Yoshizawa to diagnose the genus. *Atrichadenotecnum* is the putative sister group of *Psocus* and those species placed formerly in *Trichadenotecnum* were allocated there on wing features (namely, in having submarginal and other fore wing spots and a triangular areola postica) which are now believed to be convergent. As Yoshizawa (1998a) noted, this genus runs to *Trichadenotecnum* in most published keys but differs considerably on genitalic features, particularly in having a fringed process on the hypandrium and a pair of shallow 'trenches' near the base of the apical lobe of the subgenital plate.

Psocidus Pearman (figs. 486, 492, 514). Type species *Psocidus zanzibarensis* Pearman.
Generic distribution: Africa, Asian mainland including Hong Kong, Malaysia, Singapore, Indonesia, (?Vietnam), (?Philippines).
In the sense applied here, *Psocidus* is a relatively small group of around a dozen species, rather than the large heterogeneous assemblage implied by its former widespread use as a holding genus. This restricted use (following Thornton, 1960) is based on concordance of wing venation and male genitalic form with *P. zanzibarensis*, and was founded on two Hong Kong species. In these species the hypandrium has a setose apical lobe, the phallosome has long anterior lateral arms, lateral apophyses, a pair of long inner sclerites and the paraprocts have two trichobothrial fields, representing the main field divided by a sensilla-free zone. Thornton (1960) used these features to designate *Psocidus* s.str. and other species from the region have subsequently been allocated to this genus. It has not been reported from New Guinea, with Endang et al. (2002) regarding this absence as 'questionable'.
Several of the apparently endemic Indonesian species attributed to *Psocidus* s.str. have the trichobothrial field of the paraprocts not as markedly divided as in the Oriental mainland species, but are otherwise similar. South-east Asia at present appears to be a major centre of diversity for this genus. Although its correct placement has not been confirmed, Smithers & Thornton (1981a) noted that the New Guinea species known as *Psocidus* (s.l.) *biroi* (Enderlein) is 'probably a species of *Clematoscenea*'. It was transferred formally to this genus by Smithers & Thornton (1990).

Ptyctini

Ptyctini is much more diverse than Psocini in South-east Asia, but the integrity of some of the constituent genera is far from clear. Altogether, seven genera are represented, but several of these may be divided in the future. Thus, as examples discussed below, *Ptycta* is a 'holding genus', the distinctions between *Clematostigma, Ptycta* and *Copostigma* in the region are not always clear, and *Trichadenotecnum* comprises a number of species groups (Yoshizawa, 2001a) which might eventually be elevated to generic status. Some of the genera are widespread, but others appear to be regional endemics.

Extralimital distributions of genera as given below may be open to revision as taxa are examined more critically. The three genera listed first, *Clematostigma, Copostigma* and *Ptycta*, form part of a complex of closely-related and generally similar forms, as discussed by Smithers (1983a), together with *Maheella* Enderlein (Seychelles). The last has been regarded by some workers as a synonym of *Ptycta* (although without critical re-examination of the type species). This group of genera was designated by Badonnel (1967) as the '*Copostigma-Clematostigma-Ptycta-Maheella* complex' and the above key characters are augmented slightly below, together with notes on variation within the putative genera.

Key to the genera of Ptyctini

1 Fore wing short and elytriform (strongly thickened) ▸ *Elytropsocus**
– Fore wing of normal form, membranous and not elytriform ▸ 2

2 Fore wing with first and second portions of vein Cu1a forming almost a straight line, so areola postica appears four-sided (no spur vein to pterostigma) ▸ 3
– Fore wing with first and second portions of vein Cu1a at a distinct angle, so areola postica appears five-sided (many taxa with a short spur vein to pterostigma) ▸ 5

3 Meso- and meta-scutella of thorax with strong dorsal ridges ▸ *Kaindipsocus**
– Meso- and meta-scutella of thorax of normal form, without dorsal ridges ▸ 4

4** Fore wing usually heavily pigmented, often with numerous discrete spots ▸ *Trichadenotecnum*
– Fore wing mainly hyaline, never with numerous discrete spots ▸ *Javapsocus*

5 Fore wing veins Rs and M joined by a crossvein ▸ *Copostigma*
– Fore wing veins Rs and M fused for a length ▸ 6

6*** Female: subgenital plate with short, rounded or transverse, apical lobe; ventral gonapophysis valve about as long as dorsal gonapophysis valve. Male: phallosome frame with conspicuous outer parameres ▸ *Clematostigma*

– Female: subgenital plate with long apical lobe; ventral gonapophysis valve shorter than dorsal gonapophysis valve. Male: phallosome narrowed to slender posterior apex, sometimes bifid at apex ▸ *Ptycta*

* Genera *Elytropsocus* and *Kaindipsocus* are treated in Amphigerontiinae by Lienhard & Smithers (2002), following unpublished advice from E.L. Mockford.

** Endang et al. (2002) noted an unnamed genus from Indonesia, known from only a single damaged male from Bali (see Endang & Thornton, 1992). This specimen keys above to couplet 4, but differs from *Trichadenotecnum* and *Javapsocus* on the fore wing venational feature of the posterior border of the pterostigma being smoothly rounded and convex, rather than being basally concave or straight and apically convex. It is noted here to alert readers to the existence of this anomalous taxon in the region.

*** Couplet 6 incorporates data from non-regional taxa and is based on genitalic features summarised by Smithers (1983a, *Clematostigma*) and Thornton (1984a, *Ptycta*).

Generic information

Clematostigma Enderlein (fig. 487). Type species *Copostigma maculiceps* Enderlein. Generic distribution: Paraguay, India, China, Taiwan, Japan, Indonesia, New Guinea, Australia.

The type species of *Clematostigma* was for long the only well-characterised species in this genus (Smithers, 1983a) and the relationships of most other species are still somewhat unclear. The genus was treated as a subgenus of *Copostigma* by Roesler (1944), but its generic status was confirmed by Smithers (1983a).

The following features were noted by Smithers (1983a) in defining *Clematostigma*:

- a pterostigmal spur vein is always present,
- the first portion of vein Cu1a is longer than the second portion (also in some *Ptycta*) and at an angle to it,
- the subgenital plate lobe is long, as are the gonapophyses. The external valve is peculiar in having a dorsal tapered lobe,
- the symmetrical hypandrium lacks teeth or apophyses (cf. *Ptycta*) and the closed phallosome has strong external parameres,
- the paraproct trichobothrial field is entire and the male paraprocts do not have a basal lobe.

With this scope, only three regional species have been recognised until now.

Copostigma Enderlein (fig. 489). Type species *Copostigma dorsopunctatum* Enderlein. Generic distribution: Sri Lanka, Malaysia, New Guinea, New Caledonia, Fiji, Tonga.

Several of the New Guinea species currently placed in *Copostigma* were described in *Mecampsis* Enderlein by Smithers & Thornton (1981a) because of a general

resemblance to that genus. Following the closer generic definition provided by Smithers (1983a, 1985a), the latter genus is now believed to occur only in the New World (the type species, *M. cinctifemur* Enderlein, is from Chile), although the two seem to differ in only small venational features. Genitalic information on *Mecampsis* s.str. is not available. Nearly 20 species have been described.
A pterostigmal spur vein is present in all species, and the first portion of vein Cu1a is shorter than the second portion and at an angle to it. Based on the New Guinea species (Smithers & Thornton, 1981a), Smithers (1985a) also characterised *Copostigma* by the male paraprocts having a strongly developed (sometimes rugose) basal lobe, and the female subgenital plate with a short broad apical lobe (rarely longer). The divided trichobothrial field is also of putative generic value, but the most conspicuous generic feature has been regarded as the linking of fore wing veins Rs and M by a crossvein, rather than these being fused for a length, with the discoidal cell narrowed posteriorly.

Ptycta Enderlein (figs. 488, 493, 512). Type species *Psocus haleakalae* Perkins.
Generic distribution: North America, South America, Galapagos Islands, Hawaiian Islands, Madagascar, Indian Ocean region, Hong Kong, Japan, Malaysia, Indonesia, New Guinea, Australia, Micronesia.
There are considerable problems in interpreting the limits and distribution of *Ptycta*, largely reflecting that it has become a 'holding genus' to which a great variety of (predominantly) tropical Psocidae have been added. These include taxa variously with or without a pterostigmal spur vein and with varying forms of fore wing Rs-M connections, together with a variety of genitalic features. Smithers (1983a) helped to delimit other genera in the complex and Thornton (1984a), based on a study of more than 50 endemic Hawaiian species (including the type species of *Ptycta*), provided a far more succinct definition of *Ptycta* than available previously and also designated a number of Hawaiian species groups to reflect the major variations present. Most Hawaiian species lack a conspicuous pterostigmal spur vein, although this was long used as a reliable generic feature. Because of the considerable variation in putative *Ptycta*, Thornton's (1984a) study on the type species and its close relatives, constituting a substantial endemic radiation of Hawaiian taxa, was a necessary prerequisite to defining related genera (Smithers, 1985a). Mockford (1993a) retained two North American species in *Ptycta*, but had earlier distinguished the closely-related *Indiopsocus* Mockford from other New World genera by (in part) the phallosome being broad at the base and epiproct meeting clunium in a straight line (cf. *Ptycta*: phallosome slender and tapering to base, male epiproct extending over the clunium in the central region; Mockford, 1974a).
Thornton's genitalic diagnosis of *Ptycta* (Thornton, 1984a) notes, *inter alii*, the following features:

- ventral gonapophysis valve at least half the length of the dorsal valve, dorsal valve with subapical lobe or with blunt apex abruptly narrowing subapically, or gradually tapering,
- distal process of subgenital plate lacking lateral longitudinal sclerotisation near each edge,
- hypandrium with median tongue with teeth along each lateral margin, with or without median boss,
- phallosome closed, symmetrical, narrowed and rounded or angular anteriorly, narrowed posteriorly to apical tine which may be bifid,
- male epiproct with median basal sclerotised rugose lobe,
- male paraproct with rounded conical process (sometimes rugose) basal to trichobothrial field.

Several Indonesian taxa, amongst others, appear to be true *Ptycta*. However, a number of Indonesian species attributed to this genus by Endang et al. (2002) reflect the difficulties which remain in this complex, because an equally convincing case could be made for placing them in *Copostigma*, depending on whether genitalic or wing features are accorded greater taxonomic significance. The development and diversification of these genera in tropical South-east Asia is indeed complex but at present, and not withstanding future revisions, *Ptycta* appears to be widespread in the region.

Other ptyctine genera

Elytropsocus Smithers & Thornton (figs. 490-491). Type species *Elytropsocus coleoptratus* Smithers & Thornton. Generic distribution: New Guinea.
Elytropsocus is unique within Psocidae in its unusual elytriform fore wings, and its general appearance resembles that of a small beetle (cf. *Calocaecilius*, p. 109; Calopsocidae, p. 114). It has, so far, been found only in New Guinea (Smithers & Thornton, 1981a) and is known only from the type species. Smithers & Thornton (1990) suggested that *Elytropsocus* (together with *Kaindipsocus*, below) is 'probably related to *Copostigma*' but, because of its form, wing venation can not be used constructively to help assess relationships. Only the female has been described, and it is unknown whether males share the elytriform appearance or are of 'more normal' form. Female genitalia suggest relatedness to *Kaindipsocus*, but discovery of the male is necessary to determine even subfamily affinities more clearly. Note placement in Amphigerontiinae by Lienhard & Smithers (2002).

Javapsocus Endang, Thornton & New (figs. 494, 498, 502, 505, 513). Type species *Javapsocus intimadentatus* Endang, Thornton & New.
Generic distribution: Indonesia.
This genus was raised to contain four Indonesian species which could not be attributed easily to any existing genus or subgenus of Psocinae s.l. Fore wing veins

Rs and M meet in various ways, but M is subsequently strongly concave, so that the apical border of the discoidal cell is strongly curved. This, together with Cu1a first and second portions being in a straight line, provides the most obvious generic feature of *Javapsocus*. The hypandrium has a median pair of large 'claws' and a variety of other apophyses; the phallosome has a long apical projection; the subgenital plate has a long posterior lobe; and the ventral and dorsal gonapophysis valves are both long; the external valve is large. The genus is known only from Indonesia and seems to be widespread in Java (Endang et al., 2002).

Kaindipsocus Smithers & Thornton (figs. 495, 501, 506, 511). Type species *Kaindipsocus mixtus* Smithers & Thornton.
Generic distribution: Vietnam (unpublished), New Guinea, Australia.
See under *Elytropsocus*, above, for comment on possible affinities. *Kaindipsocus* is distinguished by the very broad discoidal cell, which is as wide as long and has a convex proximal side caused by curvature of vein Cu1a. The areola postica is tall and the fusion of Cu1a with M is very short; the Rs-M crossvein is unusually long. The highly unusual sclerotised ridges on the pterothorax are not known elsewhere in Psocidae and may be involved with sound production (Smithers & Thornton, 1981a), although this intriguing suggestion has yet to be investigated experimentally. A ventral apophysis on the female paraproct (as found also in *Elytropsocus*) is an unusual feature. Females are unusual also in lacking setae at the apex of the subgenital plate. Note placement of *Kaindiposcus* in Amphigerontiinae by Lienhard & Smithers (2002).

Trichadenotecnum Enderlein (figs. 496-497, 499-500, 503-504, 507-508, 516). Type species *Hemerobius sexpunctatus* L. Generic synonymy *Trichadenopsocus* Roesler, type species *Psocus desolatus* Chapman.
Generic distribution: North, Central and South America, Europe, Africa, India, Nepal, Bhutan, China, Taiwan, Japan, Thailand, Vietnam, Malaysia, Singapore, Indonesia, New Guinea, Australia.
Although they were listed as subgenera of *Trichadenotecnum* by Roesler (1943c), *Loensia* Enderlein and *Psocomesites* Roesler are now generally regarded as distinct genera, restricted respectively to the Holarctic region and South America. *Trichadenotecnum* was for many years accepted in Roesler's broad sense of spotted-winged Psocidae, having the first and second portions of vein Cu1a in a straight line so that the areola postica is of characteristic shape, and with veins Rs and M fused for a short length. Such a grouping is highly artificial, as revealed in a number of subsequent studies. However, even with transfer of a few anomalous species to *Atrichadenotecnum* (Yoshizawa, 1998a, p. 160) and restriction by removal of the former subgenera noted above, the genus is large, very varied and very strongly represented in South-east Asia. Thus, for example, Yoshizawa (2001a) noted that some species should be removed from *Trichadenotecnum* by lacking

generic apomorphies but '... are retained in *Trichadenotecnum* because there is no genus suitable for them'. A few species from Indonesia have been listed as *Loensia*, but this placement has not been confirmed. However, Li Fasheng (2002) placed 20 Chinese species in *Loensia* and also recognised *Trichadenopsocus* as a distinct genus, with 26 Chinese species.

The features considered by Yoshizawa to diagnose *Trichadenotecnum* predominantly involve wing pattern and male genitalia (Yoshizawa, 2001a). He noted that female genitalia 'are identical with those of other genera in the tribe Ptyctini', and therefore do not help in defining the genus. Yoshizawa (2001a) listed a diagnostic character suite, as below, but noted the difficulties of discussing the diagnostic fore wing markings (fig. 497) in taxa with extensively and/or densely spotted wings. In such species, the diagnostic marks are usually in part identifiable (thus, submarginal spots in cells R3 and R5, proximal bar visible between M+Cu fork and distal third of cell Cu2).

Trichadenotecnum s.str. was defined by Yoshizawa (2001a) on the following putative apomorphic characters, which he considered to delimit the genus as a monophyletic entity:

- fore wing markings – spots in cell C; opposing spots near base of cell r; proximal band from stigmasac passing through M+Cu fork to distal third of cell cu2; a submarginal row of six spots,
- male genitalia – presence of clunial arms; hypandrium with median tongue (unique among Psocoptera but (? secondarily) absent in some species); combination of apophyses on posterior margin of hypandrium.

Within this combination, considerable variety occurs. Yoshizawa recognised different character combinations to designate five major species groups based on the Japanese representation of 22 species. The genus appears to be particularly diverse in the Malaysia/Indonesia region but poorly represented further east in New Guinea. Of Yoshizawa's species groups, the following are represented in South-east Asia, but some species are difficult to allocate, may constitute as yet undefined groups and further revision is needed. See also New (1978c) and Endang & New (2005).

- '*majus*-group'. This group includes *T. apertum* Thornton (Hong Kong) in addition to species from Europe, China, Japan and Indonesia. The major autapomorphy is the presence of two or three dorsal processes on the male epiproct, and the fore wing is not extensively covered with small spots;
- '*spiniserrulum*-group'. This was raised to contain species from India, China and Japan. Other possible inclusions are species from Malaysia (*T. malayense* New, *T. gombakense* New & Lee), as well as species from Indonesia (Endang et al., 2002). Species of this group are united by three apomorphies (Yoshizawa, 2001a), namely (1) the male epiproct being conical, (2) the male paraproct having a basal lateral process and (3) the ventral gonapophysis valve being very short – the latter feature found also in some members of the *majus*-group. Many, although not all, species of the *spiniserrulum*-group

have a very weakly defined posterior lobe to the external gonapophysis valve (in contrast to the well developed lobe of other groups) and this is not found in any *majus*-group species. Note that *T. spiniserrulum* Datta was transferred to *Trichadenopsocus* by Li Fasheng (2002), so that his concept of *Trichadenotecnum* differs widely from that of Yoshizawa and of most other authors;

– '*medium*-group'. This group is based on *T. medium* Thornton (Hong Kong) and also contains ten Japanese species of *Trichadenotecnum*.

TABLE 6
Recognition features for some species groups of *Trichadenotecnum* (Yoshizawa, 2001a). Apomorphic features are underlined.

CHARACTER	SPECIES GROUP				
	sex-punctatum	*medium*	*alexanderae*	*majus*	*spini-serrulum*
fore wing covered with small spots	no	yes/no	yes	no	yes
male					
– epiproct	chair-shaped	chair-shaped	chair-shaped	chair-shaped	conical
– epiproct lobe	long/wide	long/wide	long/wide	2-3 posterior processes	(none)
– paraproct with latero-basal process	no	no	no	no	yes
– hypandrium with					
· right process	yes	yes	no	no	no
· median tongue	yes	yes	yes	reduced	yes
· keel-like process	yes	no	no	no	no
female					
– subgenital plate with median sclerotised band	no	yes	no	no	no
– gonapophyses					
· ventral valve	long	long	long	long/short	short
· external valve posterior lobe well-developed	yes	yes	yes	yes	no

The character matrix used by Yoshizawa (2001a) to designate species groups is shown in table 6, to provide a guide to help appraisal of the variety of undiscovered *Trichadenotecnum*-species in South-east Asia. An additional species group was designated by Yoshizawa & Lienhard (2004) for two species from Hong Kong and Indonesia. This, the '*vaughani*-group', has apomorphies of the median tongue of the hypandrium and phallosome, and the ventral gonapophysis valve is short.

Thyrsophorinae

Formerly regarded as a separate family, Thyrsophoridae, these narrow-winged Psocidae are the largest of all Psocoptera and until recently were believed to be found exclusively in the Neotropics (New, 1978a). The pterostigma is strongly broadened in the basal half and narrowed apically and (as in Cycetini, Psocinae, p. 156) vein R4+5 separates from the apparent vein M rather than forking directly with vein R2+3. The recent discovery of a member of this group in Vietnam (Mockford, 2004) is an intriguing geographical puzzle in representing a remarkable disjunct incidence far from the group's documented range.

Generic information

Thyrsopsocopsis Mockford (figs. 517-518). Type species: *Thyrsopsocopsis thorntoni* Mockford.
Generic range: Vietnam.
This genus is known from a single male specimen of the type species captured in a Malaise trap. It shares with the Neotropical *Thyrsopsocus* Enderlein the fore wing feature of vein R 4+5 joined to M by a crossvein, but differs clearly in the relatively narrower fore wing, the pterostigma being shallower, and in hypandrial features. The large size (fore wing longer than 8 mm) renders this by far the largest psocid known from the region.

Chapter 6

Figures

FIGURE 1

Map of South-east Asia and nearby areas, to indicate major political boundaries.

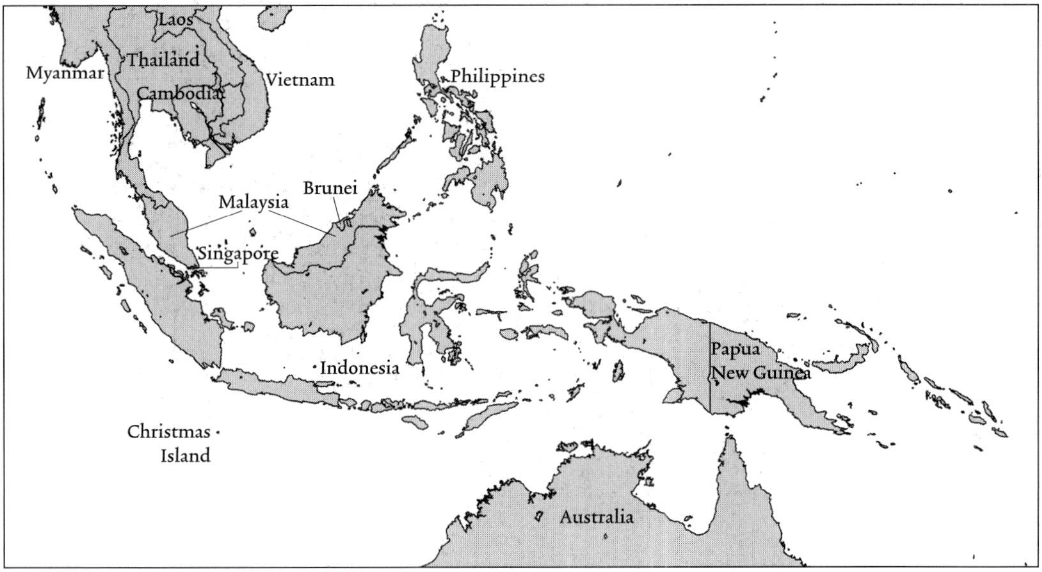

FIGURES 2 – 12

Immature stages of Psocoptera: oviruptors of 2 *Stenopsocus* – 3 a typical bark-frequenting psocid – 4, 5 *Ectopsocus* – 6 *Graphopsocus* – 7, 8 *Caecilius* (2, 3, 4, 6, 7 profile, anterior to right; 5, 8 ventral aspect, anterior toward top of page) – 9 typical first instar nymph (no trace of wing pads, reduced number of antennal segments) – 10 last instar nymph – 11 egg batch, with covering webbing, of foliage-frequenting psocid – 12 two eggs, in profile, of bark-frequenting psocid, covered with faecal material (2-10 after New, 1969d).

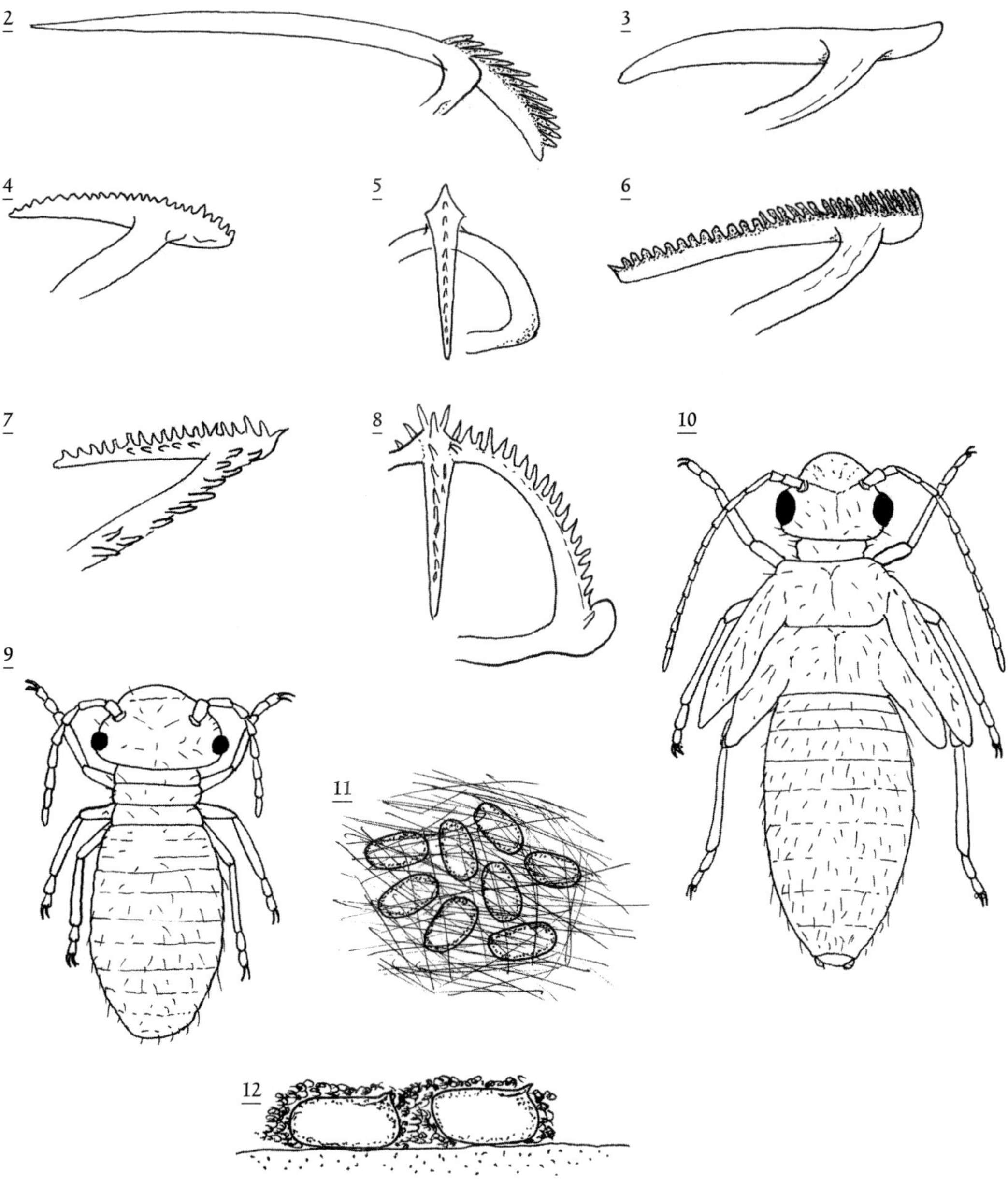

FIGURES 13 – 16

Diagnostic measurements for Psocoptera, used in taxonomic description. **13** IO/D, Pearman's method – **14** IO/D and PO/D, Badonnel's method – **15** wing length, point of articulation with thorax to apex of membrane – **16** tarsal segment lengths, condyle to condyle.

13

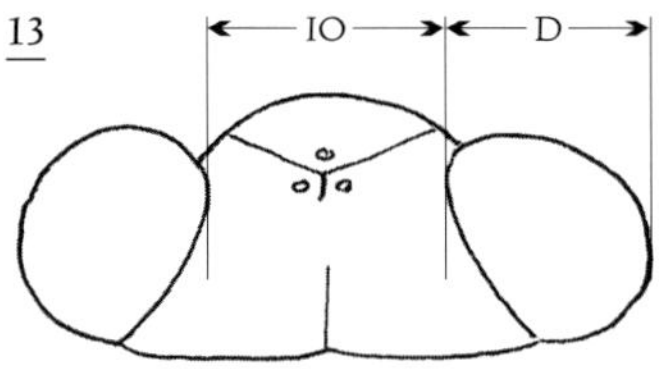

14

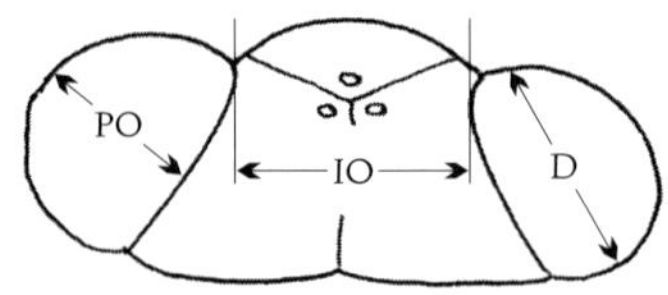

15

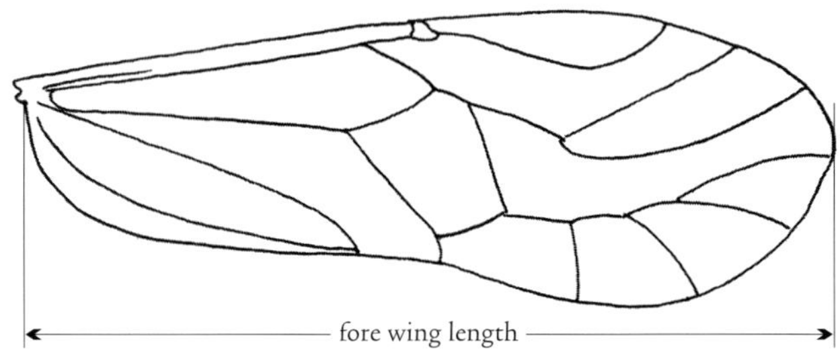

16

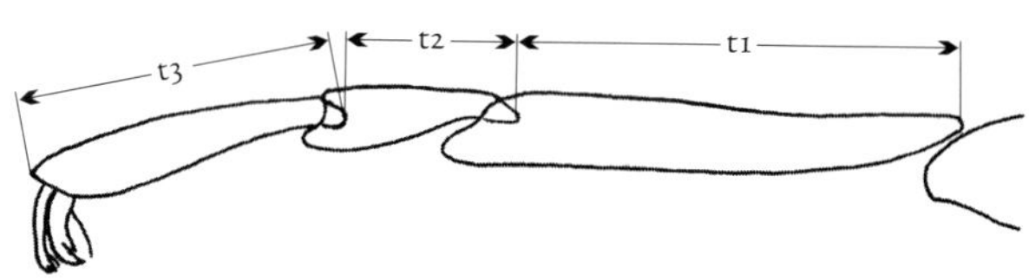

FIGURES 17 – 19

Representative adult Psocoptera, general body form in profile (details of wing venation not shown). **17** Lepidopsocidae (numerous antennal segments, wings held in shallow 'tent' over body) – **18** Liposcelididae (apterous, body depressed) – **19** Psocidae (no more than thirteen antennal segments, wings held in steep 'tent' over body) (wing venation not shown).

17

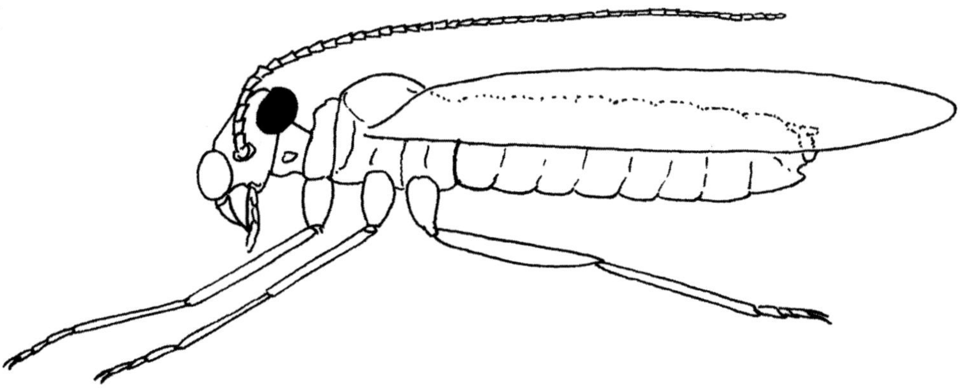

18

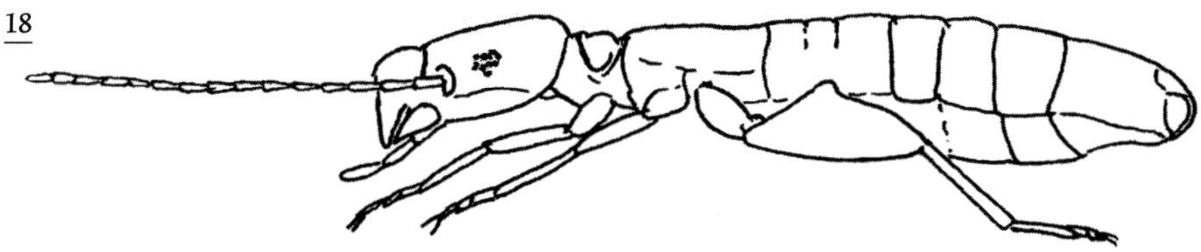

19

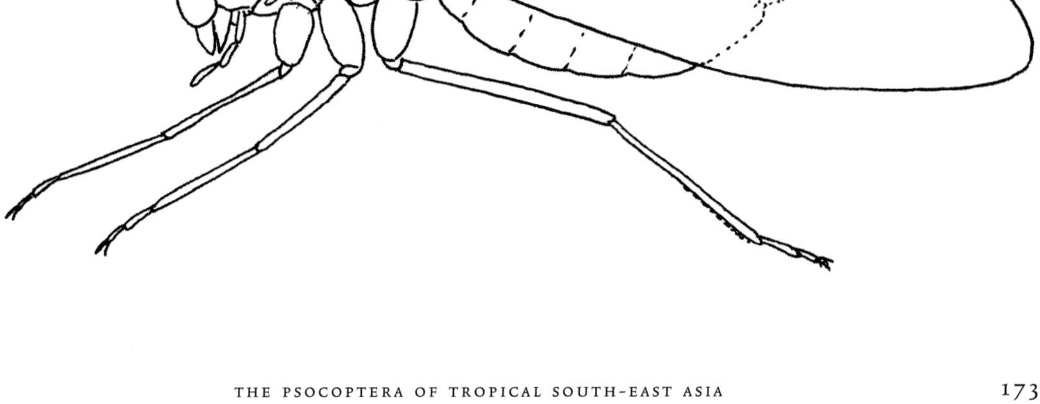

FIGURES 20 – 32 [FACING PAGE]
Morphology, head. Basic structure of head as seen in **20** facial and **21** lateral aspect; forms of mandible in **22** *Liposcelis* – **23** *Psyllipsocus* – **24** *Caecilius* – **25** *Peripsocus*; labium with **26** two-segmented palp – **27** single segmented palp – **28** maxilla, with palp (always four-segmented: P1-P4) and position of lacinia shown; representative laciniae of **29** Epipsocidae, apex expanded with several teeth and **30** Caeciliusidae, apex narrow with two shallow tines; representative antennal flagellar segments of **31** *Archipsocopsis* (as in all Psocomorpha, lacking secondary annulations) with characteristic discoidal sensillum with short cone on flagellar segment vi, and **32** Liposcelididae (note secondary annulations on surface).

20

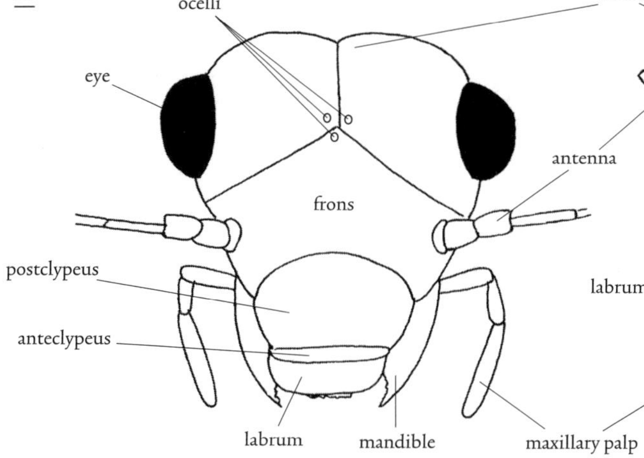

21

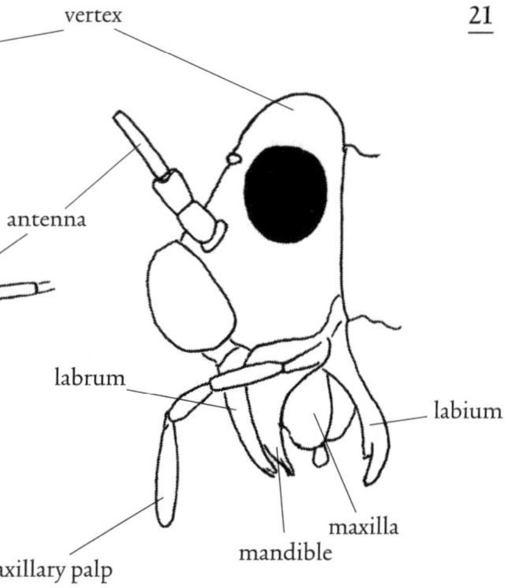

22

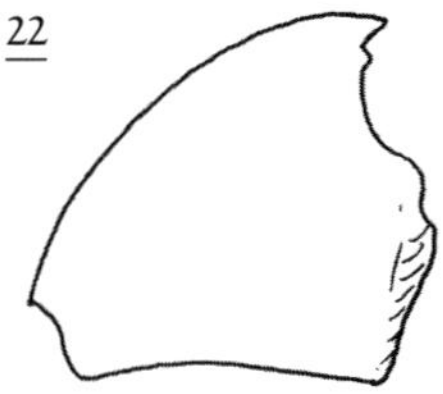

23

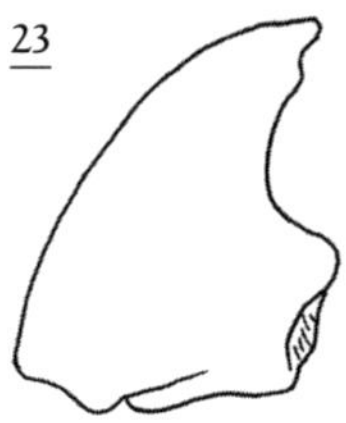

24

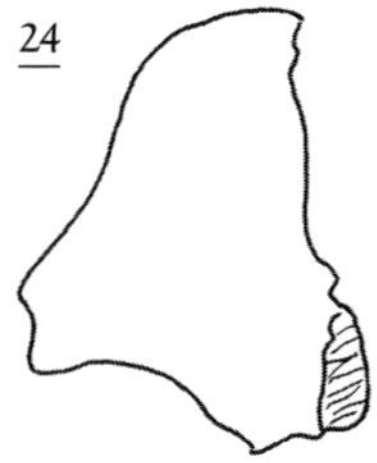

25

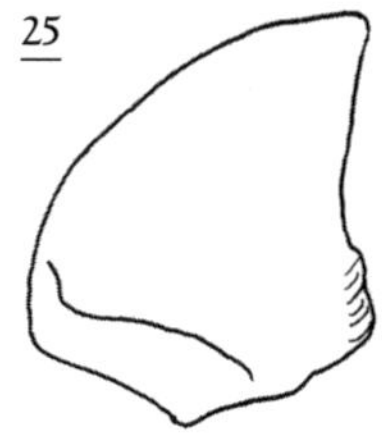

26

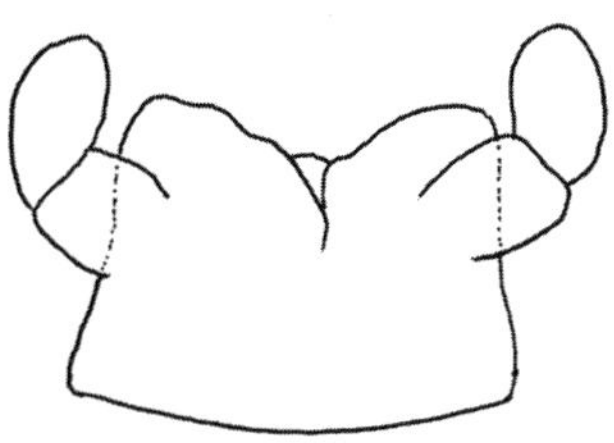

27

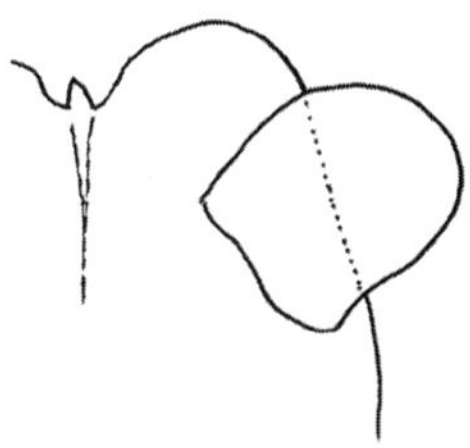

28

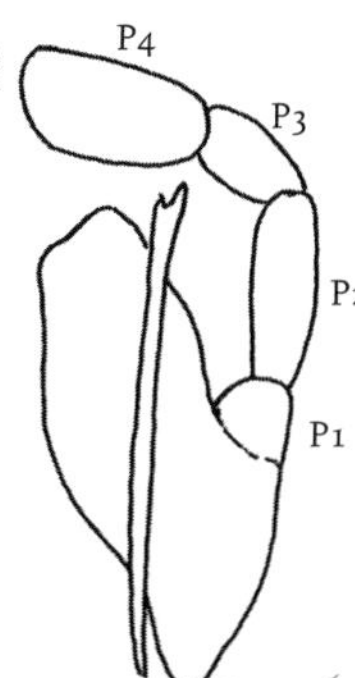

29

30

31

32

FIGURES 33 – 37

Morphology. **33** Dorsum of pterothorax to indicate main structures – **34** hind leg, with parts indicated: coxa with Pearman's organ, enlargement of ctenidiobothrium inserted; tarsal claws of **35** typical bark-frequenting psocid (subapical tooth present, pulvillus slender) and **36** typical foliage-frequenting psocid (claw smooth ventrally, pulvillus expanded) – **37** profile of abdomen to indicate position of eversible ventral vesicles between central segments in foliage-frequenting psocids.

33

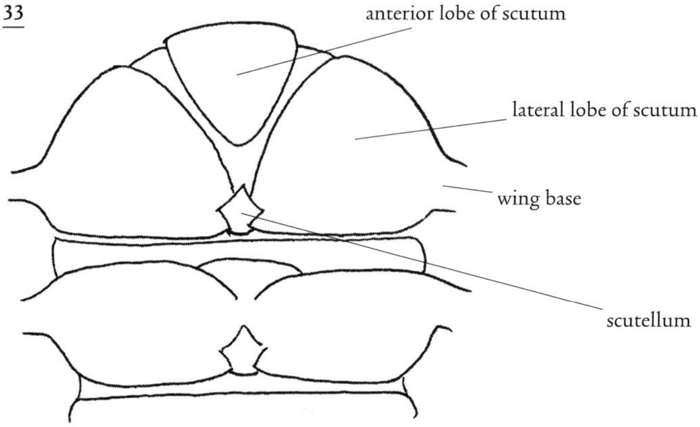

34

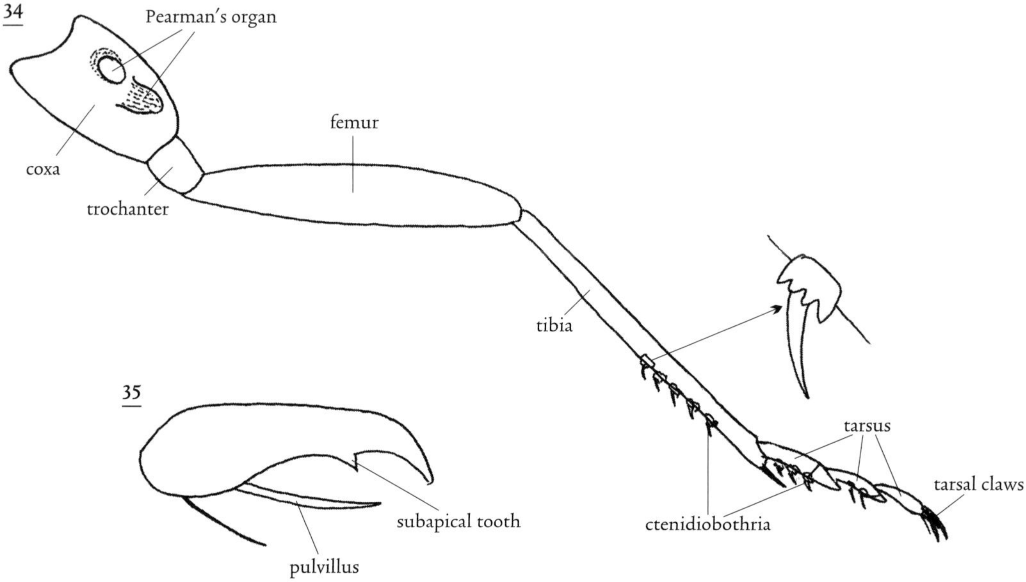

36

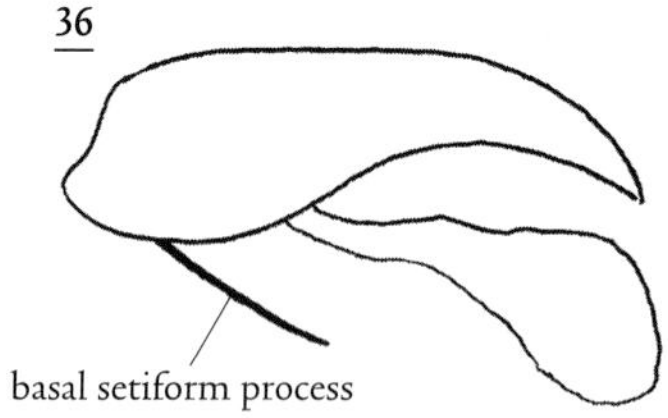

37

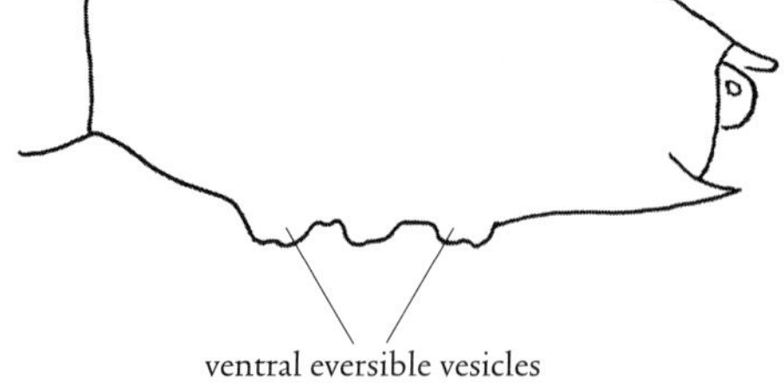

FIGURES 38 – 45
Wing venation. Basic pattern of venation of **38** fore wing and **39** hind wing; representative variations on basic pattern (see also later illustrations for each family): – **40, 41** Liposcelididae, *Embidopsocus*, venation strongly reduced and evanescent; fore wings of **42** *Stenopsocus* (pterostigma and areola postica linked by crossveins to other veins) – **43** Psocidae (areola postica fused to M by joint length to produce closed discoidal cell in centre of wing – **44** *Ectopsocus* (areola postica absent, as Cu1 unforked) – **45** Lepidopsocidae.

38

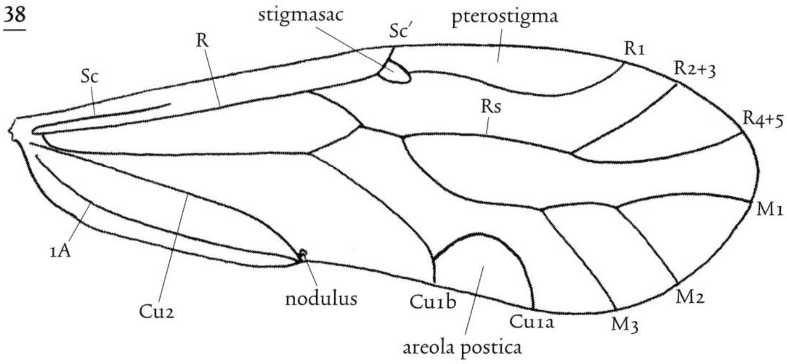

39

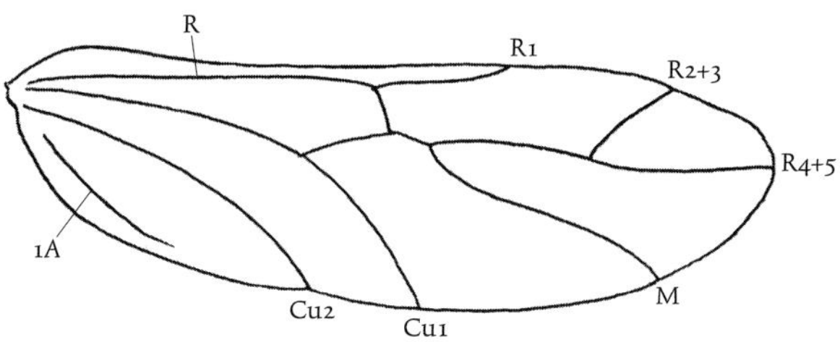

40

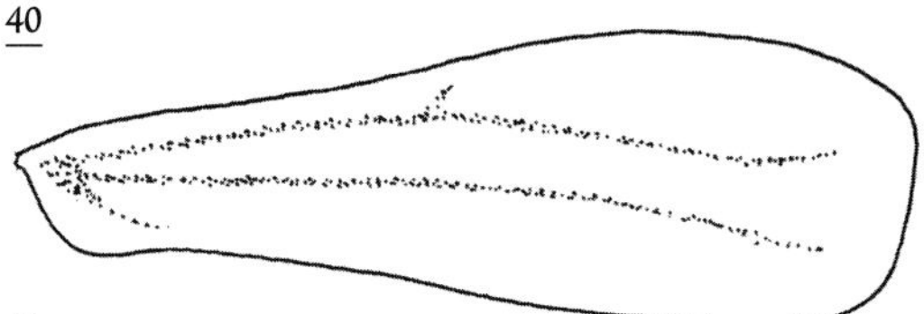

41

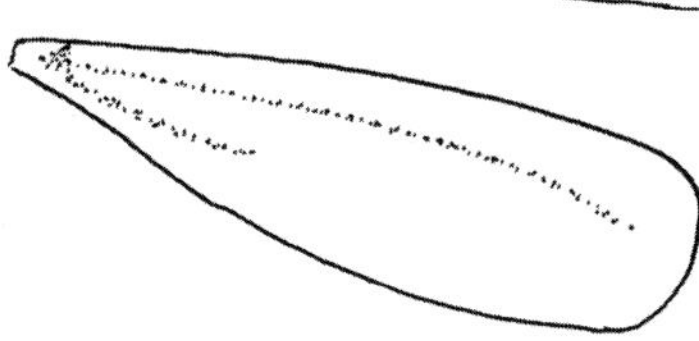

42

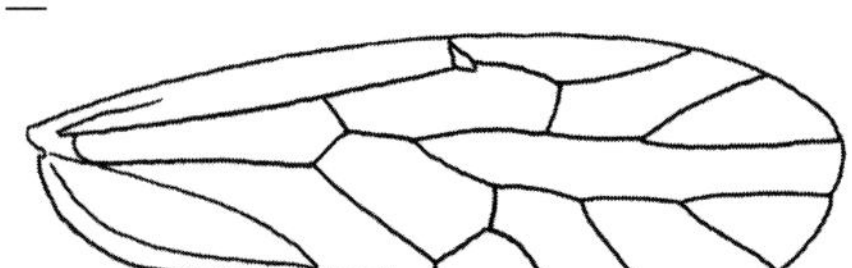

43

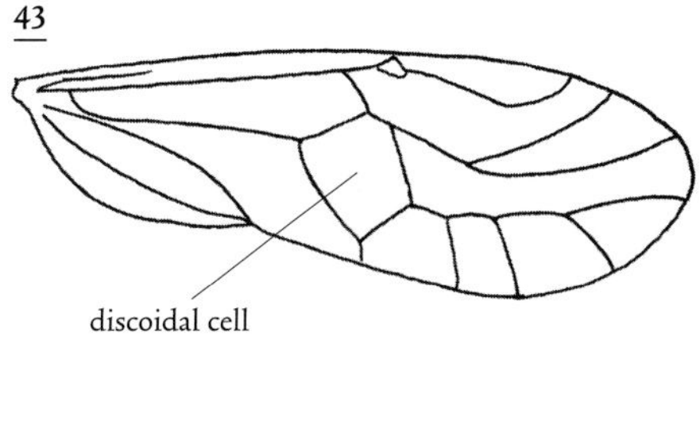

44

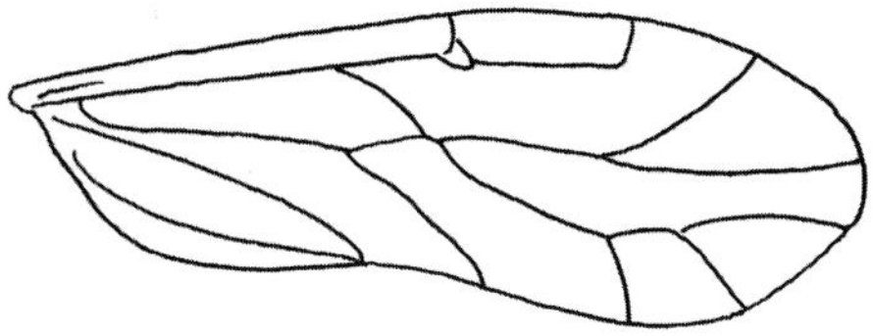

45

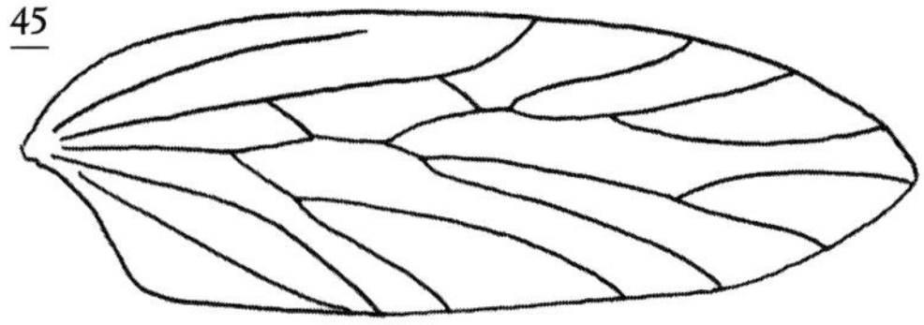

FIGURES 46 – 64 [FACING PAGE]

Abdomen, basic structure and patterns of genitalic structure – **46** profile of apex of female abdomen (posterior to right) to indicate position of genitalic structures – **47** complete gonapophyses, illustrating the three pairs of valves, each of which may be modified; forms of subgenital plate **48** simple, rounded apex (many taxa) – **49** short apical lobe, sometimes bilobed (Peripsocidae) – **50** margin with slight lobes (Elipsocidae) – **51** long, tapered apical lobe (some Psocidae); spermatheca – **52** with duct surrounded by thickened glandular area as in many Caeciliusidae – **53** large with internal maculae, as in Trogiidae; paraproct – **54** rounded with marginal hyaline cones and rounded field of trichobothria (Caeciliusidae) – **55** narrowed with basal stem and apical hooked process (Psocidae) – **56** long preapical spine and loosely grouped trichobothria (Lepidopsocidae) – **57** male clunium expanded and overlying base of epiproct (Peripsocidae); phallosome **58** simple closed frame – **59** frame with external parameres prominent – **60** reduced to parameres, associated basally (some Psocidae); hypandrium **61, 62** simple, rounded and more or less convex – **63, 64** elaborate sclerotised, highly convex (some Psocidae) – **62, 64** ventral – **61, 63** lateral aspect (subgenital plates, apex toward bottom of page; male structures (62, 64), apex toward top of page).

46

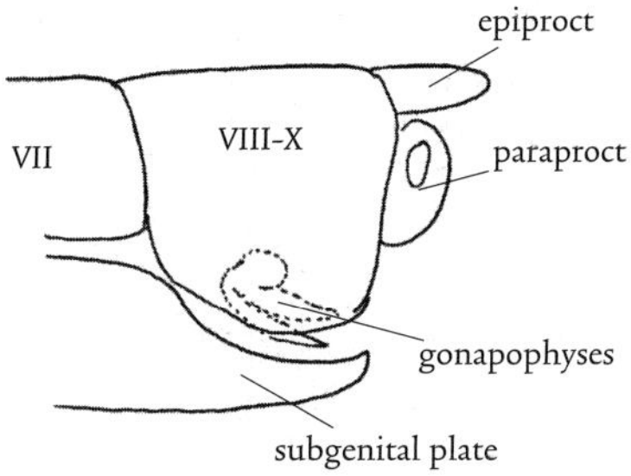

48

49

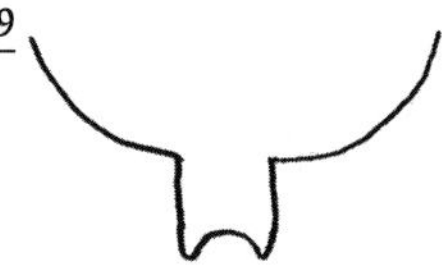

50

51

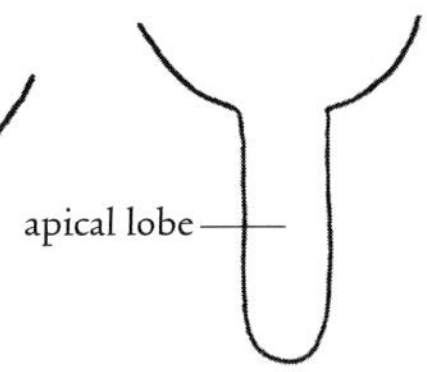

47

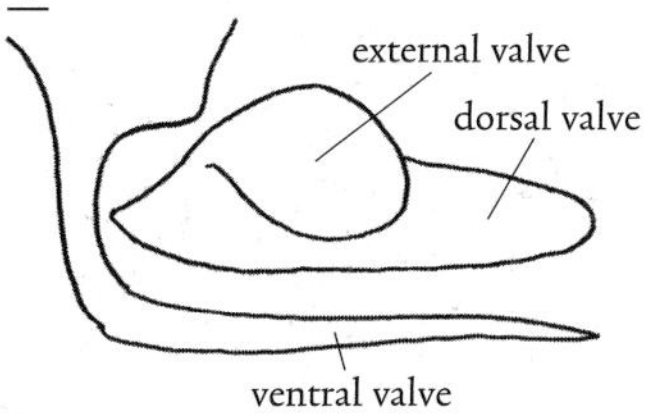

54

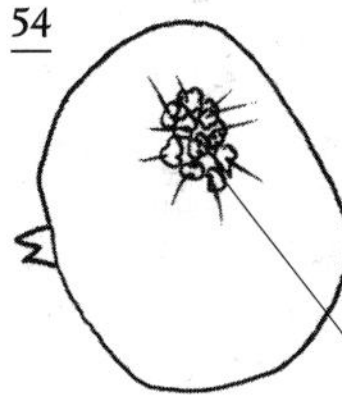

55

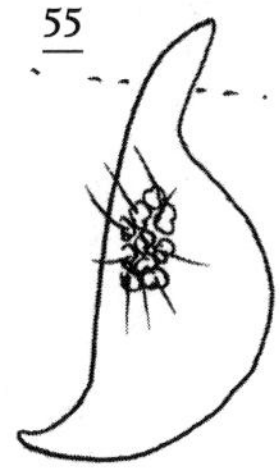

52

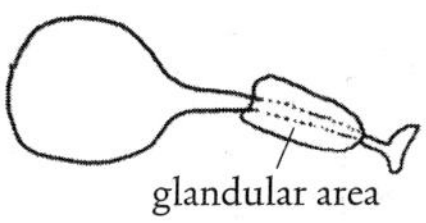

58

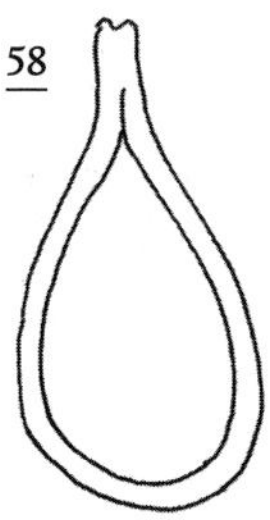

59

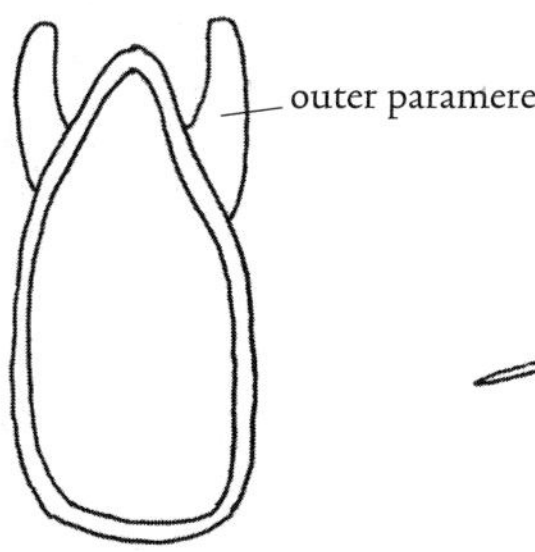

56

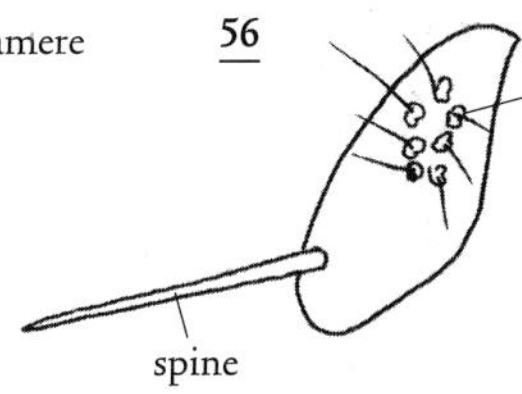

57

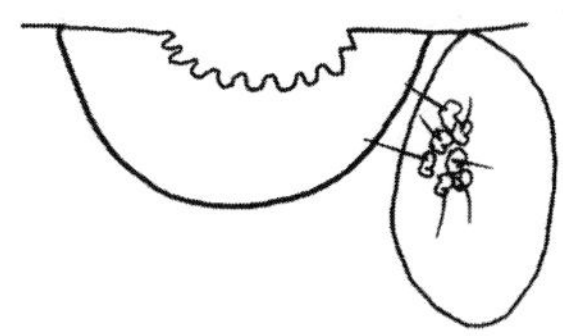

60

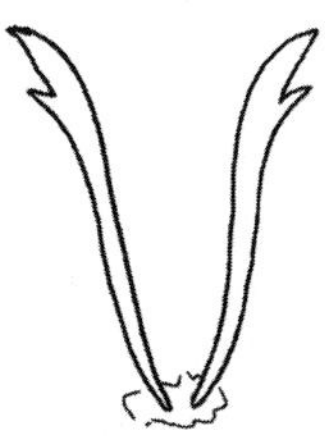

64

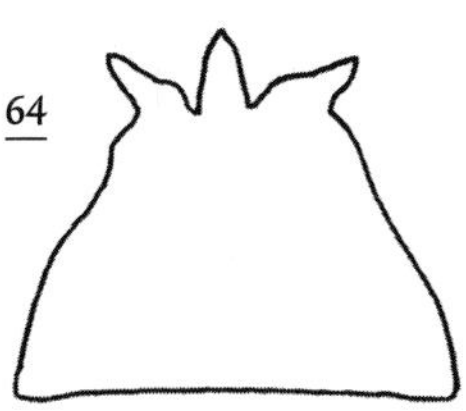

53

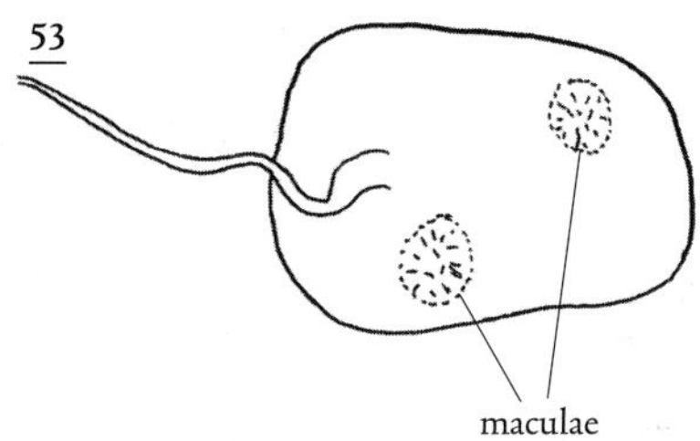

63

61

62

FIGURES 65 – 67
Hypopharynx, degrees of separation of filaments, as used for subordinal classification – 65 filaments separate throughout their length – 66 filaments separate only distally – 67 filaments joined for central length.

65

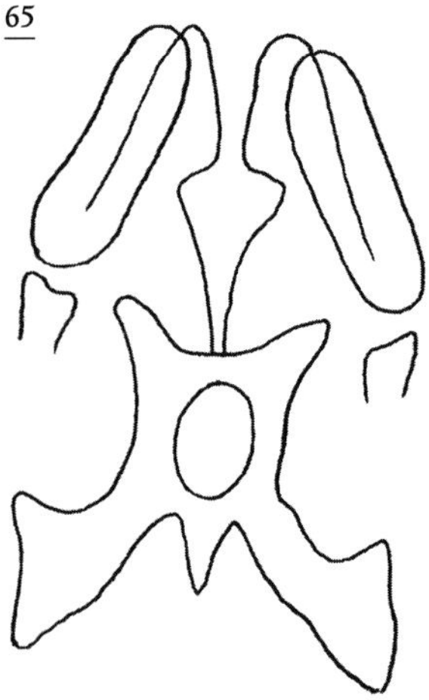

66

67

FIGURE 68
Map of Malesia to indicate demarcations of botanical provinces (after van Steenis, 1950).

68

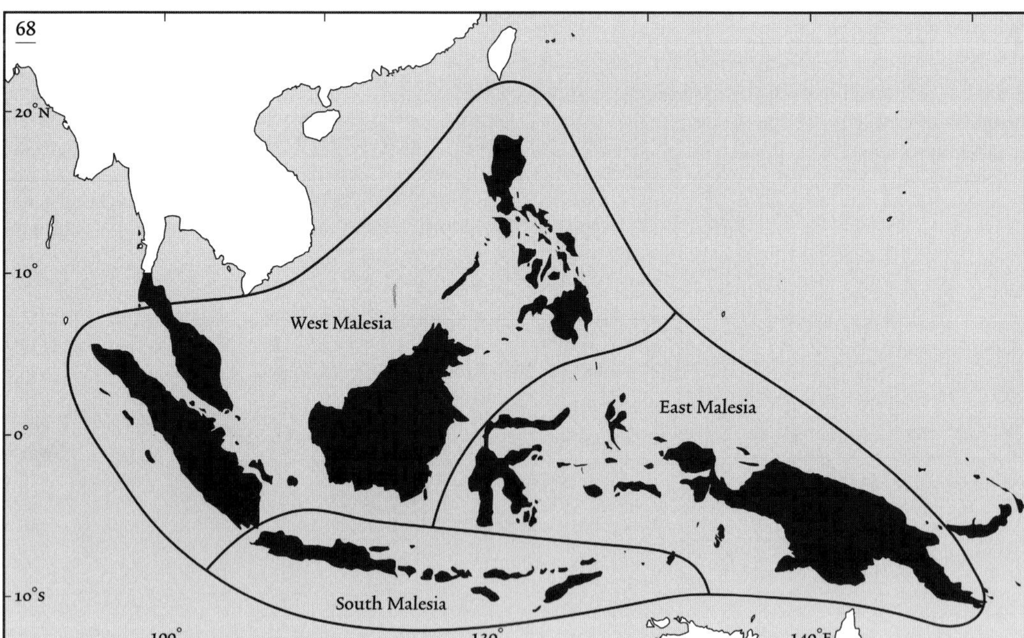

FIGURE 69
Biogeographical subdivisions of Malesia, showing positions of various 'lines' proposed for biotic delimitation, and extent of Sahul and Sunda shelves.

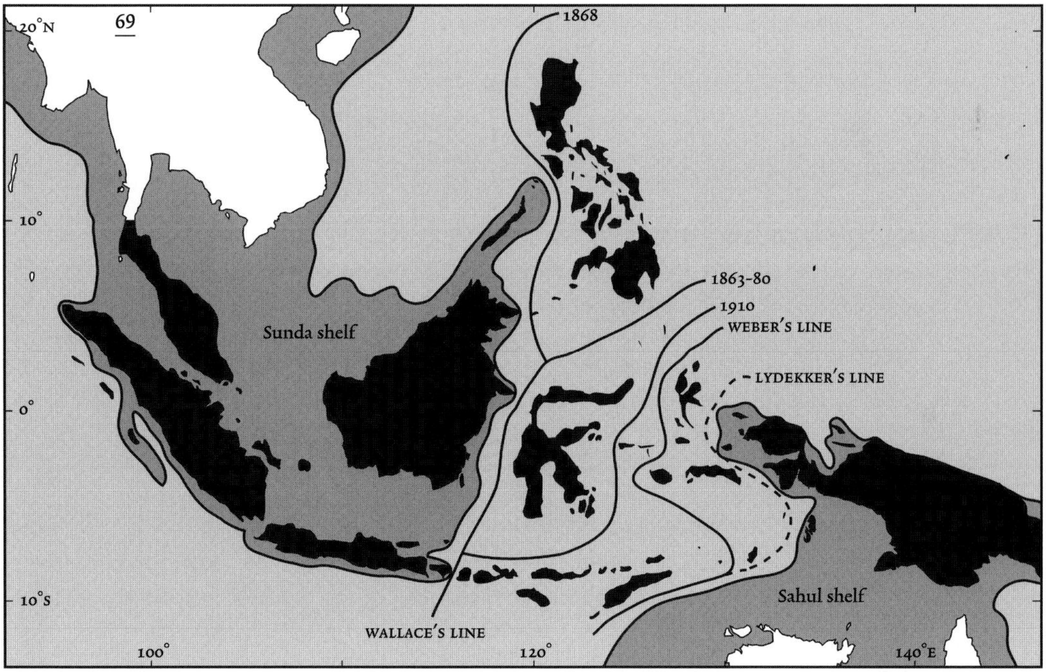

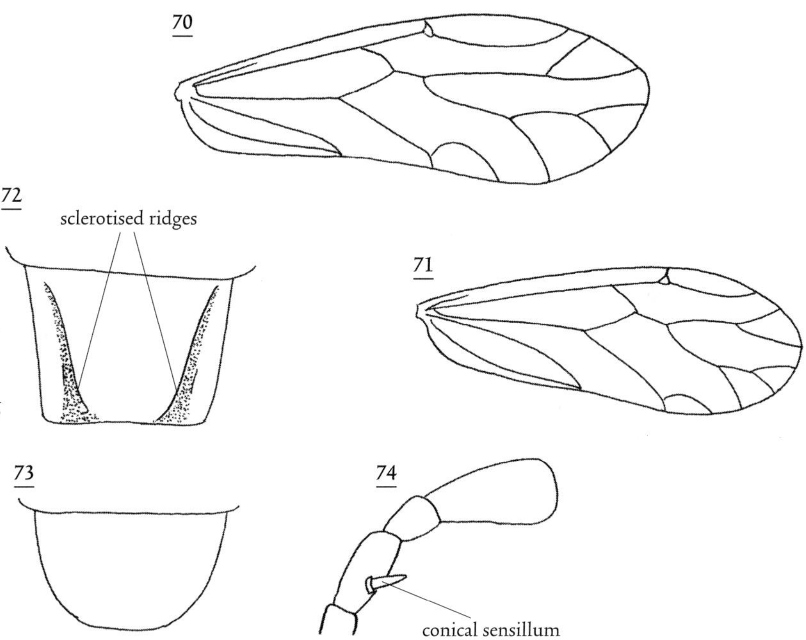

FIGURES 70 – 74
Character states used in psocid identification. Form of fore wing 70 normal, macropterous – 71 brachypterous (relative shortening of apical part of wing, venation distinct); labrum 72 with (Epipsocidae) or 73 without two longitudinal sclerotised ridges – 74 maxillary palp with conical sensillum on second segment.

FIGURES 75 – 84

Lepidopsocidae. Wing venation of **75** *Cyptophania*, fore wing – **76** *C. marginata* Thornton, Lee & Chui, elytriform, almost veinless fore wing – **77, 78** *Echmepteryx*, fore and hind wing – **79** *Echinopsocus*, fore wing – **80** *Lepidopsocus*, fore wing – **81** *Papuapsocus*, fore wing – **82** representative shapes of lepidopsocid wing scales – **83** typical gonapophyses, reduced to setose external valve – **84** *Echmepteryx*, phallosome, widened anteriorly, open posteriorly.

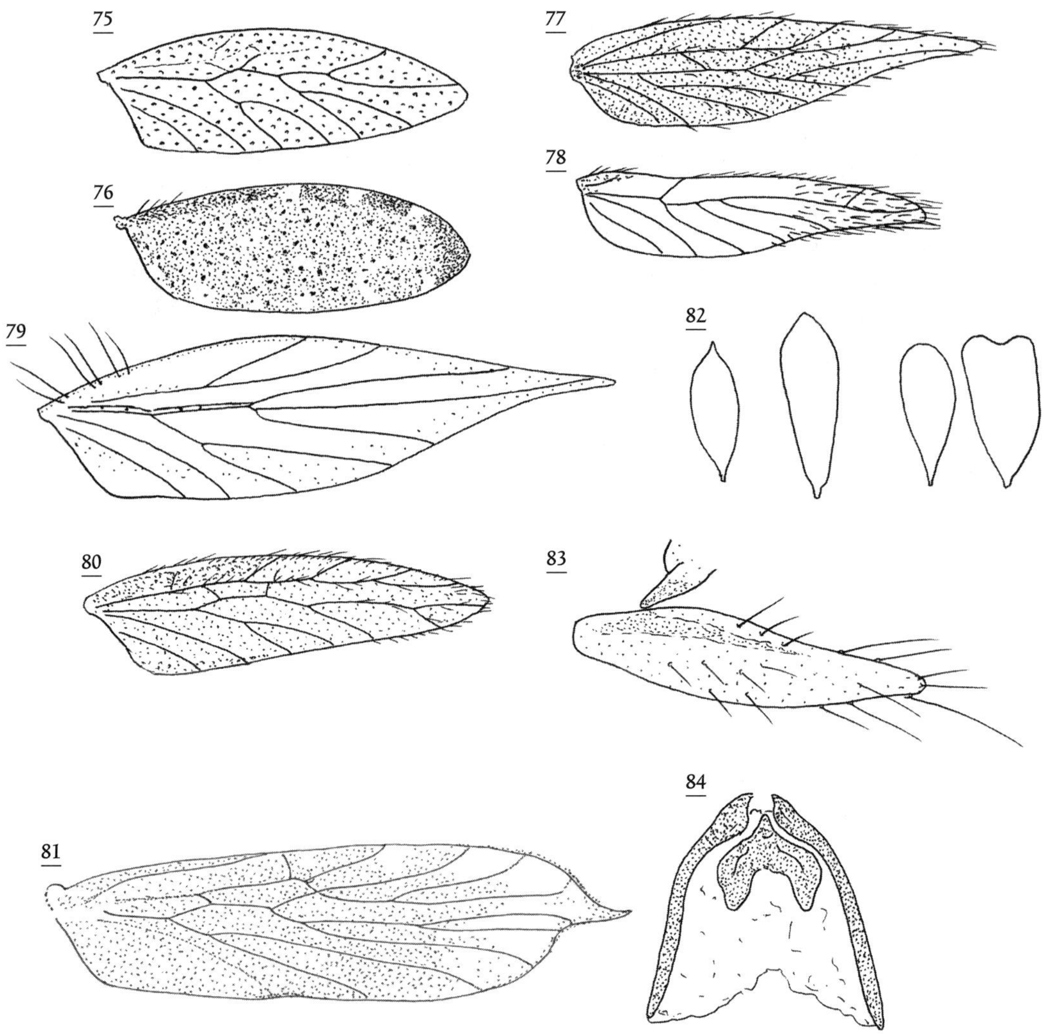

FIGURES 85 – 95

Lepidopsocidae. Wing venation (fore and hind wings) of **85, 86** *Nepticulomima* – **87, 88** *Parasoa* – **89, 90** *Perientomum* – **91, 92** *Soa*; gonapophyses of **93** *Nepticulomima* – **94** *Soa* – **95** *Parasoa*, closely associated along ventral midline.

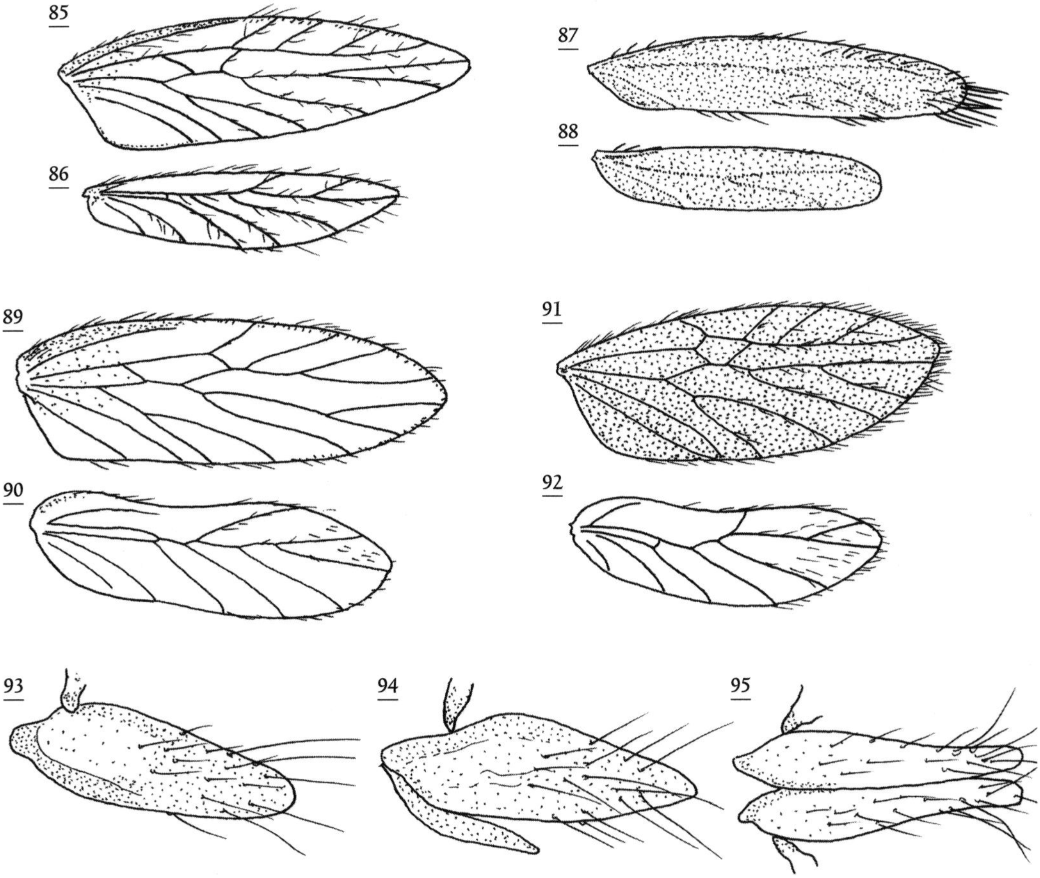

FIGURES 96 – 100

Trogiidae. Scale-like fore wing of **96** *Lepinotus patruelis* and **97** *L. reticulatus*; maxillary palp of **98** *Lepinotus* and **99** *Trogium* – **100** phallosome of *Trogium*.

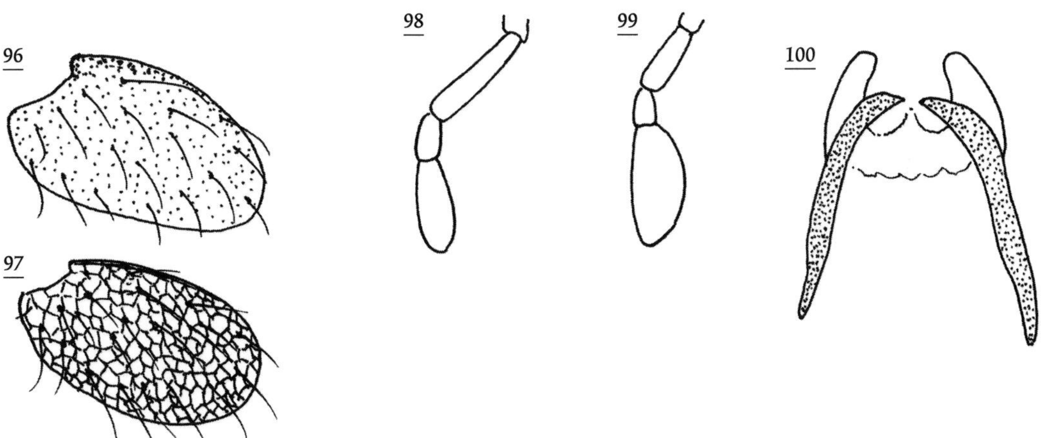

FIGURES 101 – 109

Psoquillidae, Psyllipsocidae. **101** *Eosilla*, fore wing – **102** *Rhyopsocus*, fore wing – **103** *Psocathropos*, brachypterous, fore wing – **104, 105** *Psocathropos*, macropterous, fore and hind wings; gonapophyses of **106** *Psocathropos* – **107** *Rhyopsocus* – **108** phallosome, *Rhyopsocus* – **109** *Dorypteryx*, fore wing.

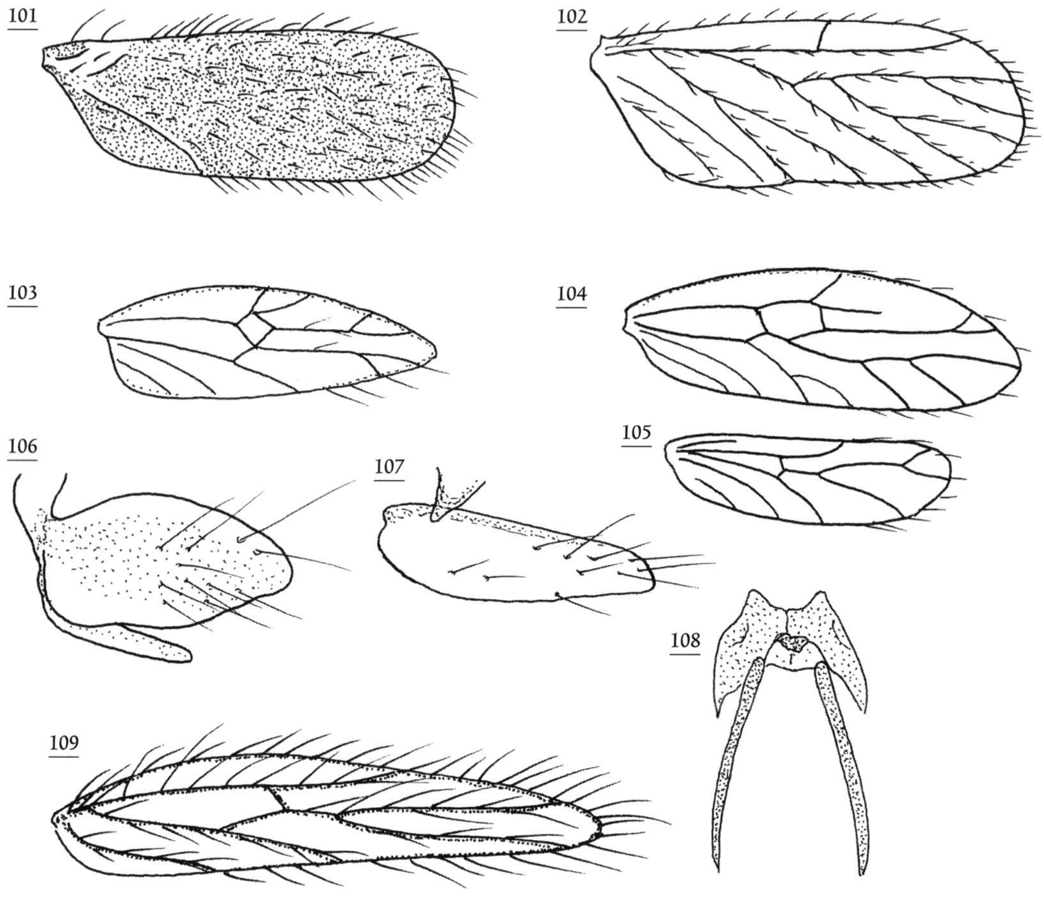

FIGURES 110 – 120
Psoquillidae, Psyllipsocidae. Wings (fore and hind wing of each) – **110, 111** *Psyllipsocus*, macropterous – **112, 113** *Psyllipsocus*, brachypterous – **114, 115** *Psoquilla* – **116, 117** *Psoquilla marginepunctata*, pattern; gonapophyses of **118** *Psyllipsocus* – **119** *Psoquilla* – **120** phallosome of *Psoquilla*.

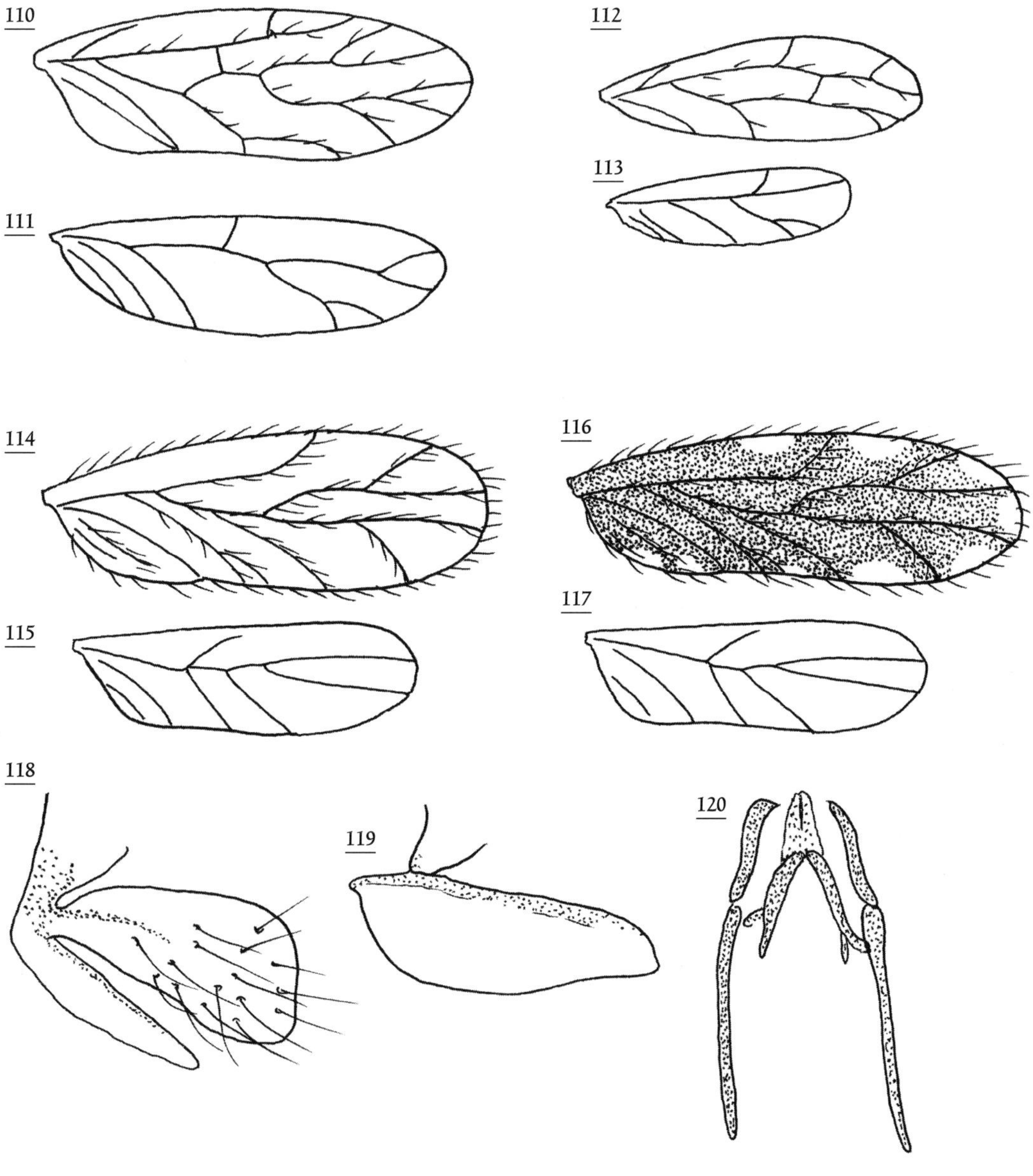

FIGURES 121 – 123
Prionoglarididae. *Siamoglaris zebrina* – 121 fore wing and 122 hind wing – 123 mandibles (after Lienhard, 2004a).

121

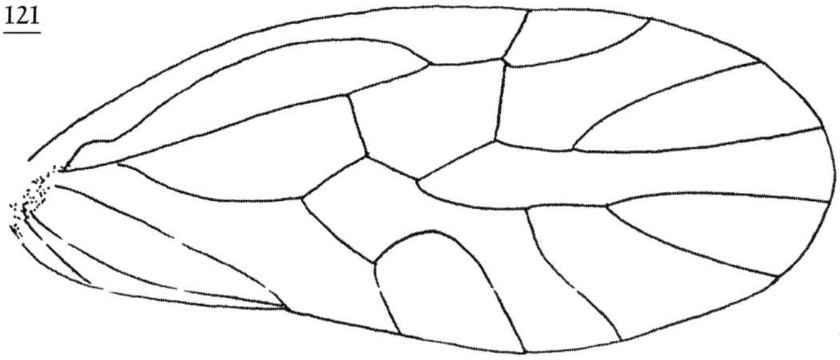

123

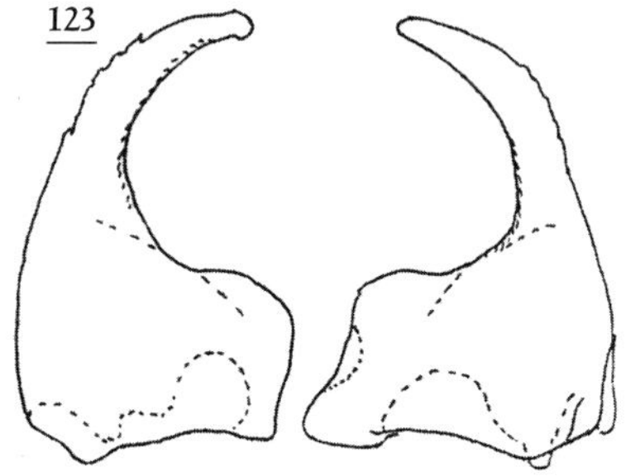

122

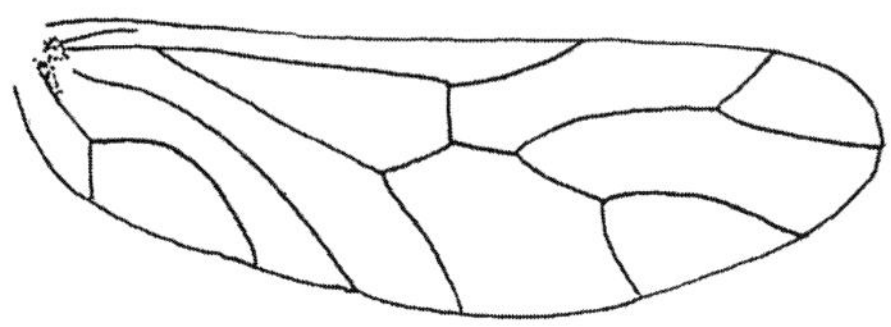

FIGURES 124 – 135

Amphientomidae. Wing venation (fore and hind wings) of **124, 125** *Paramphientomum* – **126, 127** *Stigmatopathus* – **128, 129** *Seopsis* – **130** *Stimulopalpus* (fore wing) – **134, 135** *Pseudoseopsis*; *Paramphientomum* – **131** subgenital plate, indicating position of internal sclerite – **132** gonapophyses – **133** phallosome.

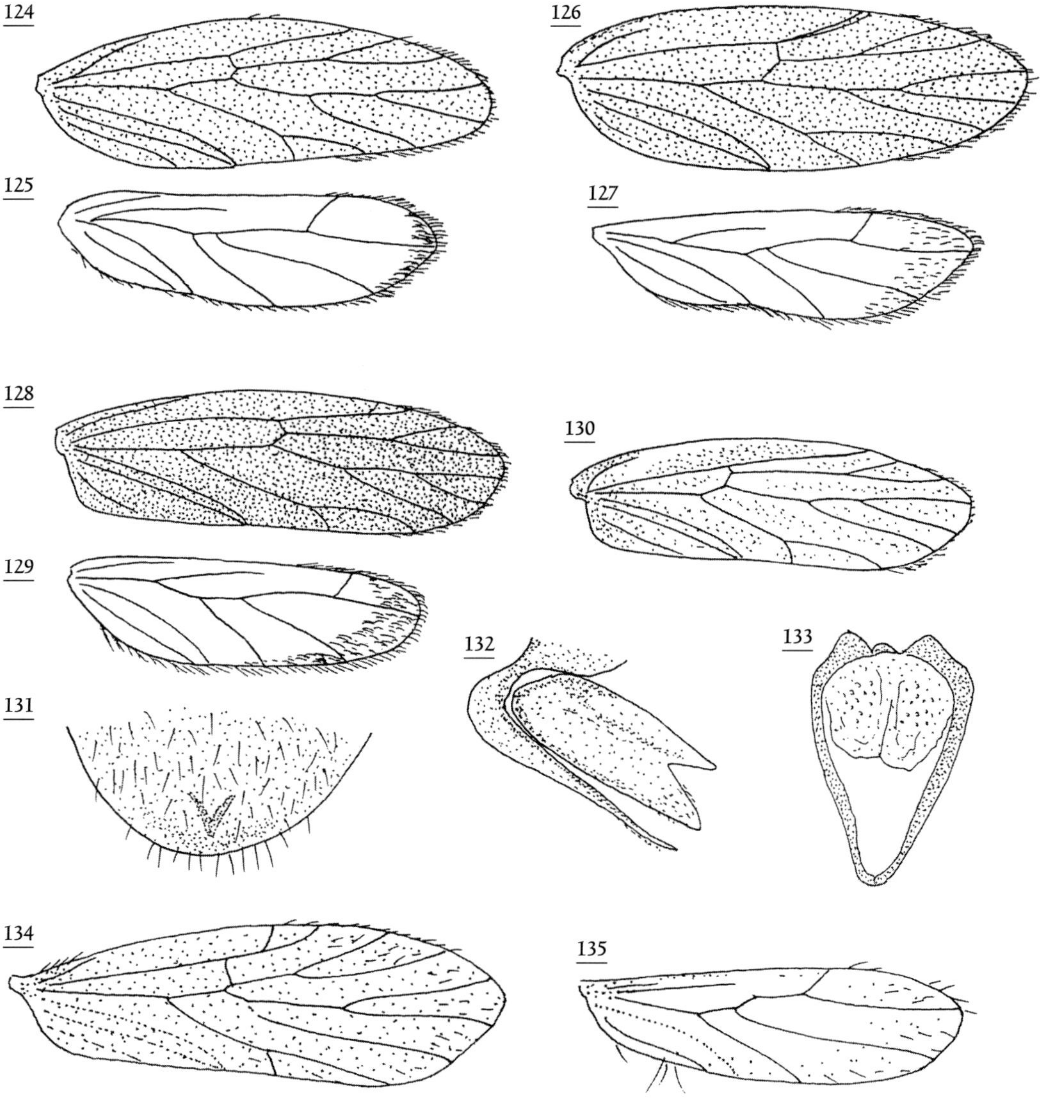

FIGURES 136 – 146

Amphientomidae. Wing venation (fore and hind wings) of **136, 137** *Syllysis* – **138, 139** *Cymatopsocus* – **140, 141** *Tineomorpha* – **142** subgenital plate of *Cymatopsocus* – **143, 144** gonapophyses of **143** *Tineomorpha* – **144** *Cymatopsocus*; phallosome of **145** *Tineomorpha* – **146** *Cymatopsocus*.

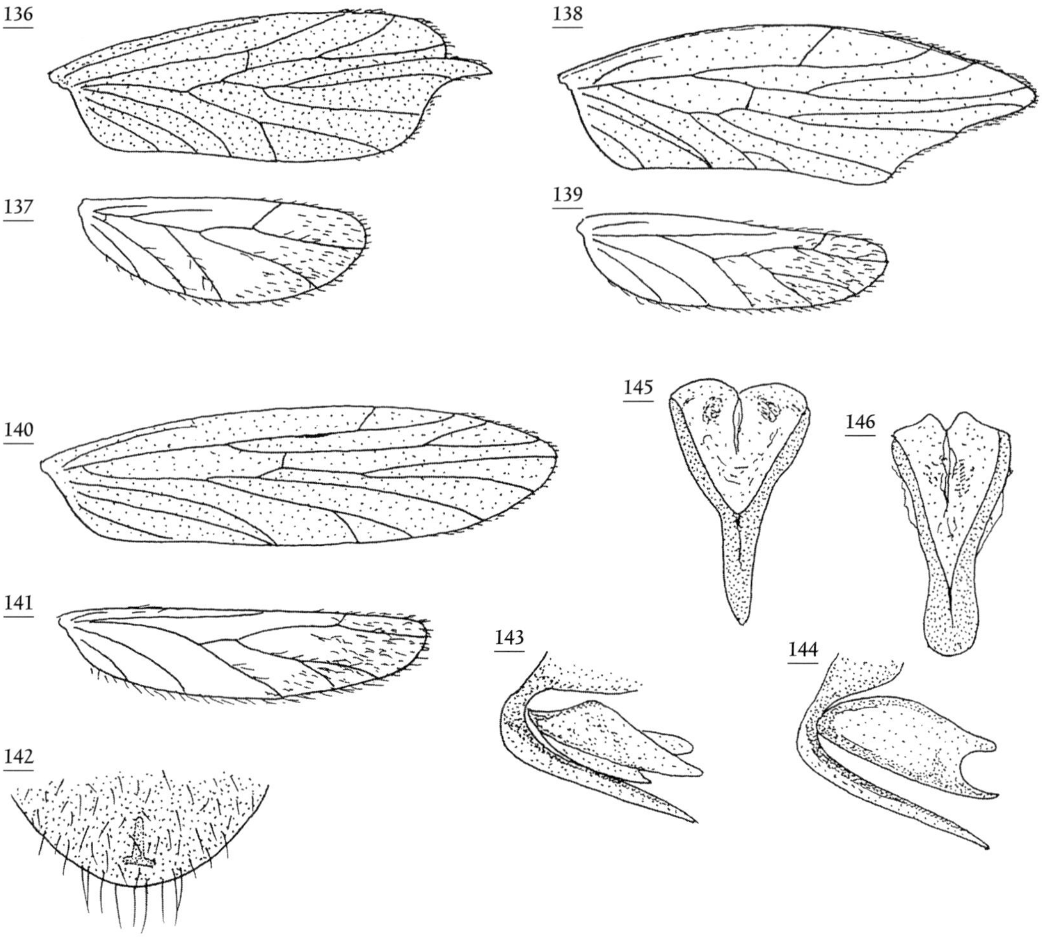

FIGURES 147 – 163 [FACING PAGE]

Troctopsocidae. Wing venation (fore and hind wings) of **147, 148** *Thaipsocus* – **149, 150** *Selenopsocus* – **151, 152** *Coleotroctellus* macropterous; fore wing, brachypterous morphs of **153** *Coleotroctellus venosus* and **154** *C. burckhardti* (after Lienhard & Mockford 1997); tarsal claw of **155** *Thaipsocus* – **156** *Selenopsocus*; lacinial apex of **157** *Thaipsocus* – **158** *Selenopsocus*; subgenital plate of **159** *Thaipsocus* – **160** *Selenopsocus*; gonapophyses of **161** *Thaipsocus* – **162** *Selenopsocus* – **163** phallosome, *Thaipsocus*.

147

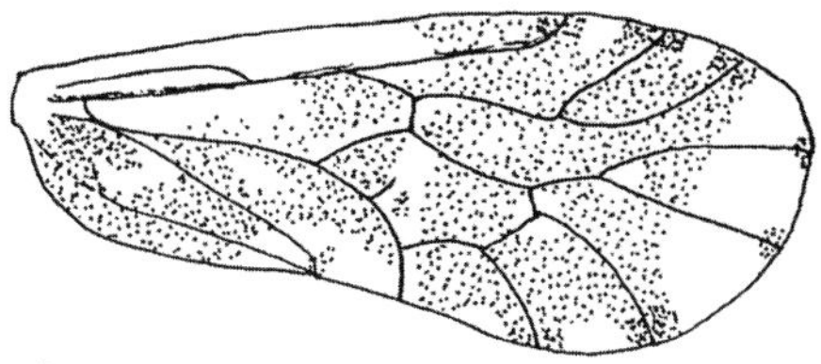

148

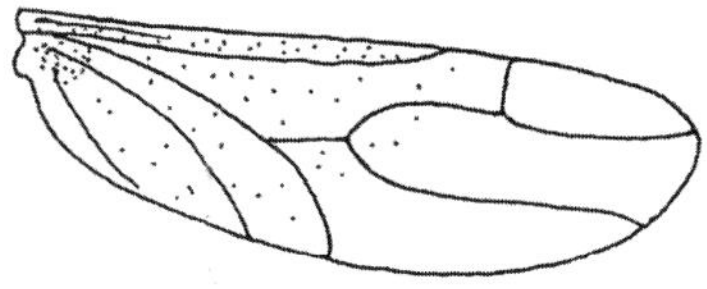

149

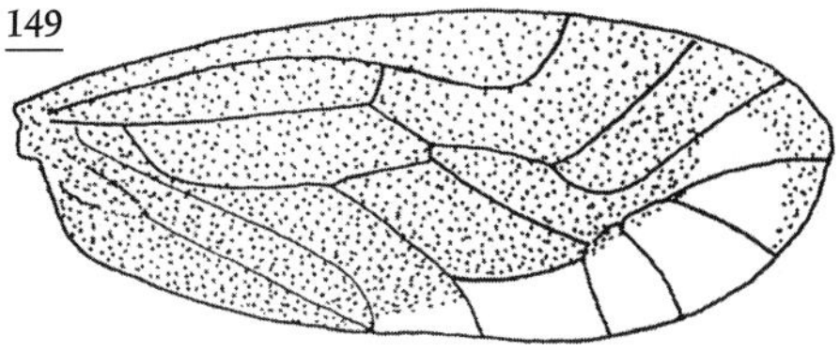

150

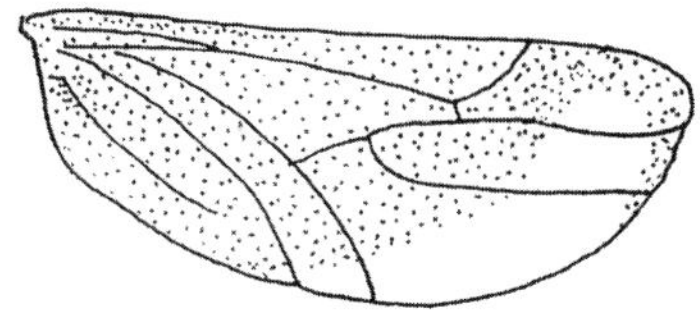

151

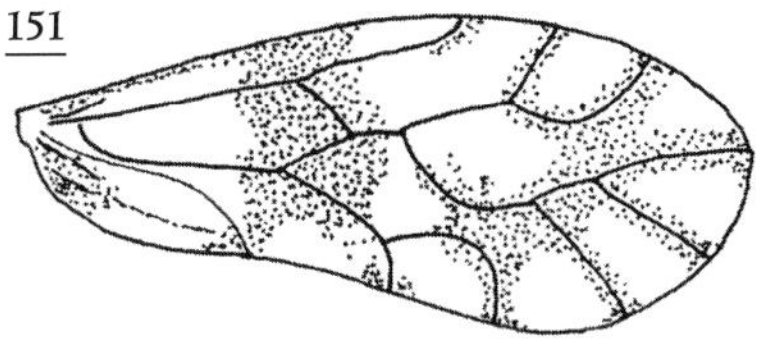

152

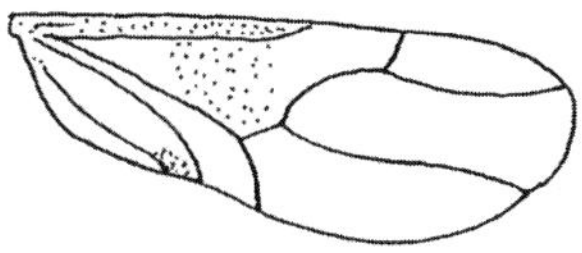

153

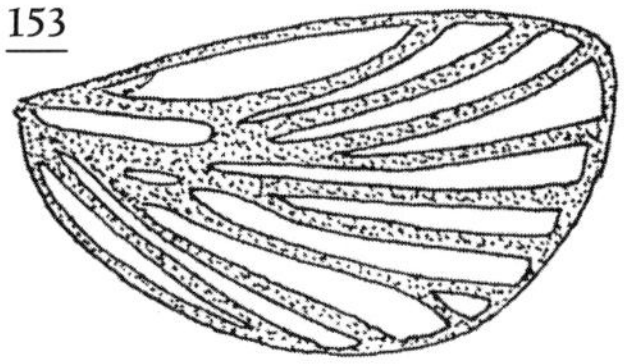

154

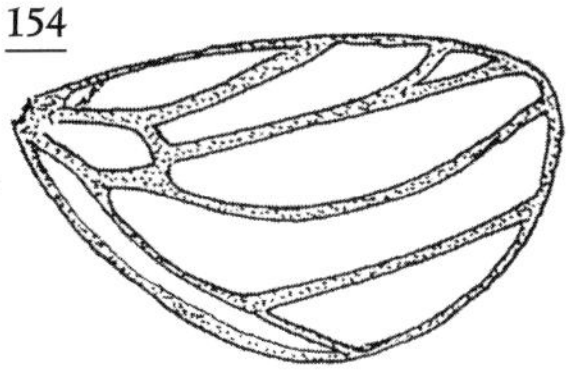

155

157

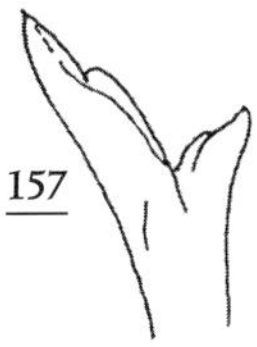

159

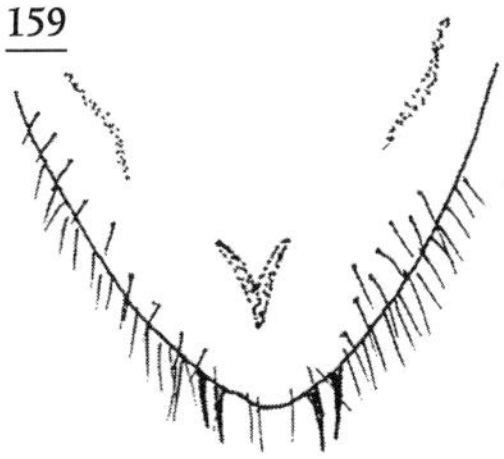

156

158

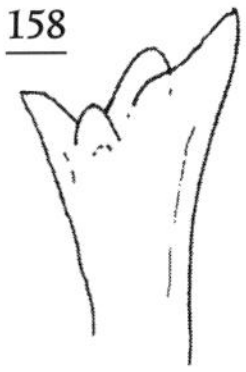

161

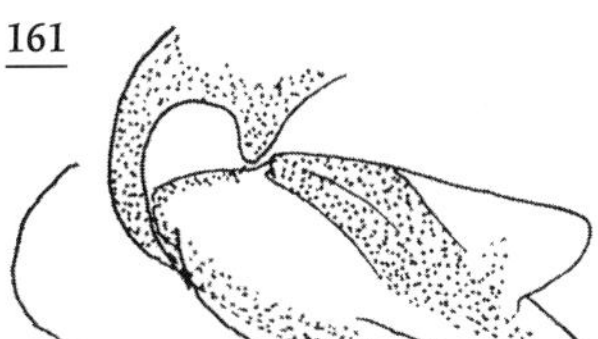

160

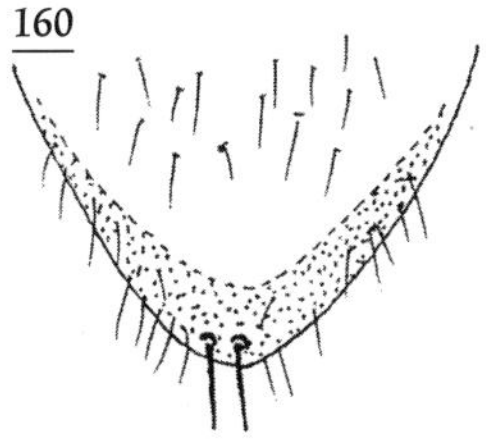

162

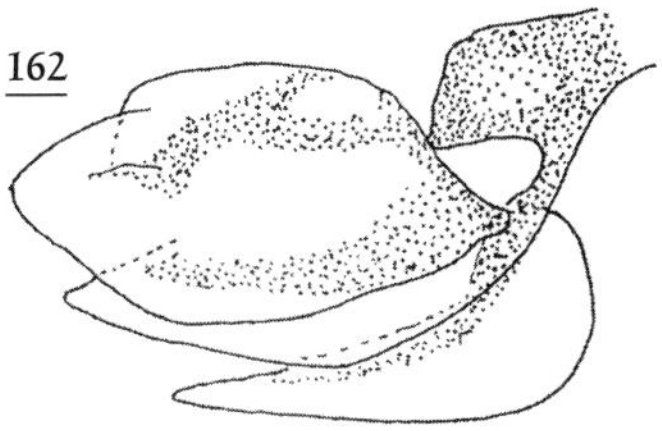

163

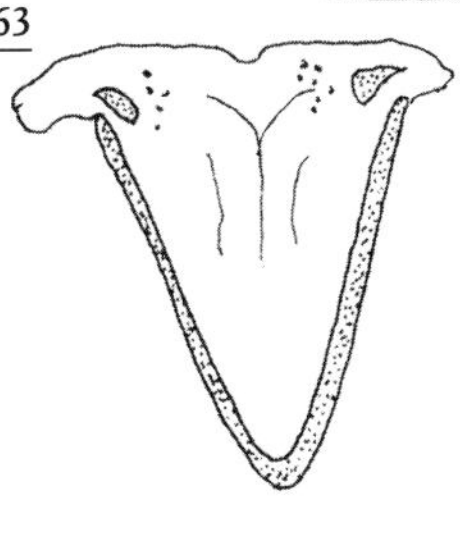

164

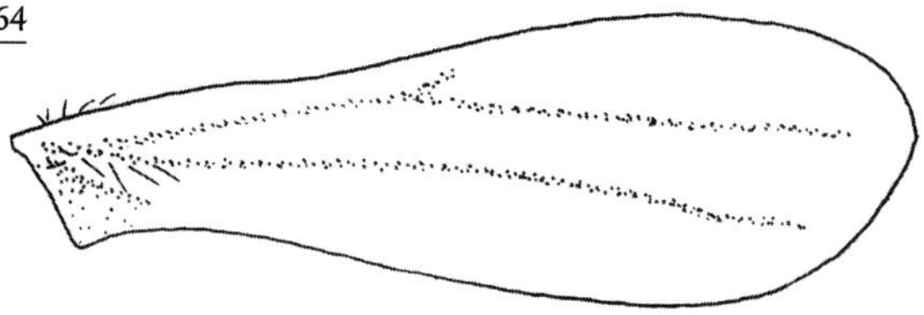

165

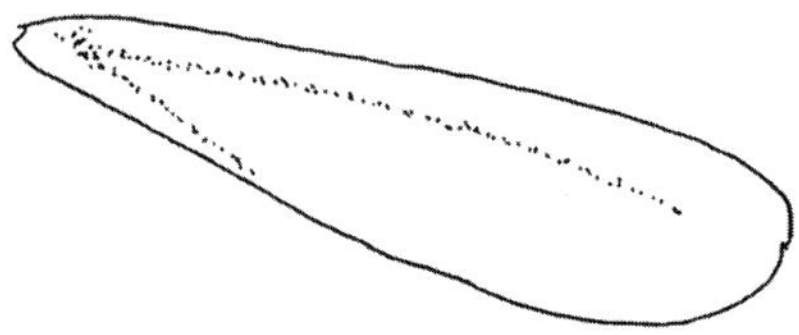

176

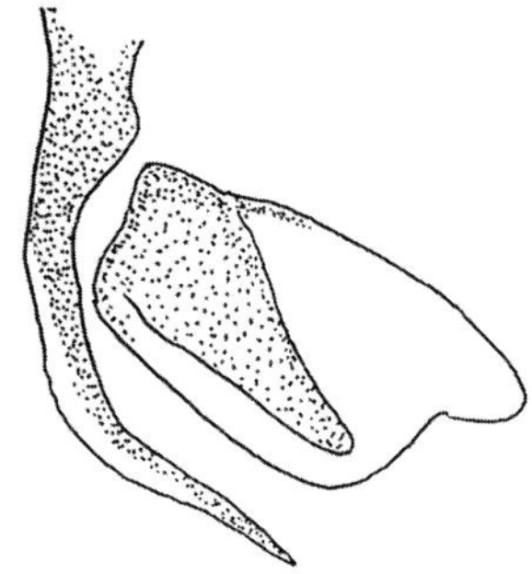

168

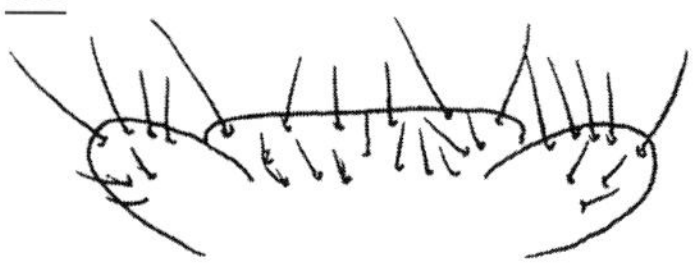

169

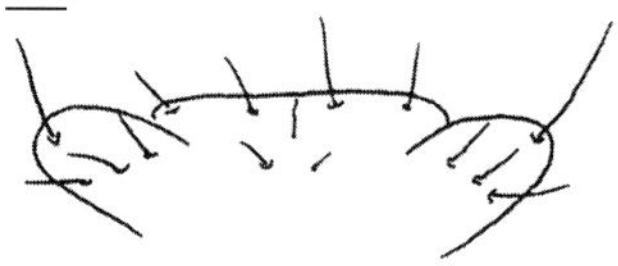

166

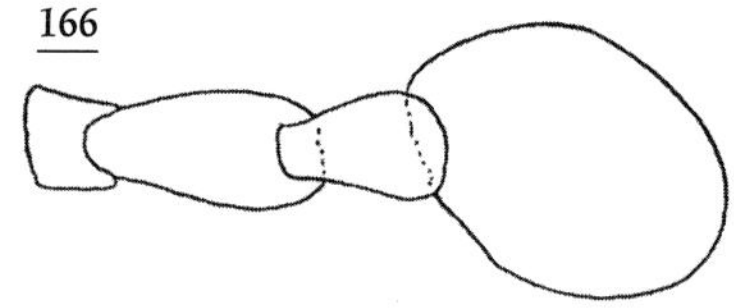

170

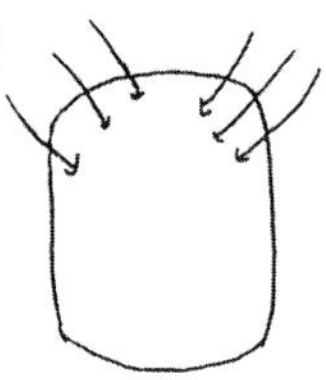

177

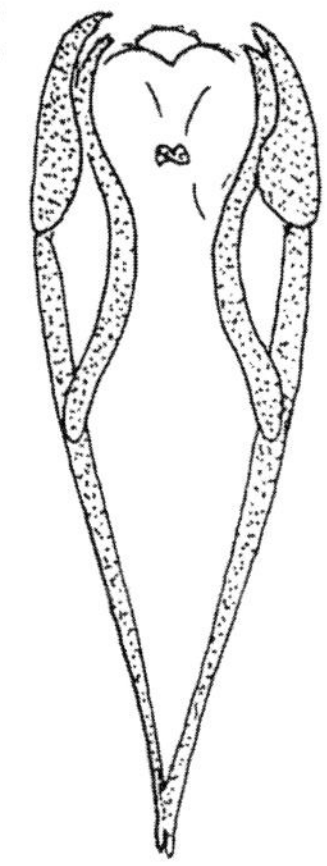

175

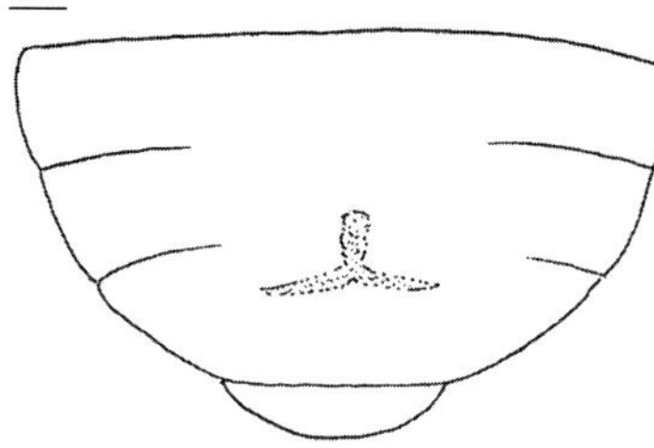

171

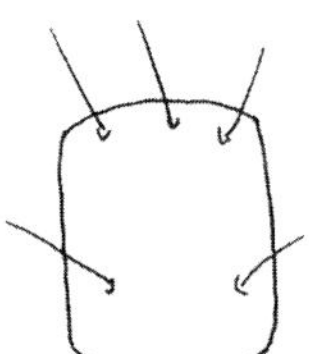

174

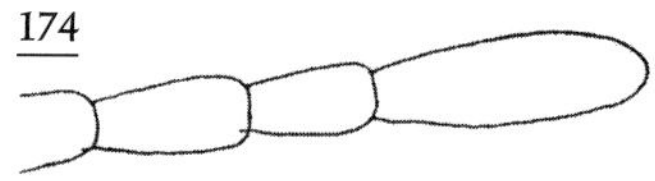

167

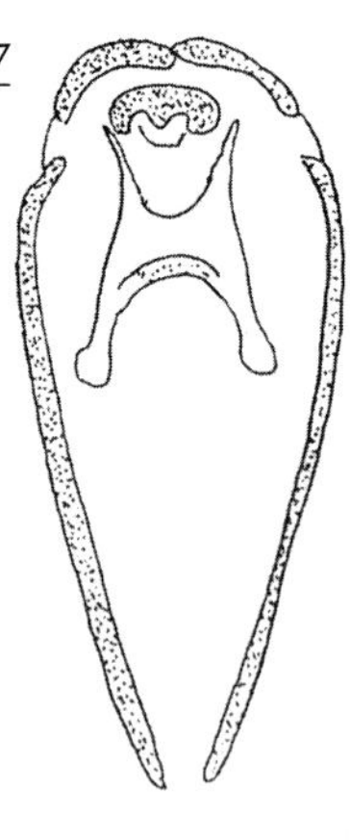

172

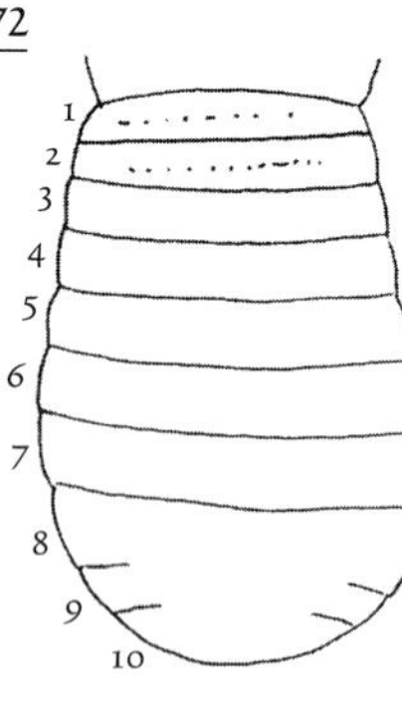

173

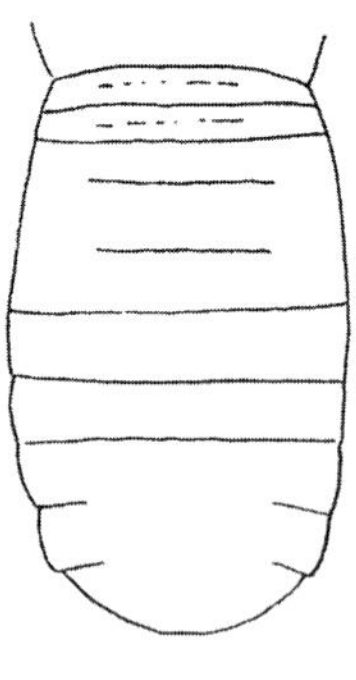

FIGURES 164 – 177 [FACING PAGE]
Liposcelididae. **164** fore and **165** hind wing of *Embidopsocus*, venation reduced and evanescent; *Belaphopsocus*: **166** maxillary palp, with globular apical segment – **167** phallosome; *Liposcelis* (168-177): pronotum representative of **168** group A – **169** group B; prosternum, bristle distribution in **170** groups A, B, C and **171** group D; dorsum of abdomen to indicate **172** normal and **173** compact morph, with hind margins of tergites 3-4 either distinct or incompletely defined – **174** normal maxillary palp of *Embidopsocus* and *Liposcelis* (cf. *Belaphopsocus*, figure 166) – **175** subgenital plate with broad internal T-sclerite (dorsal view of abdominal apex) – **176** gonapophyses – **177** phallosome.

FIGURES 178 – 188
Pachytroctidae. Wing venation of fore and hind wing of **178, 179** *Tapinella* – **180, 181** *Pachytroctes* – **182, 183** *Nanopsocus* – **184** subgenital plate of *Tapinella*, with T-sclerite; gonapophyses of **185** *Tapinella* – **186** *Pachytroctes* – **187** phallosome of *Nanopsocus* – **188** asymmetrical tarsal claws of *Nanopsocus*.

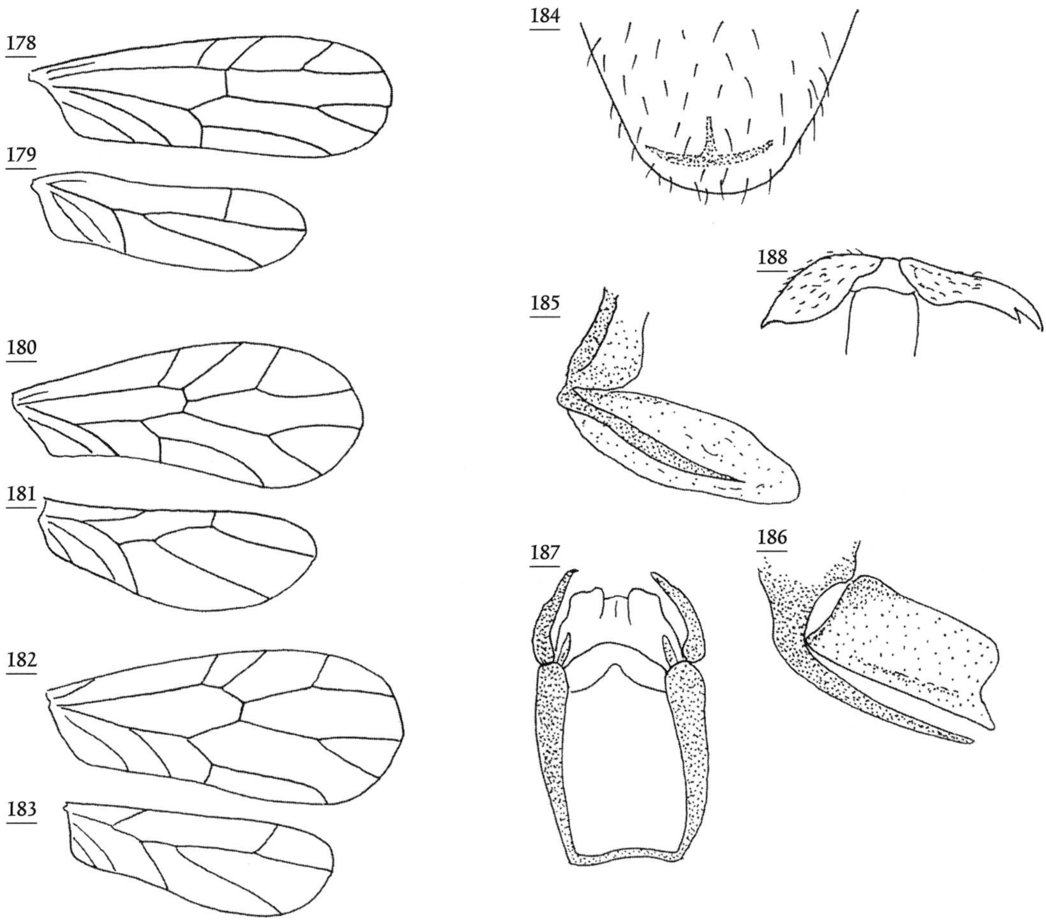

FIGURES 189 – 197

Epipsocidae, Dolabellopsocidae. Fore wing venation of **189** *Epipsocus* s.l. (with hind wing, figure **190**) – **191** *Epipsocopsis* – **192** *Dicropsocus* – **193** *Auroropsocus* – **194** lacinial apex of *Epipsocopsis*; gonapophyses of **195** *Epipsocus* s.l., ventral valve present – **196** *Epipsocopsis*, ventral valve absent – **197** phallosome, *Epipsocopsis*.

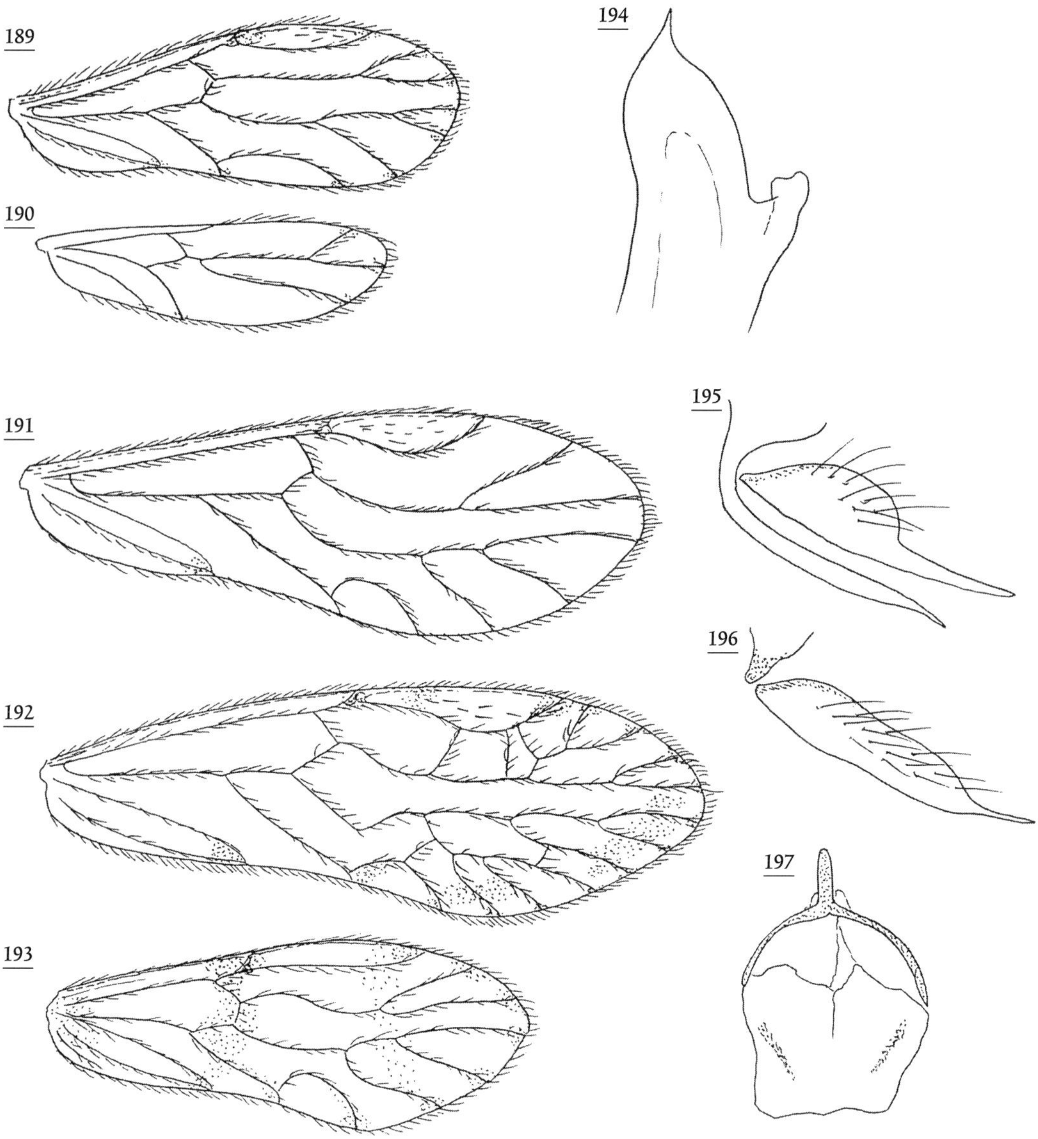

FIGURES 198 – 213 [FACING PAGE]

Caeciliusidae. Character states (198-202): (198-200) distal inner labral sensilla of **198** type 1 – **199** type 2 – **200** type 3 (see text); facial aspect to indicate position and width of clypeal shelf (201-203, dashed lines): **201** wide – **202** narrow – **203** intermediate (after Mockford, 2000); anterolateral margin of labrum **204** with or **205** without labral stylet; lacinial apex of **206** *Caecilius* – **207** *Dypsocus* – **208** *Thorntoniella* – **209** *Maoripsocopsis* – **210** *Valenzuela* – **211** *Aphyopsocus* – **212** *Enderleinella* – **213** *Protodypsocus*.

198

199

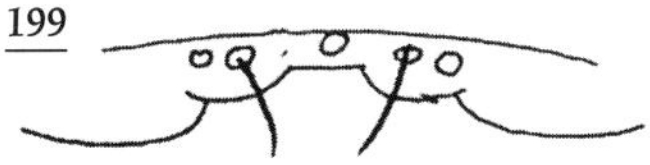

200

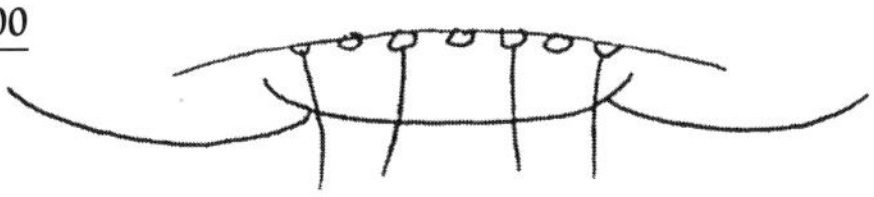

201

206

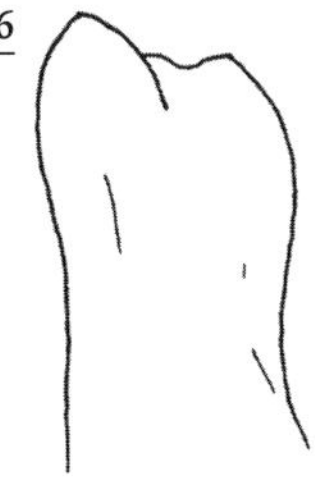

207

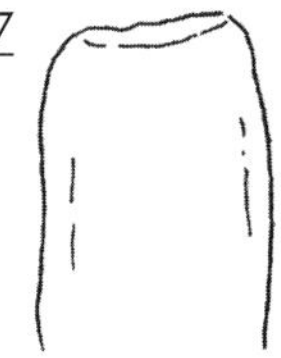

208

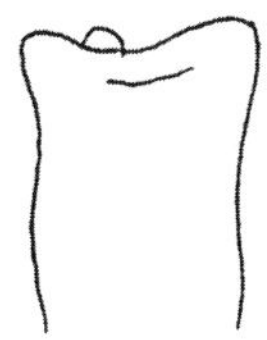

209

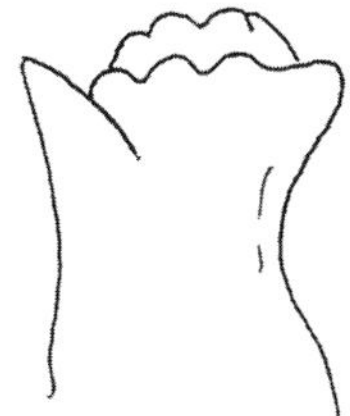

202

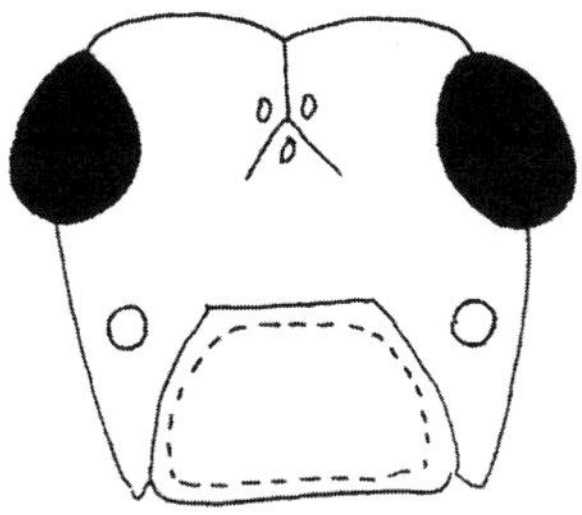

210

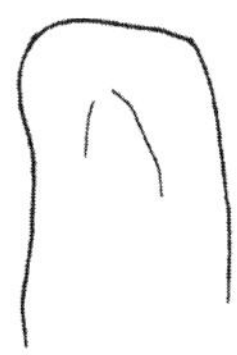

211

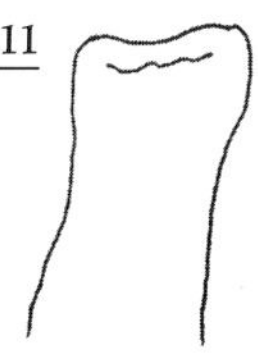

203

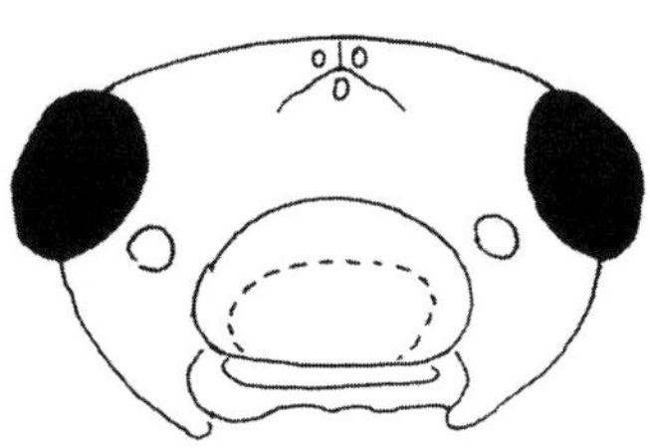

212

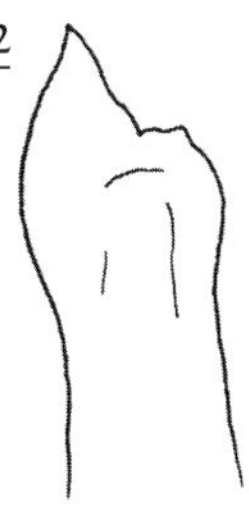

213

204

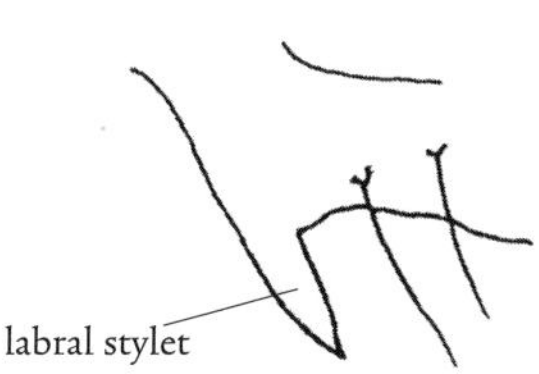

205

FIGURES 214 – 229

Caeciliusidae. Wing venation, fore wing – **214** *Aphyopsocus* – **215** *Dypsocus* – **216** *Caecilius* – **218** *Asiocaecilius* – **219** *Coryphosmila* – **217** hind wing of *Caecilius* – **220** *Dypsocus*, base of antenna to show incrassate basal flagellar segments – **221, 222** spermatheca – **223, 224** subgenital plate – **225, 226** gonapophyses of (221, 223, 225) *Caecilius*, (222, 224, 226) *Dypsocus* – **227** gonapophyses of *Aphyopsocus*; phallosome of **228** *Caecilius*, – **229** *Aphyopsocus*.

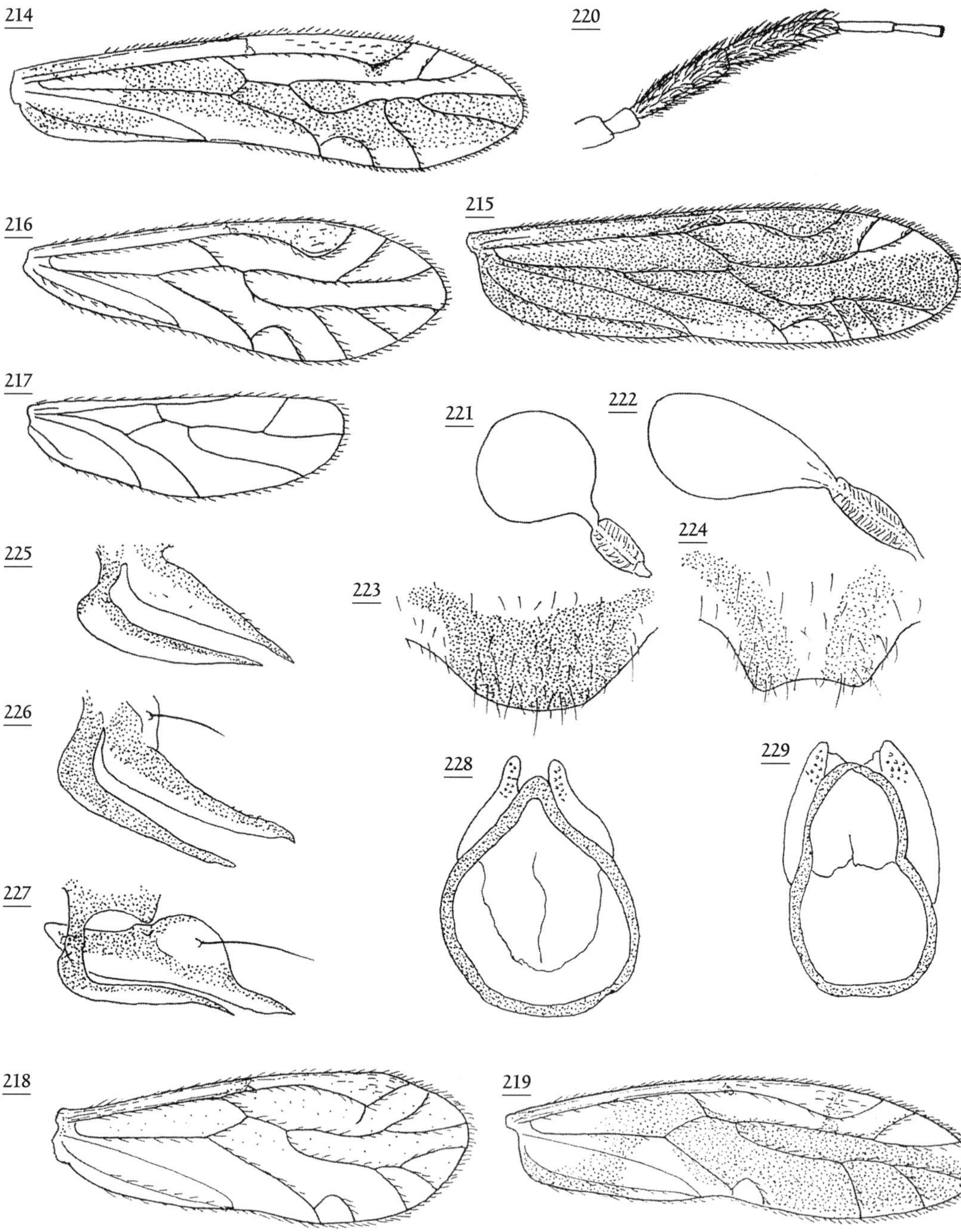

FIGURES 230 – 239

Caeciliusidae. Fore wing venation of 230 *Maoripsocopsis* – 231 *Pericaecilius* – 232 *Protodypsocus* – 233 *Thorntoniella*; spermatheca of 234 *Thorntoniella* – 235 *Maoripsocopsis* – 236 *Protodypsocus*; gonapophyses of 237 *Maoripsocopsis* – 238 *Protodypsocus* – 239 subgenital plate of *Protodypsocus*.

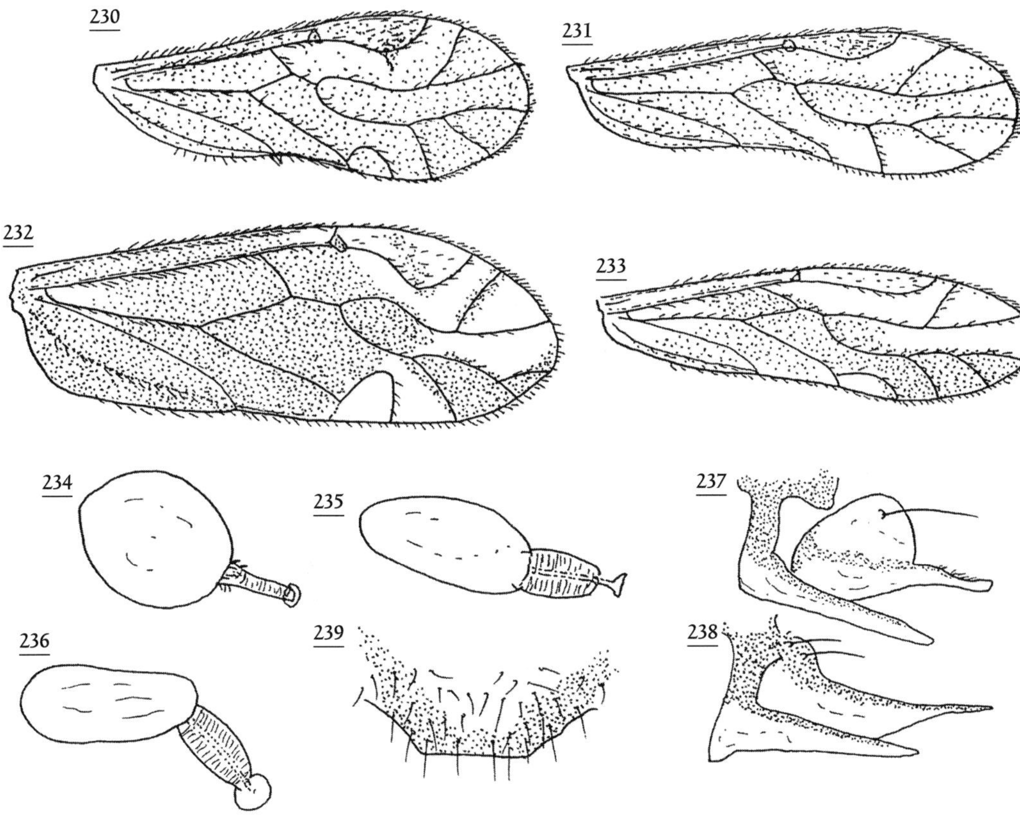

FIGURES 240 – 248

Caeciliusidae. Fore wing venation of 240 *Valenzuela* – 241 *Enderleinella* – 242 *Paracaecilius*; spermatheca of 243 *Valenzuela* – 244 *Enderleinella* – 245 *Paracaecilius*; gonapophyses of 246 *Paracaecilius* – 247 *Enderleinella* – 248 phallosome of *Paracaecilius*.

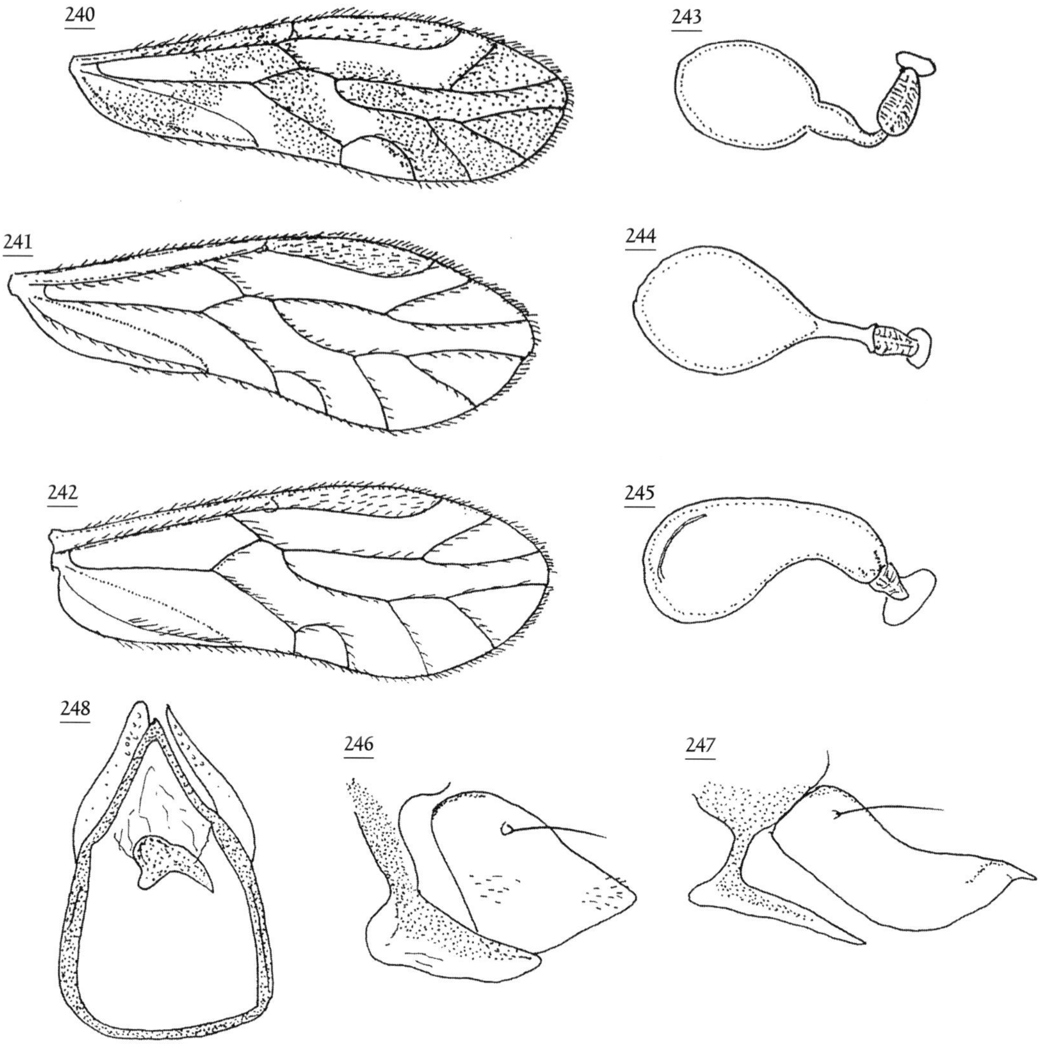

FIGURES 249 – 256
Amphipsocidae. Fore wing venation of **249** *Amphipsocus* – **250** *Kodamaius* – **251** *Dasypsocus* – **252** *Tagalopsocus* – **253** idem hind wing – **254** gonapophyses of *Dasypsocus* – **255** subgenital plate of *Tagalopsocus* – **256** phallosome of *Amphipsocus*.

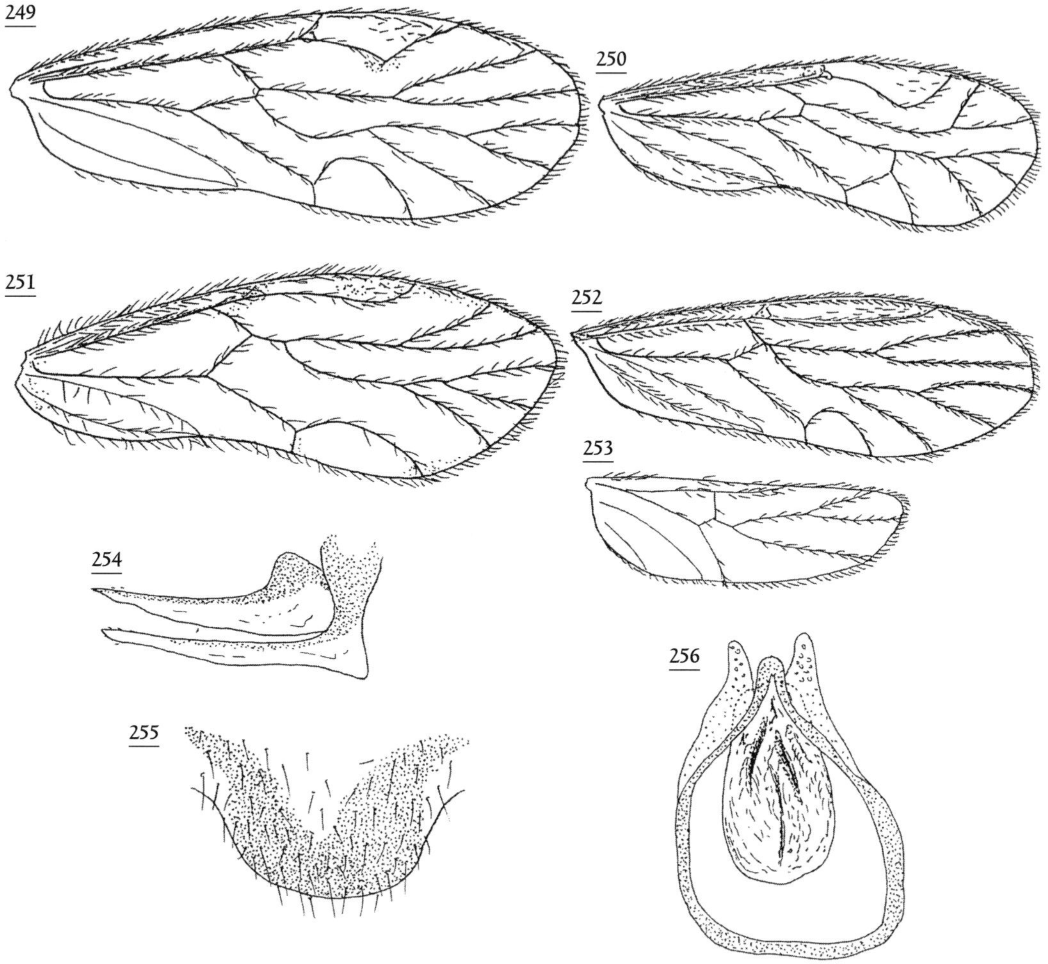

FIGURES 257 – 267

Amphipsocidae, Stenopsocidae. *Calocaecilius*: **257** fore and **258** hind wing – **259** gonapophyses – **260** phallosome; fore wing venation of **261** *Taeniostigma* – **262** *Stenopsocus* – **263** *Graphopsocus*; gonapophyses of **264** *Taeniostigma* – **265** *Stenopsocus* – **266** *Graphopsocus* – **267** phallosome of *Taeniostigma*.

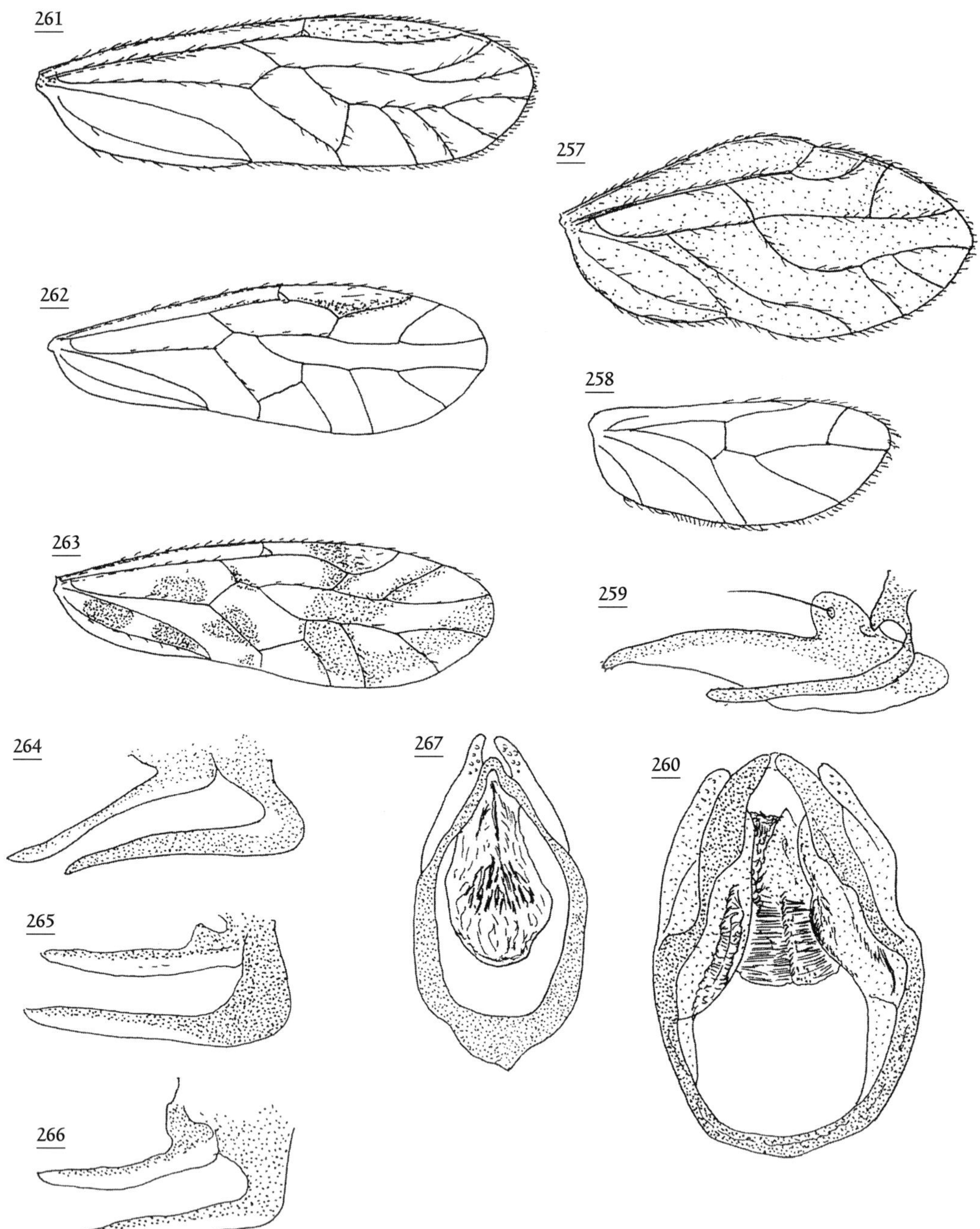

FIGURES 268 – 273
Archipsocidae. **268, 269** Fore and hind wing venation of *Archipsocus* – *Archipsocus* **270** gonapophyses – **271** subgenital plate – **272** phallosome – **273** *Archipsocopsis*, phallosome.

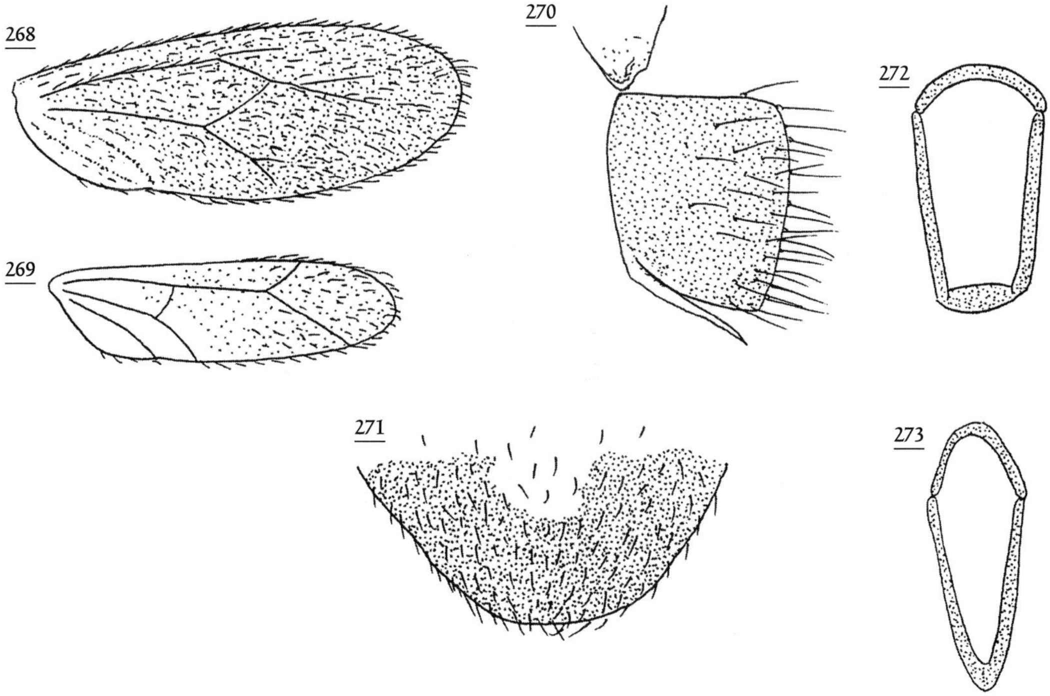

274

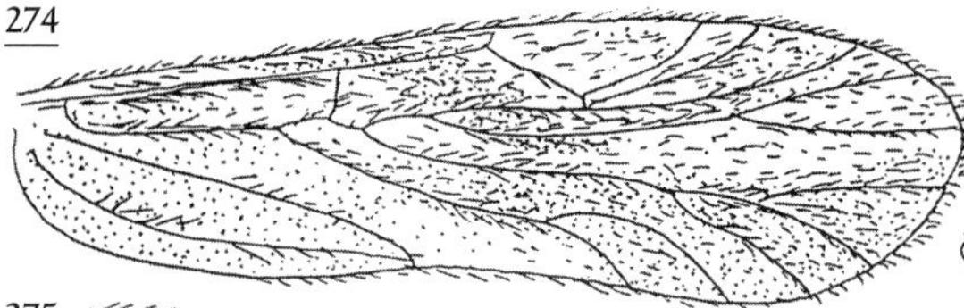

275

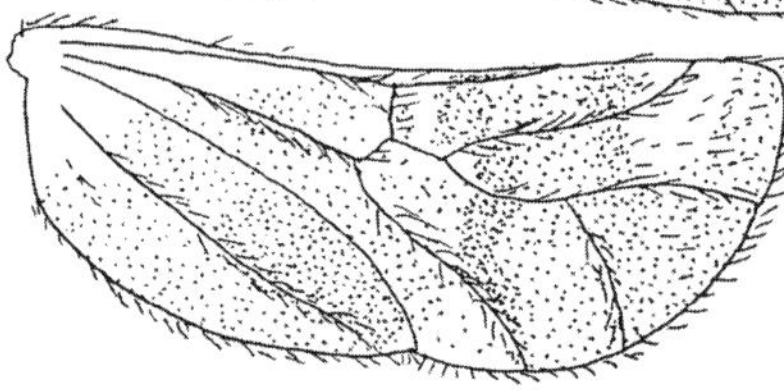

276

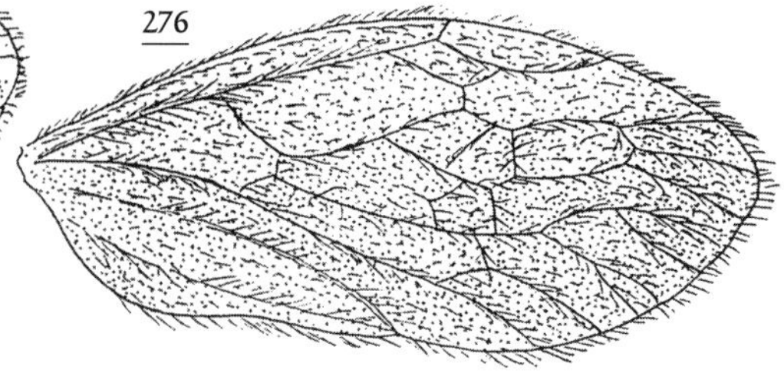

277

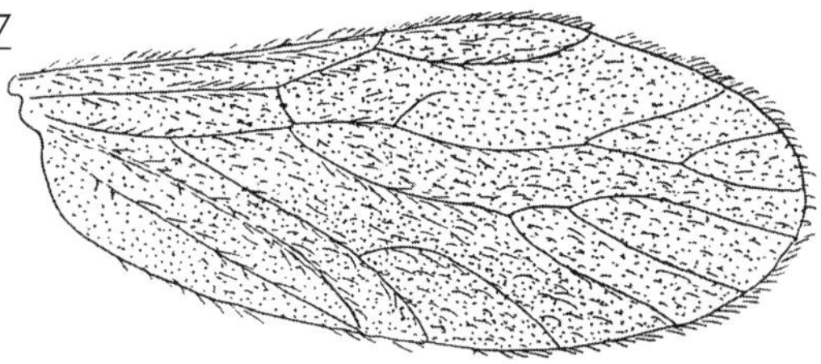

279

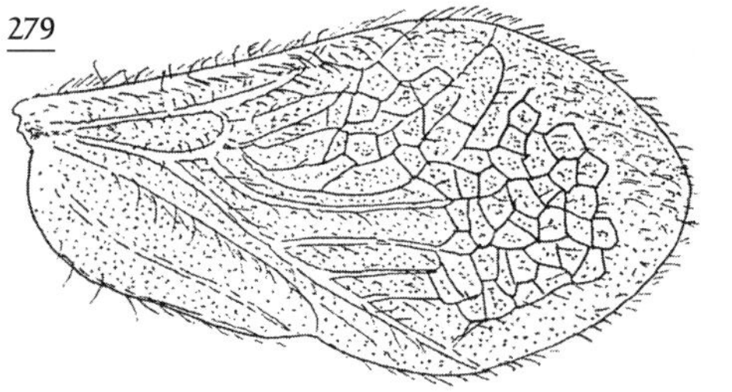

278

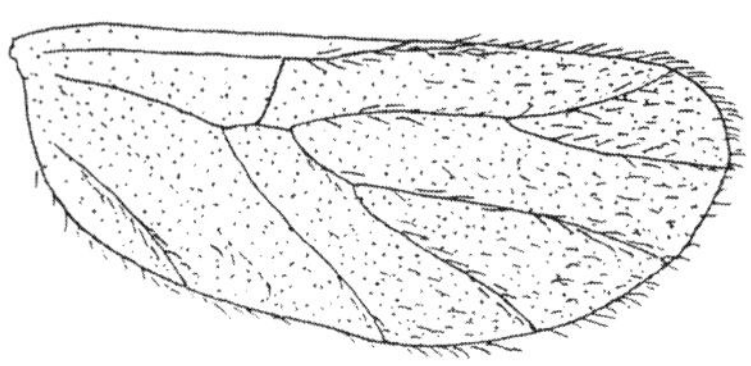

280

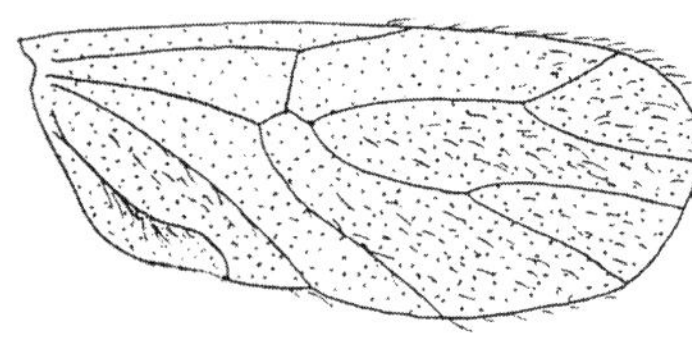

281

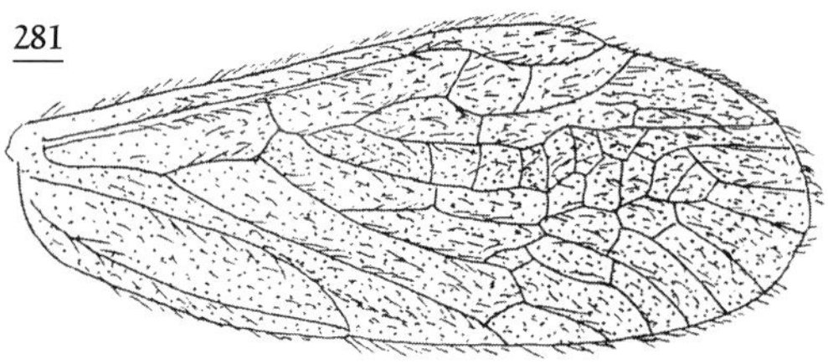

282

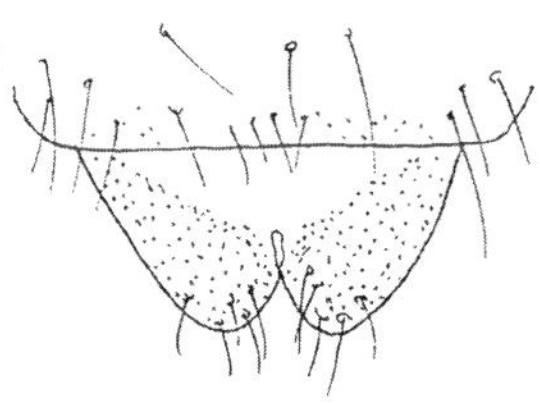

285

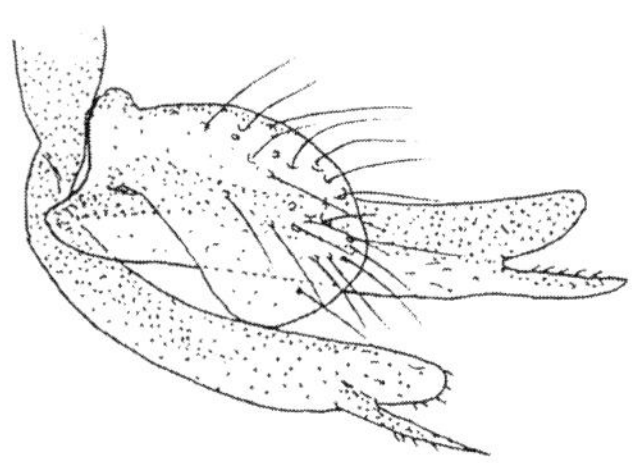

284

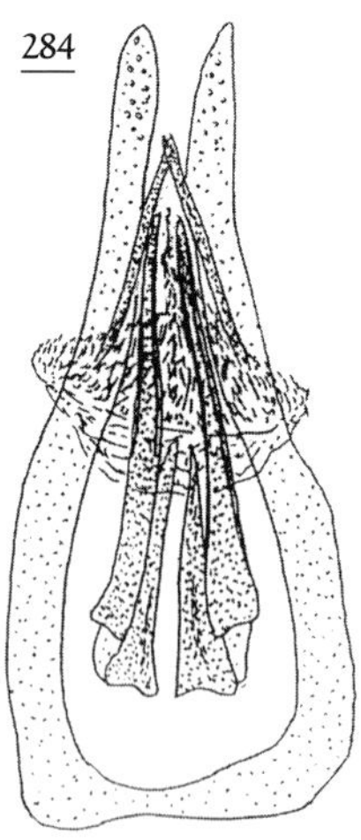

283

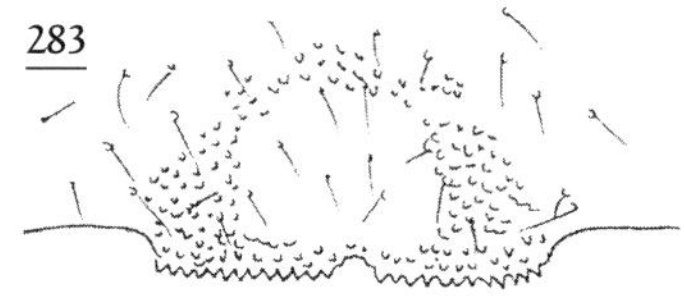

286

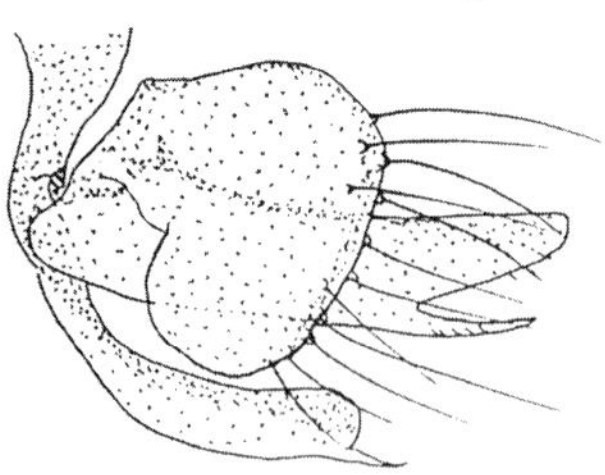

FIGURES 274 – 286 [FACING PAGE]

Calopsocidae. Wing venation: **274, 275** fore and hind wing of *Callistoptera*; *Calopsocus*, various taxa: **276** *Calopsocus* sp., fore wing – **277, 278** *C. marginatus* – **279, 280** *C. reticulatus* – **281** *C. rizali* (fore wing) – **282** subgenital plate – **283** male clunium – **284** phallosome; gonapophyses of **285** *Callistoptera* and **286** *Calopsocus*.

FIGURES 287 – 294

Calopsocidae. Fore wing venation of **287** *Cyclopsocus* – **288** *Calosema* – **289** *Dendropsocus* – **290** *Torrepsocus* – **291** *Nemupsocus* – **292** *Neurosema* – **293** gonapophyses of *Torrepsocus* – **294** subgenital plate apex of *Neurosema*.

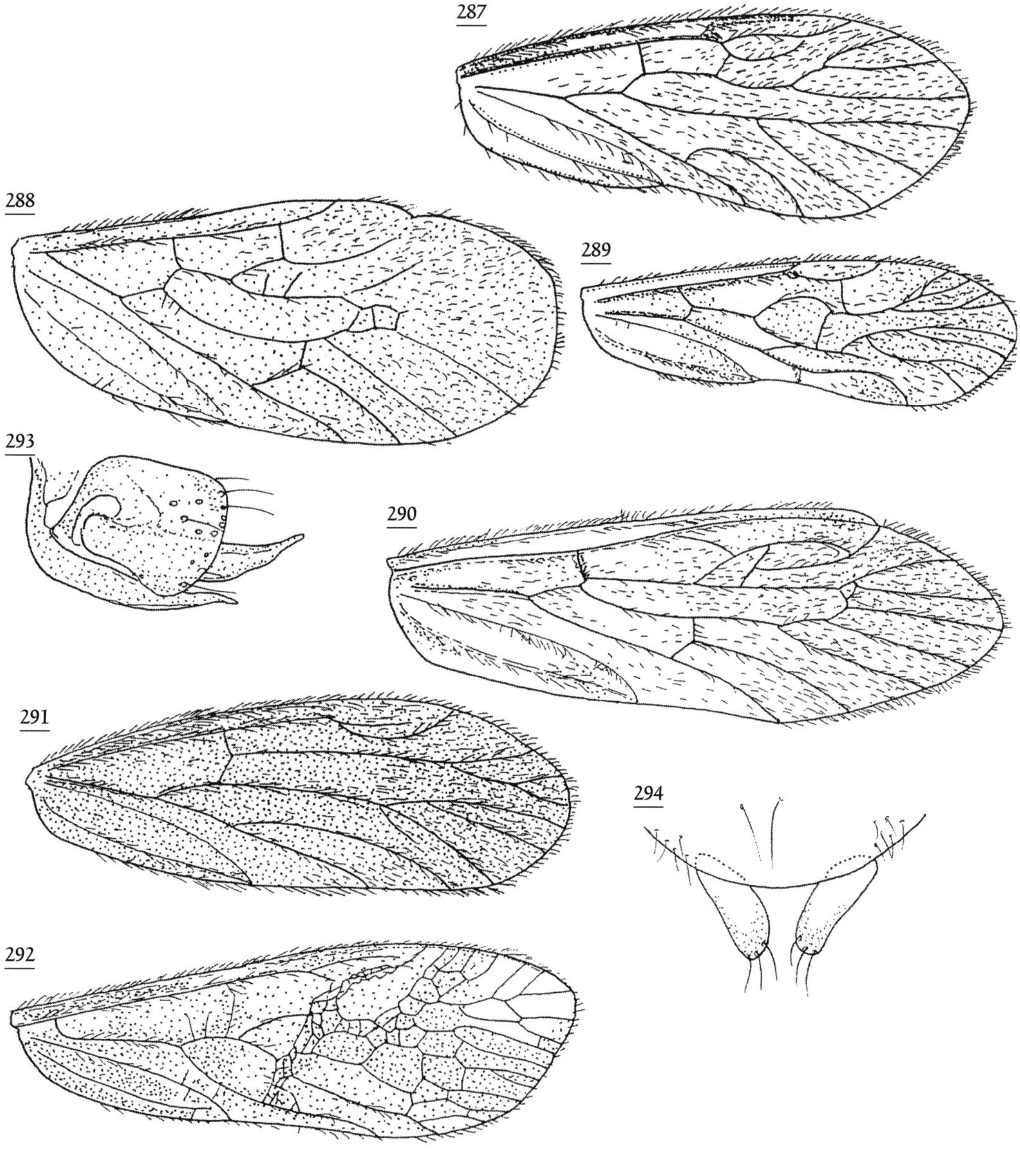

FIGURES 295 – 304

Ectopsocidae. Wing venation: **295, 296** fore and hind wings of typical *Ectopsocus* – **297, 298** same of *E. venosus* Lienhard; gonapophyses of **299** *Ectopsocus* and **300** *Ectopsocopsis* (reduced to small lobes); subgenital plate of **301** *Ectopsocus* – **302** *Ectopsocopsis* – **303** clunial margin and epiproct of *Ectopsocus* male – **304** phallosome of *Ectopsocus*.

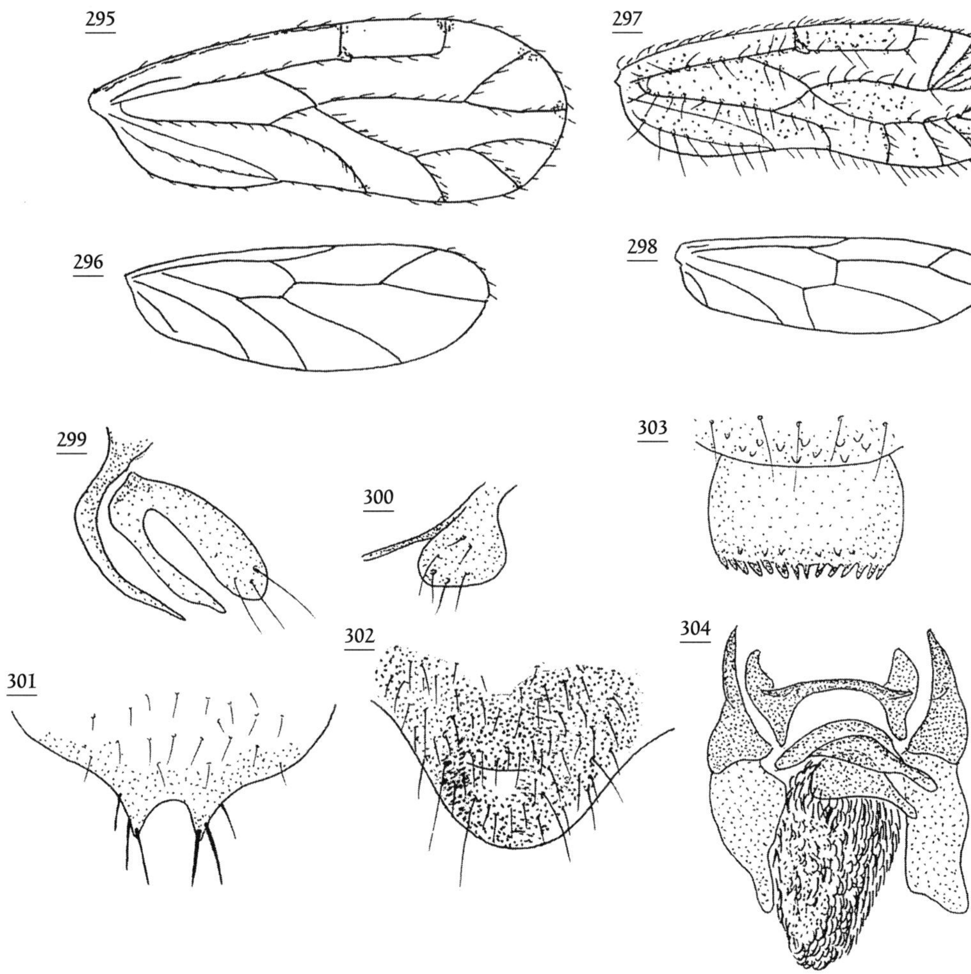

FIGURES 305 – 316 [FACING PAGE]

Elipsocidae, Lachesillidae, Mesopsocidae. (305-308) *Elipsocus*: **305** fore wing – **306** gonapophyses – **307** subgenital plate – **308** phallosome; (309-312) *Lachesilla*: **309** fore wing – **310** gonapophyses – **311** subgenital plate – **312** hypandrium and closely associated claspers; (313-316) *Mesopsocus*: **313** fore wing – **314** gonapophyses – **315** subgenital plate – **316** phallosome.

305

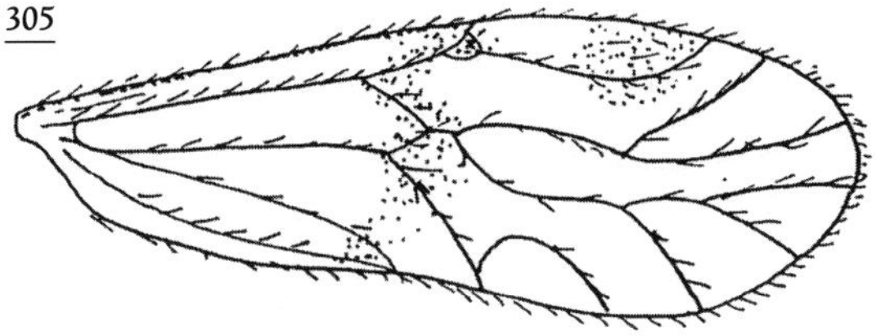

306

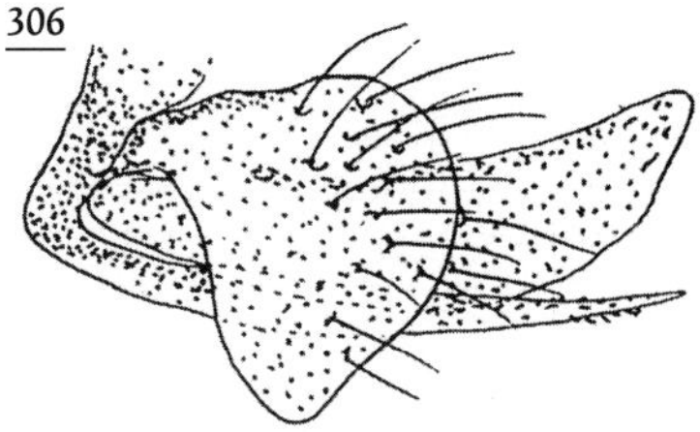

309

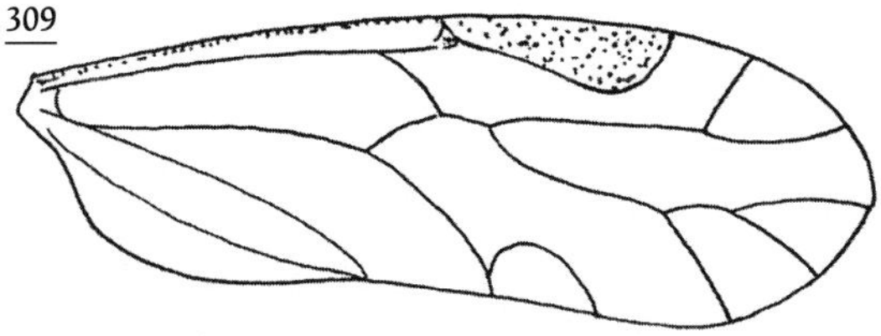

310

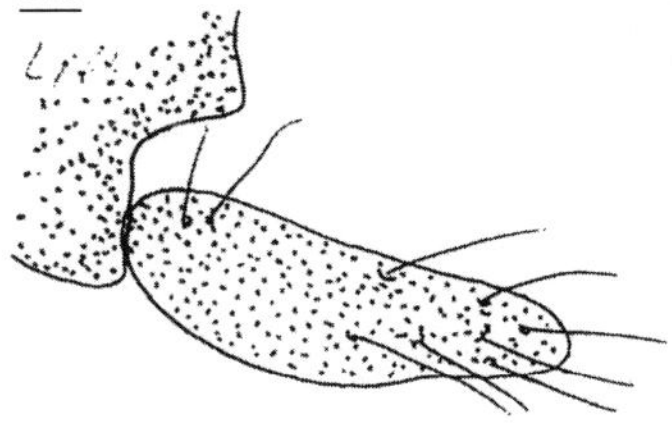

313

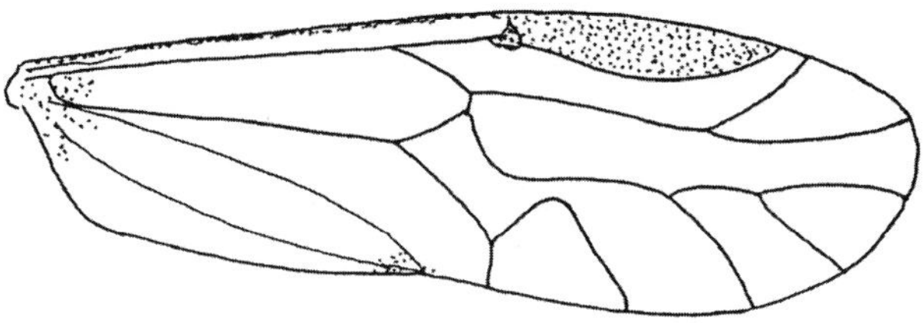

314

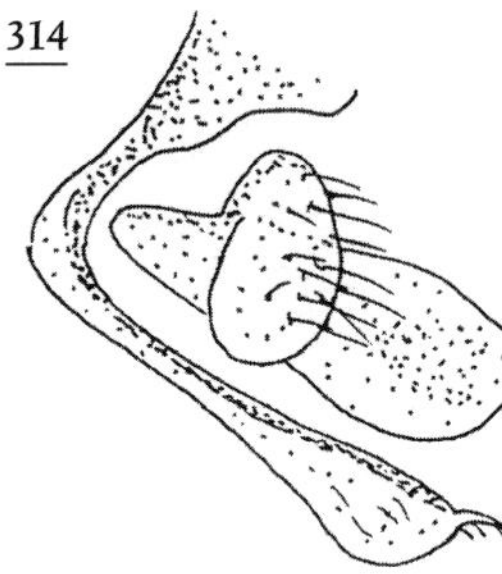

307

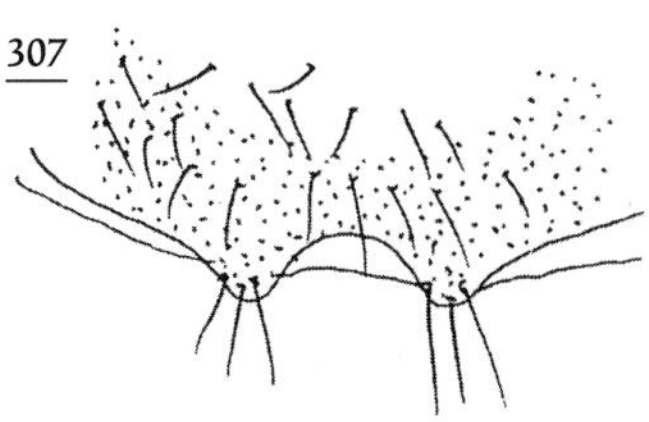

311

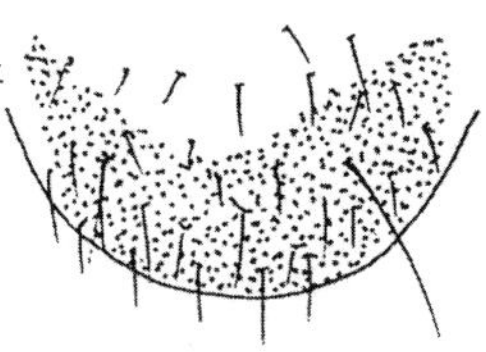

315

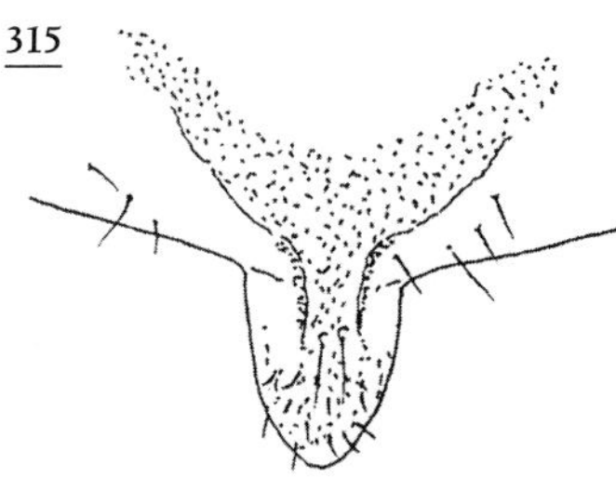

308

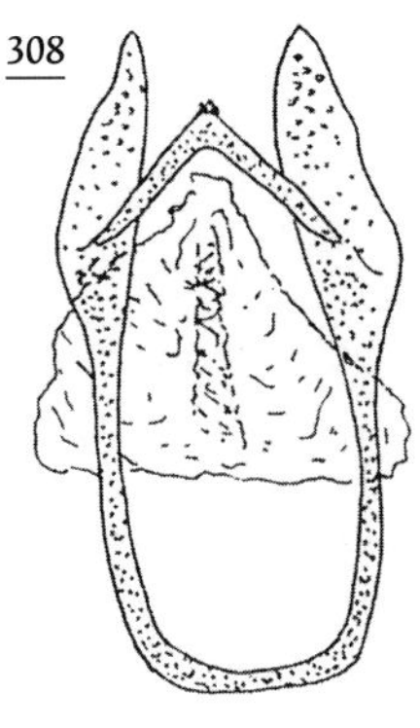

312

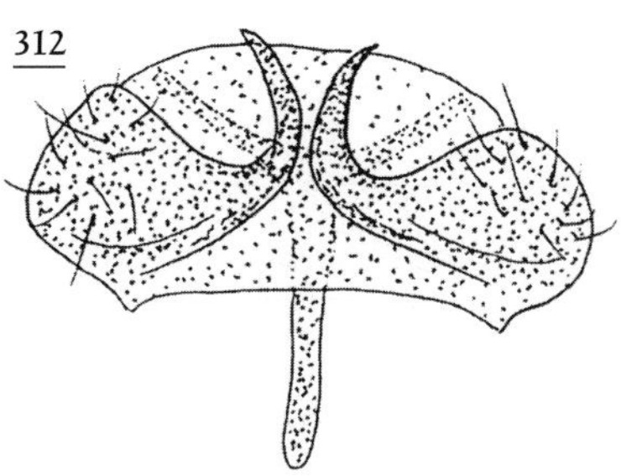

316

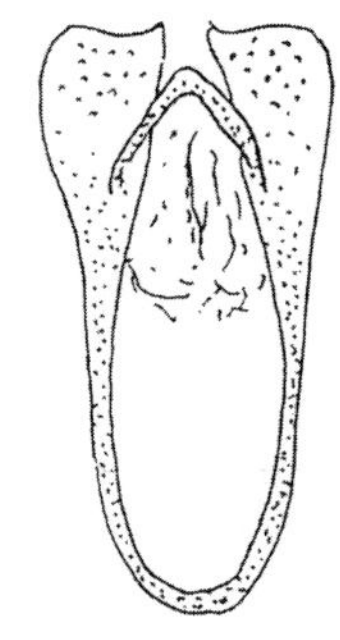

FIGURES 317 – 327

Peripsocidae. (317-320) Fore and hind wing venation of **317, 318** *Peripsocus* – **319, 320** *Diplopsocus* – **321** subgenital plate apex of *Peripsocus* – **322** subgenital plate apex of *Diplopsocus* – **323** gonapophyses of *Peripsocus* – **324** gonapophyses of *Diplopsocus* – **325** phallosome of *Peripsocus*, with enlarged insert of central sclerites – **326** phallosome of *Diplopsocus* – **327** male clunial ornamentation of *Peripsocus*.

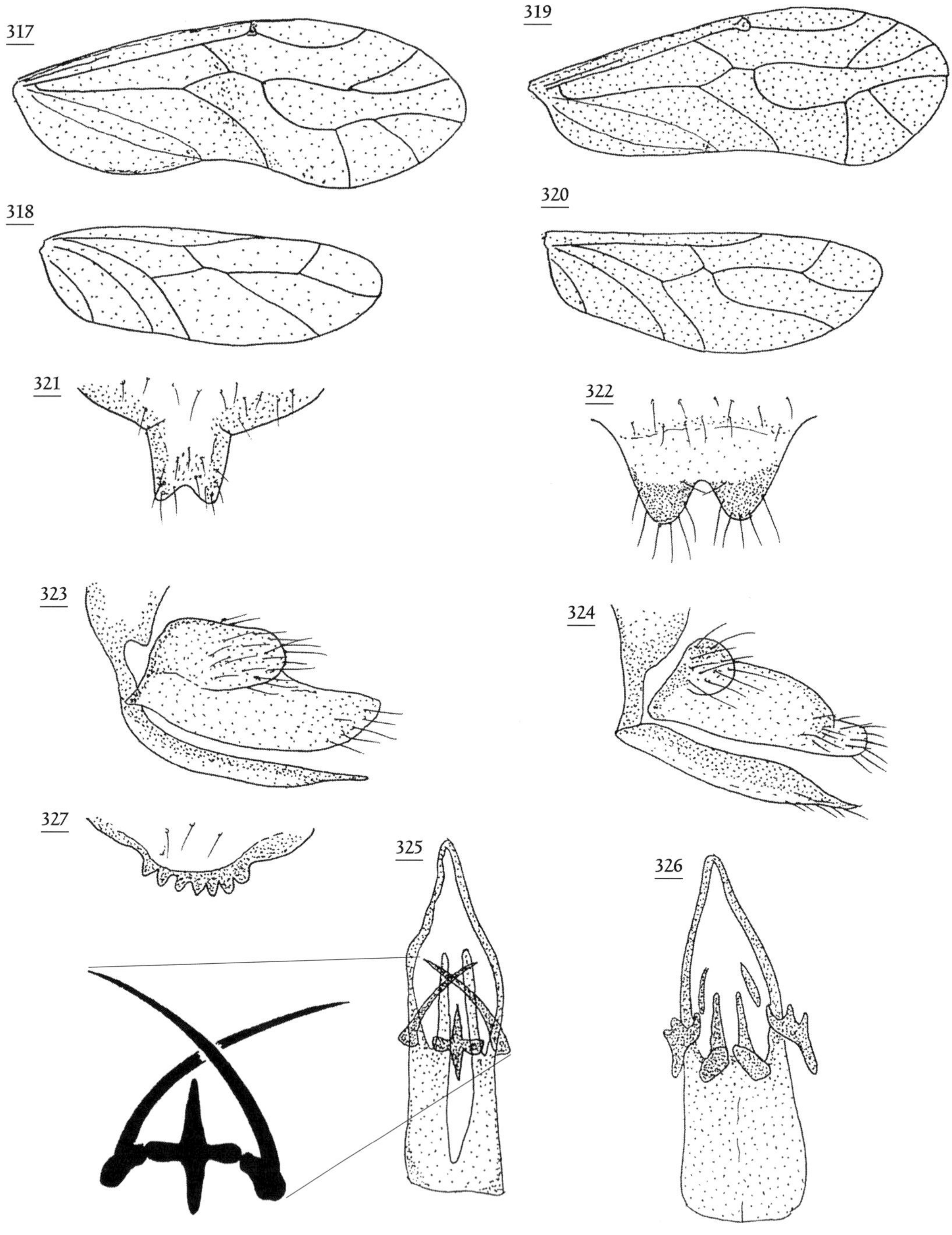

FIGURES 328 – 339

Philotarsidae. Fore wing venation of 328 *Haplophallus* – 329 *Aaroniella* – 330 *Philotarsopsis* – 331 apical flagellar segment of *Aaroniella*, to indicate long apical seta; gonapophyses of 332 *Haplophallus* – 333 *Aaroniella*; subgenital plate of 334 *Aaroniella* – 335 *Haplophallus* – 336 *Philotarsopsis*; phallosome of 337 *Aaroniella* – 338 *Haplophallus* – 339 *Philotarsopsis*.

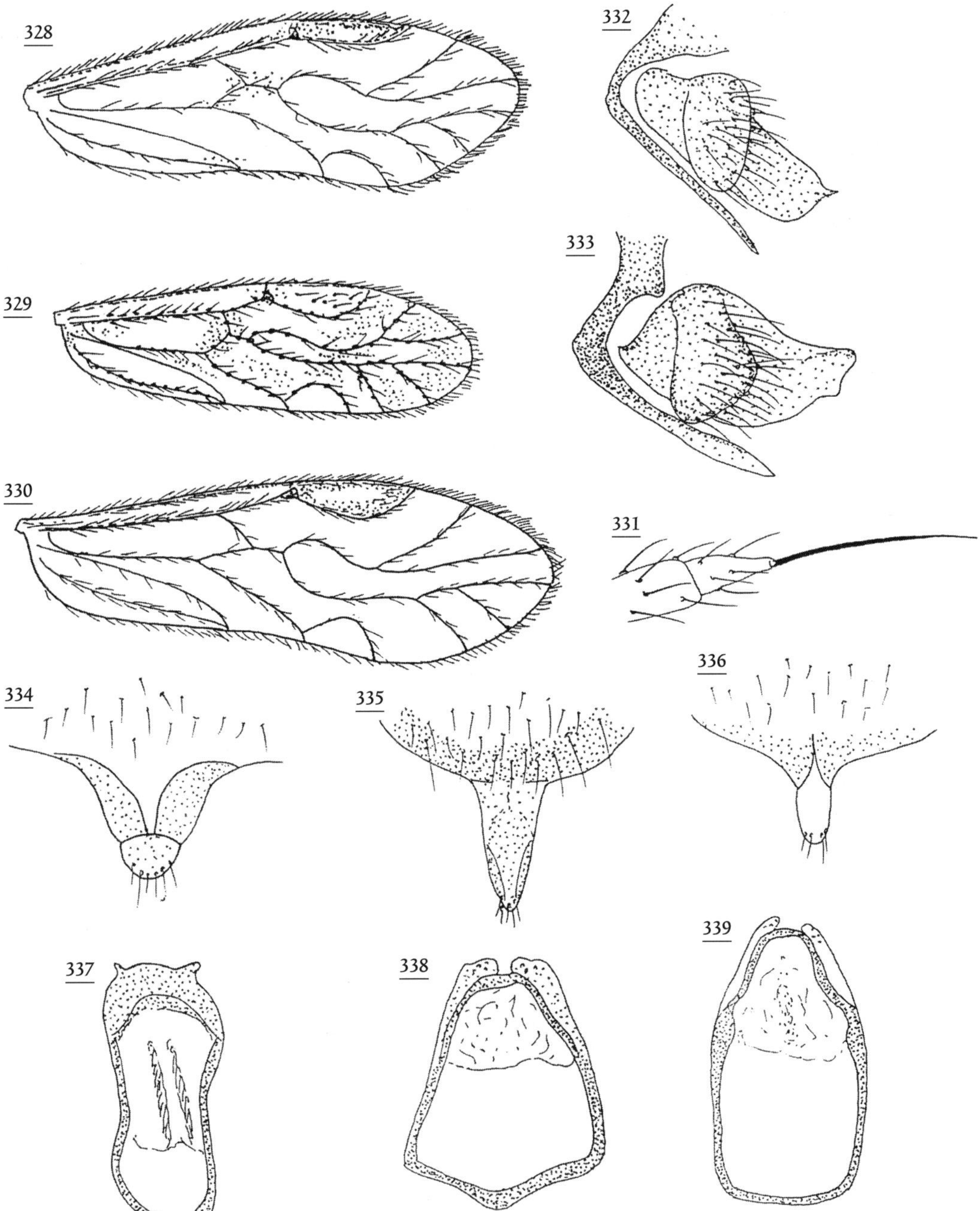

340

342

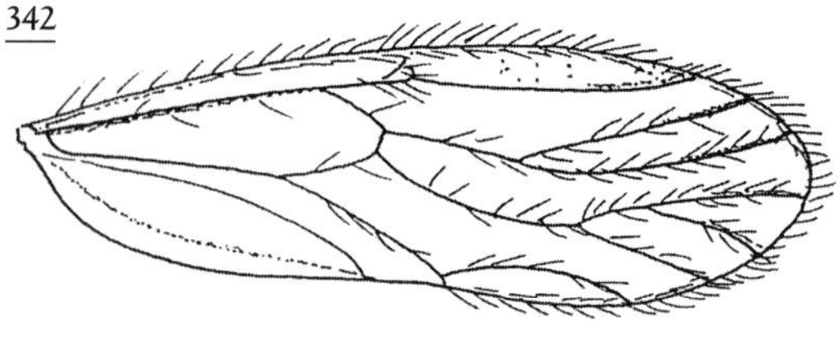

341

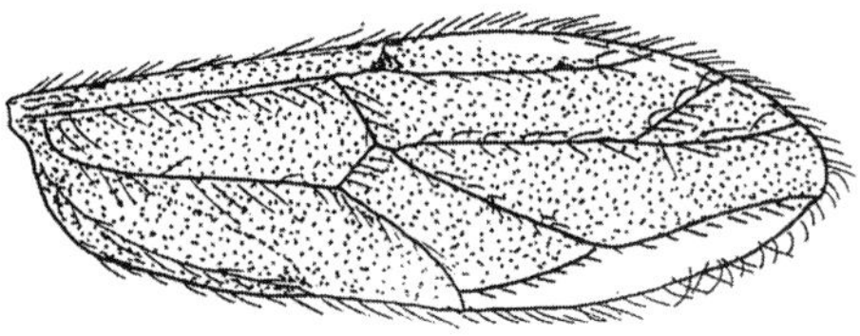

343

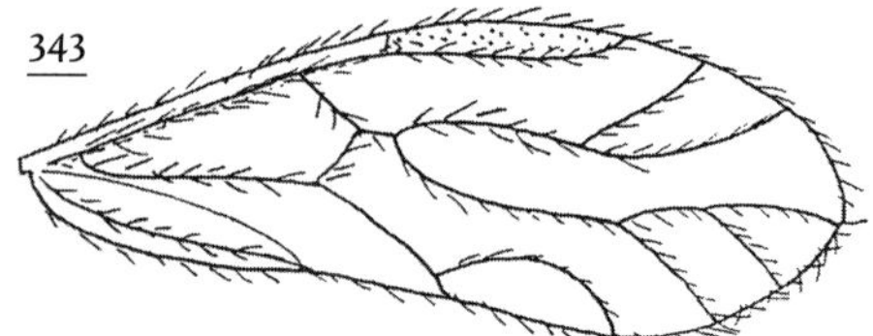

344

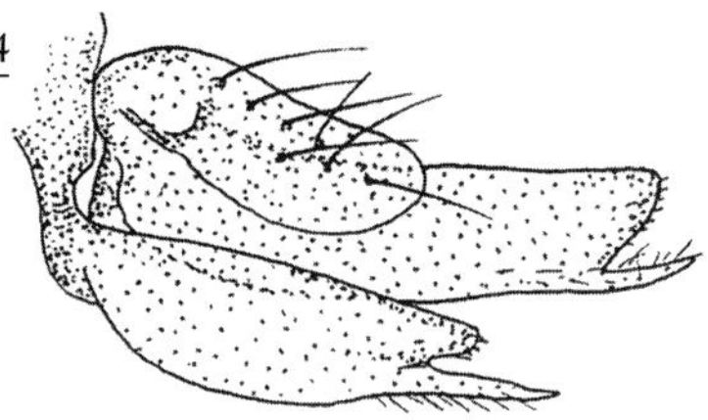

348

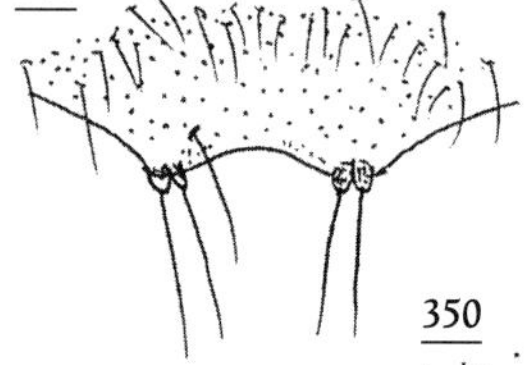

350

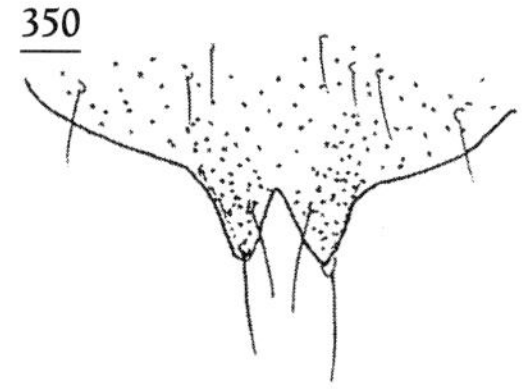

345

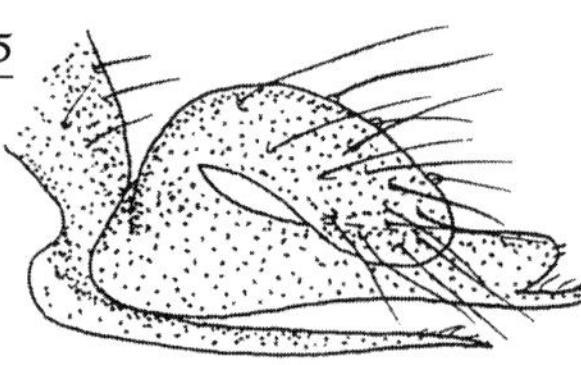

349

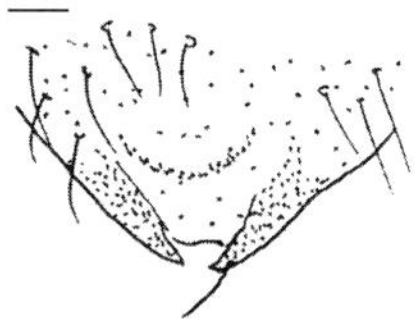

346

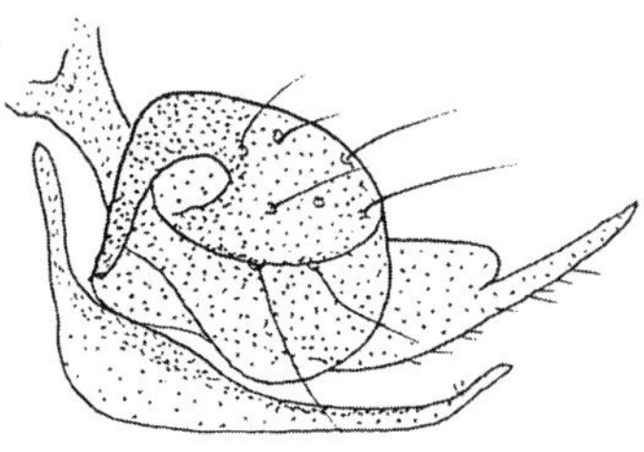

351

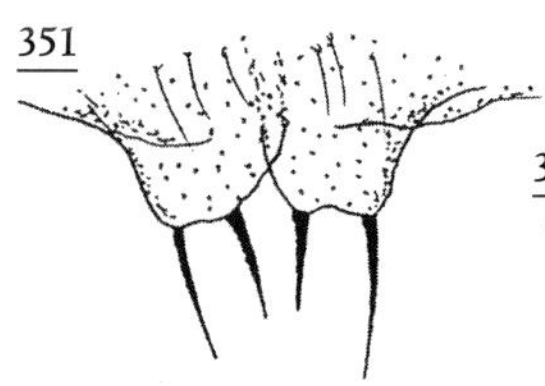

352

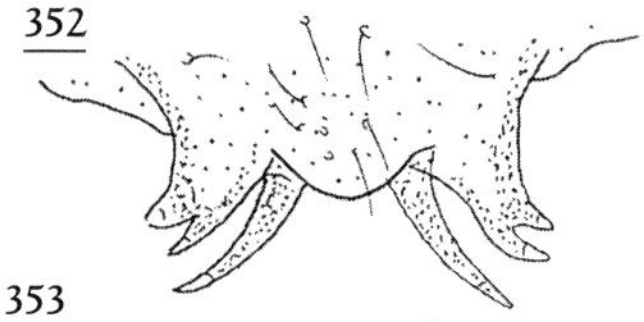

353

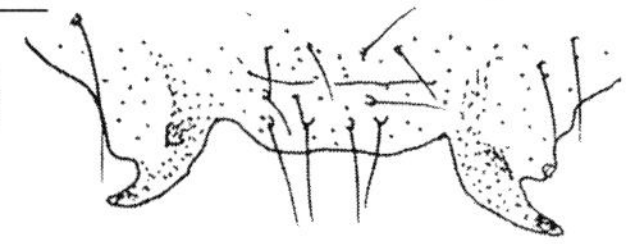

354

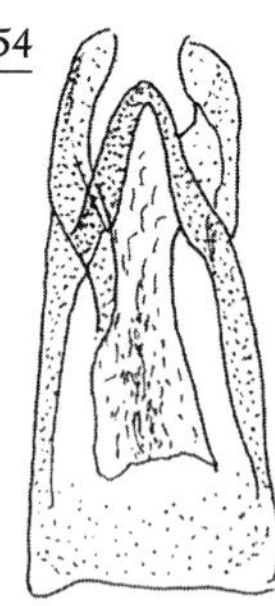

355

347

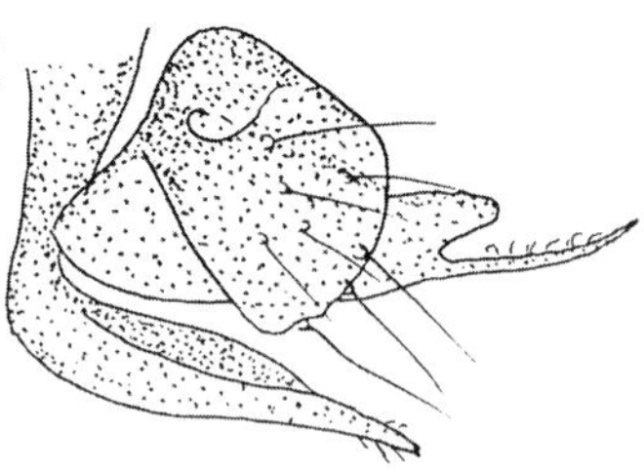

FIGURES 340 – 355 [FACING PAGE]
Pseudocaeciliidae. Fore wing venation of **340** *Allocaecilius* – **341** *Allopsocus* – **342** *Lobocaecilius* – **343** *Levucaecilius*; (344-347) gonapophyses, (348-351) subgenital plates: **344, 348** *Allocaecilius* – **345, 349** *Allopsocus* – **346, 350** *Levucaecilius* – **347, 351** *Lobocaecilius*; (352, 353) hypandrial margin and (354, 355) phallosome: **352, 354** *Lobocaecilius* – **353, 355** *Allocaecilius*.

FIGURES 356 – 360
Pseudocaeciliidae. *Heterocaecilius minotus* Lee & Thornton: **356** fore wing – **357** subgenital plate apex – **358** gonapophyses – **359** apex of hypandrium – **360** phallosome.

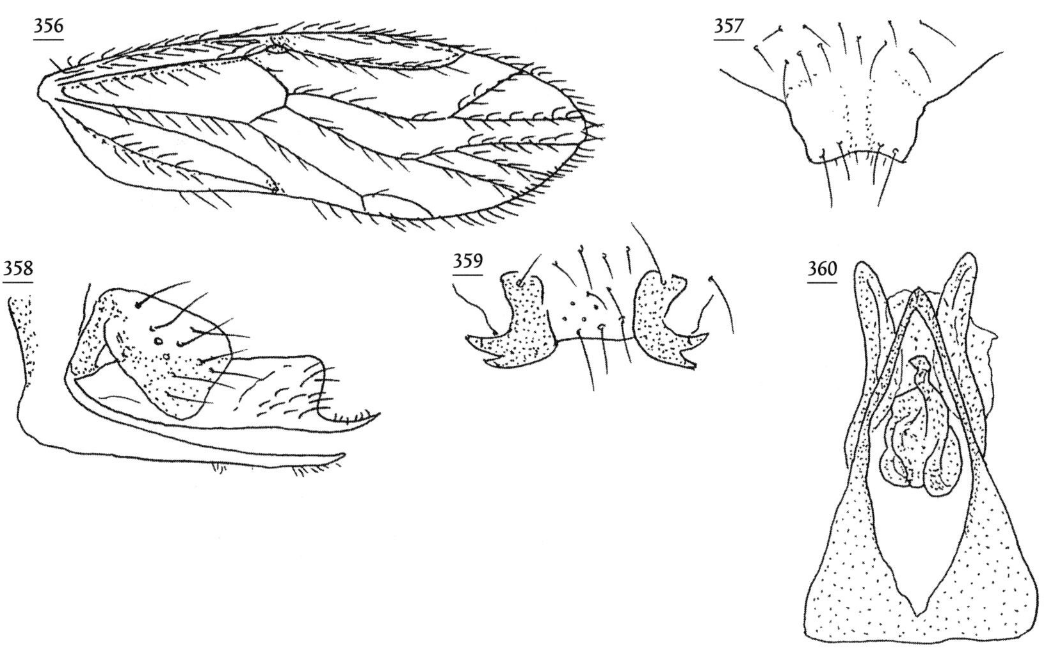

FIGURES 361 – 373

Pseudocaeciliidae. Fore wing venation of **361** *Mepleres* – **362** *Ophiodopelma* – **363** *Phallocaecilius*, with enlargement of anal region papillae – **364** *Pseudocaecilius*; gonapophyses of **365** *Mepleres* – **366** *Ophiodopelma* – **367** *Phallocaecilius* – **368** *Pseudocaecilius*; subgenital plate apex of **369** *Mepleres* – **370** *Phallocaecilius*; phallosome of **371** *Mepleres* – **372** *Phallocaecilius* – **373** *Pseudocaecilius*.

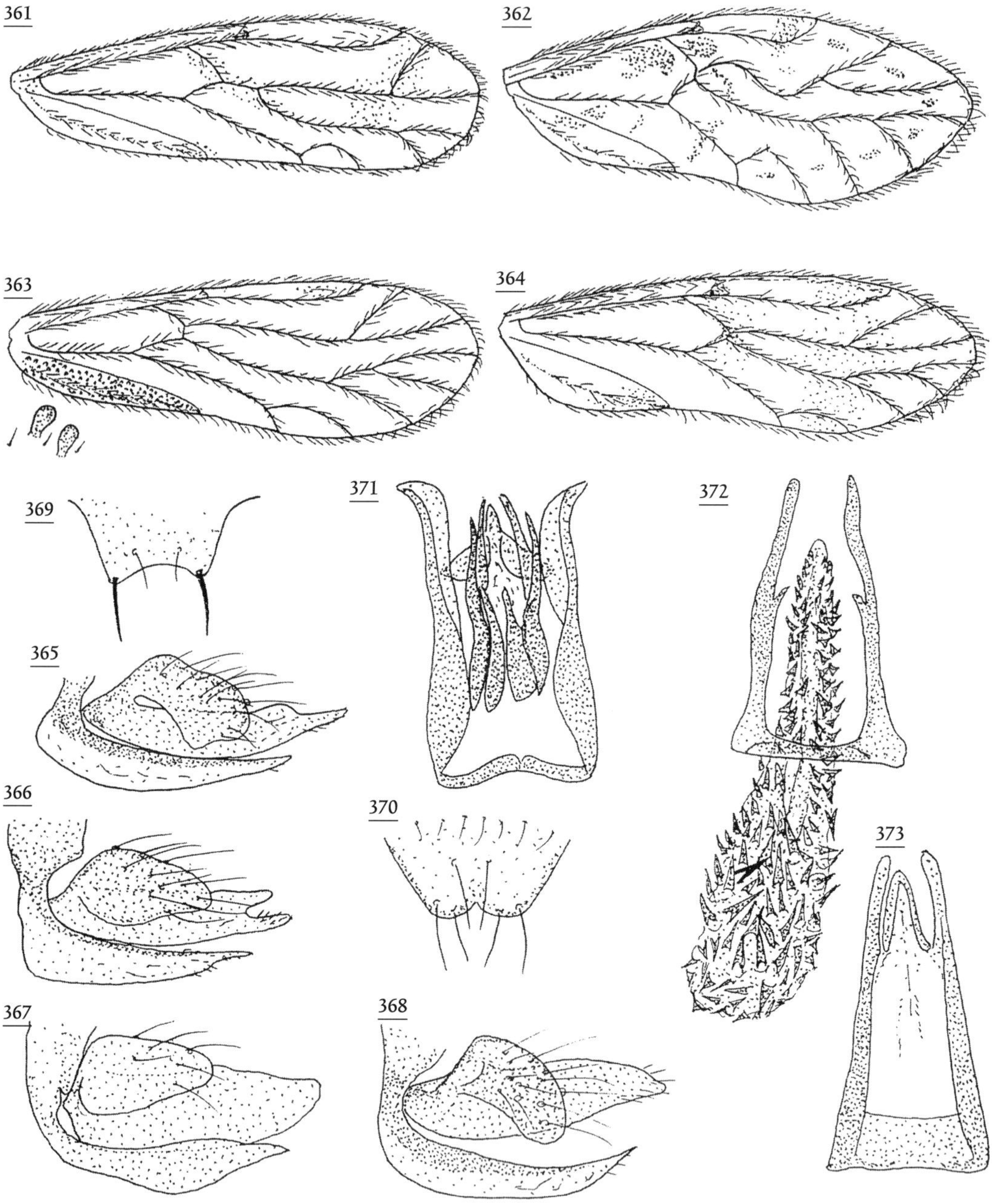

FIGURES 374 – 385

Pseudocaeciliidae: Zelandopsocinae. Fore wing venation of **374** *Novopsocus* – **375** *Austropsocus* – **376** *Zelandopsocus* – **377** vertex of *Novopsocus*, profile (eye black, antennal socket open, anterior to left); gonapophyses of **378** *Novopsocus* – **379** *Austropsocus* – **380** *Zelandopsocus*; subgenital plate apex of **381** *Austropsocus* – **382** *Zelandopsocus* – **383** *Novopsocus*; phallosome of **384** *Austropsocus* – **385** *Zelandopsocus*.

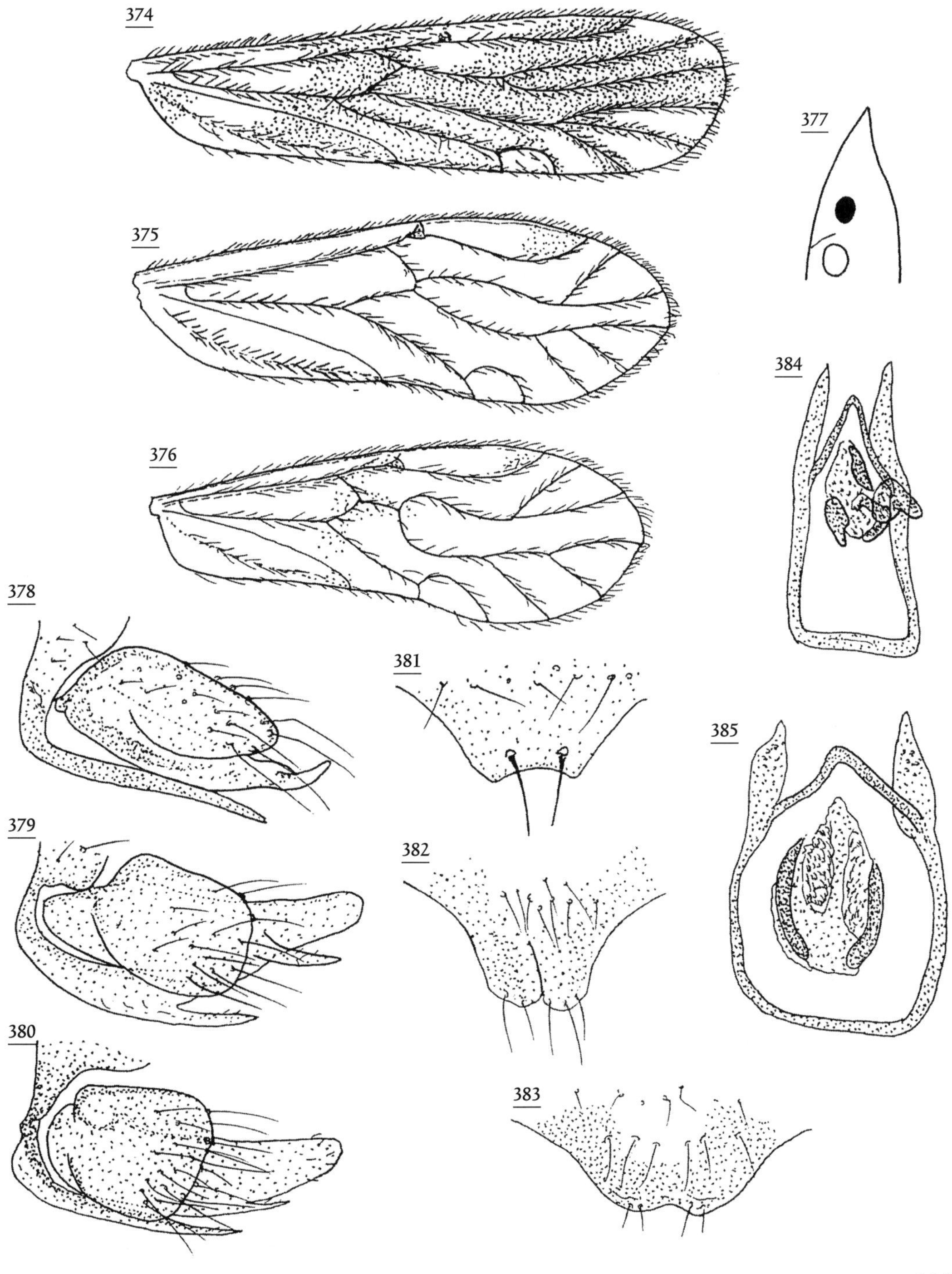

FIGURES 386 – 391

Hemipsocidae, *Hemipsocus*. Fore wing venation and appearance of **386** maculate winged and **387** pale-winged species – **388** gonapophyses – **389** subgenital plate – **390** phallosome – **391** male, clunium and epiproct.

386

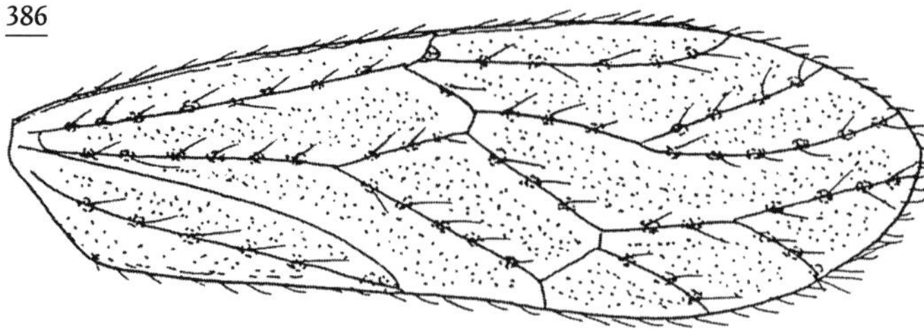

387

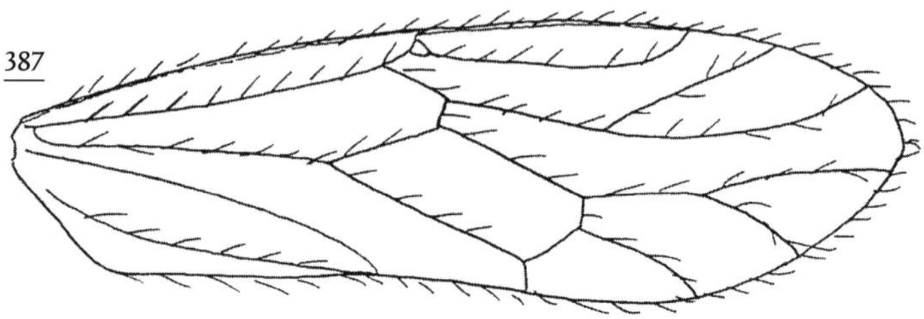

388

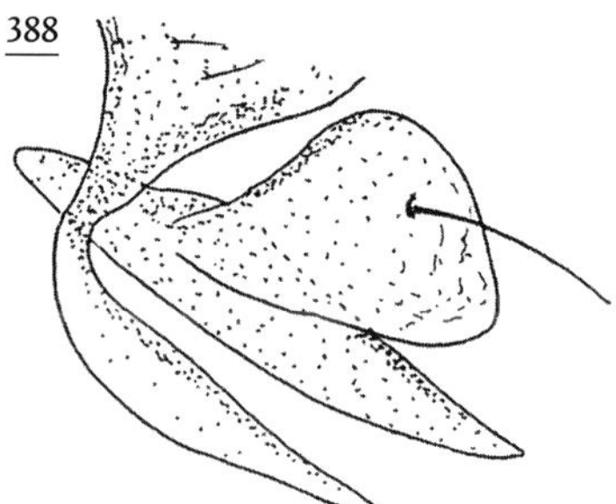

389

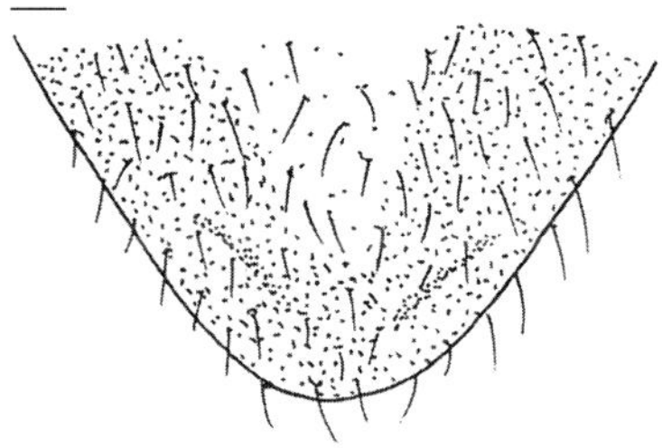

390

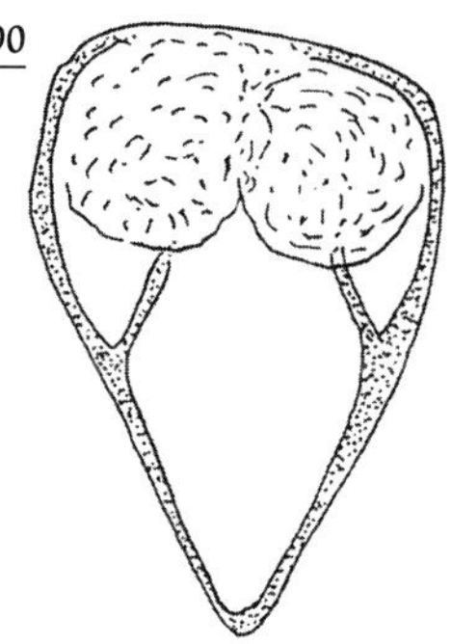

391

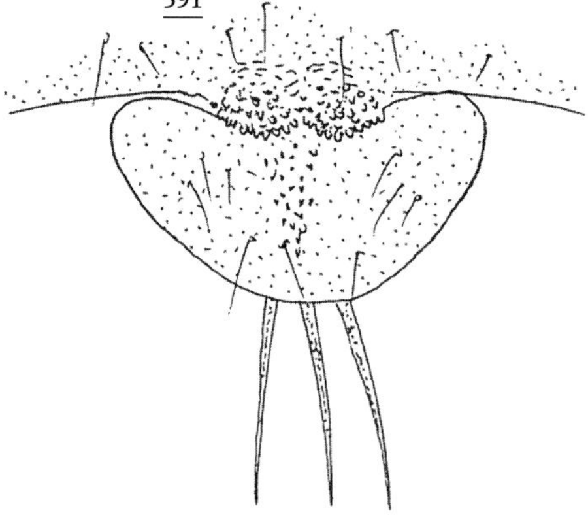

FIGURES 392 – 405

Myopsocidae. Wing venation: fore wing of **392** *Myopsocus* – **393** *Smithersia* – **394** *Lophopterygella* – **395** *Mouldsia*; hind wing of **396** *Smithersia* – **397** *Lichenomima*; gonapophyses of **398** *Smithersia* – **399** *Myopsocus*; subgenital plate apex of **400** *Lichenomima* – **401** *Myopsocus* – **402** *Smithersia*; phallosome of *Myopsocus* spp.: **403** rounded, with radular plates – **404** narrowed anteriorly, with central rod – **405** elongate, no central rod.

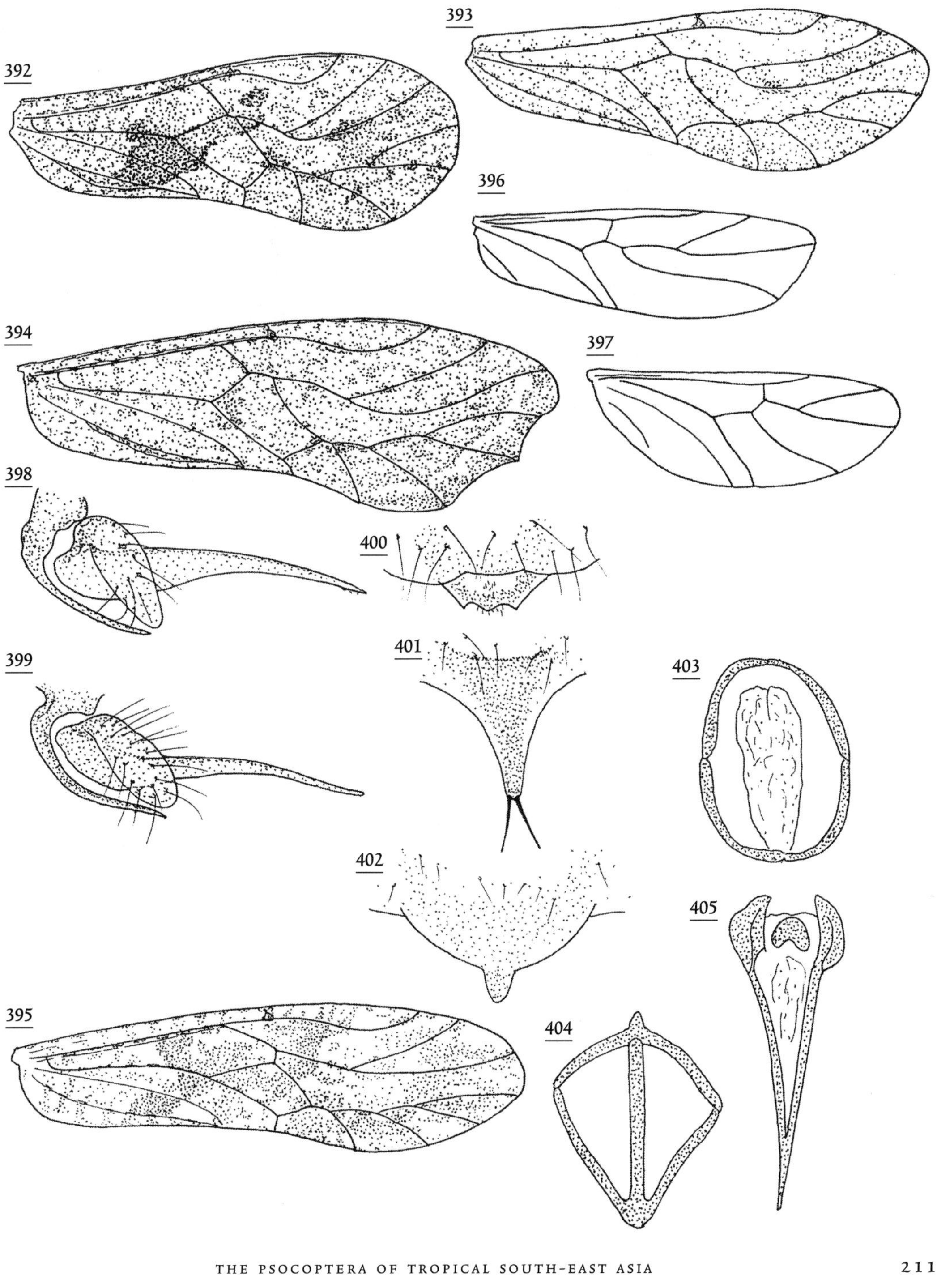

FIGURES 406 – 410

Myopsocidae. **406** Phallosome of *Nimbopsocus* – (407-410) hypandrium of **407, 408** *Lophopterygella camelina* – **409, 410** *Thorntonodes lobatus* (407, 409, ventral, posterior toward bottom of page; 408, 410, lateral, posterior toward left of page. After Lienhard, 2004b).

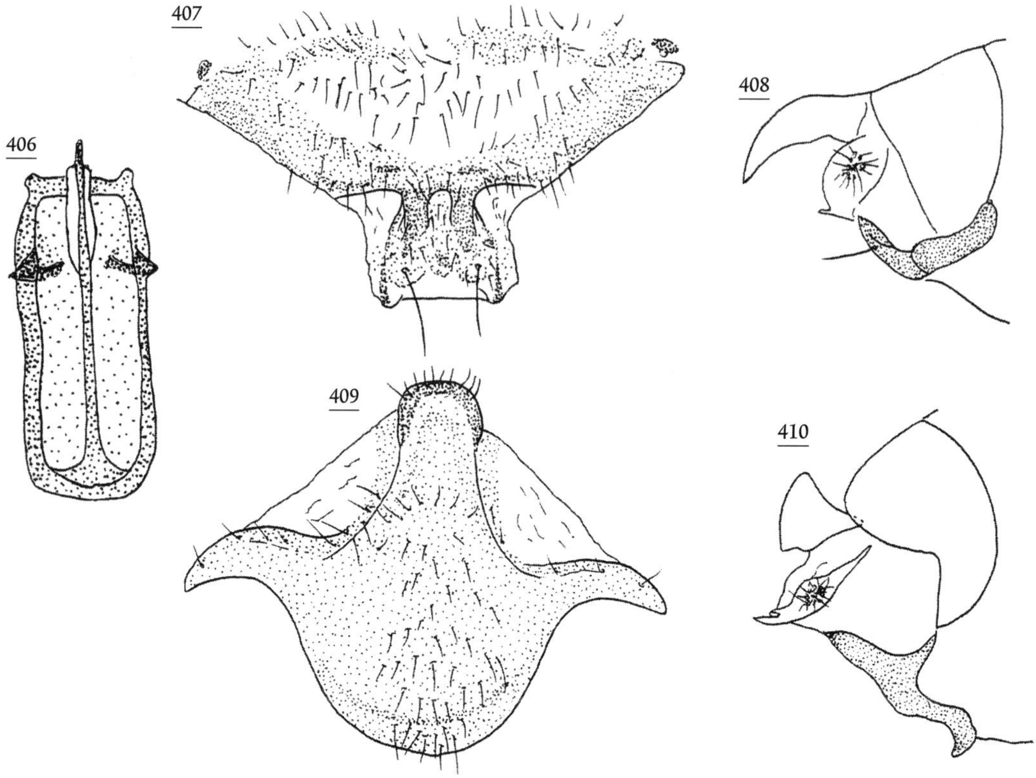

FIGURES 411 – 417

Psilopsocidae, *Psilopsocus*. **411, 412** Fore wing venation, indicating variety in form of areola postica – **413** lacinial apex – **414** tarsal claw – **415** gonapophyses – **416** subgenital plate – **417** phallosome.

411

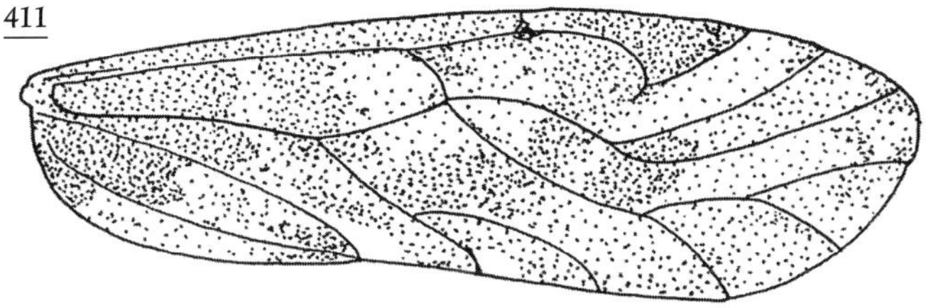

413

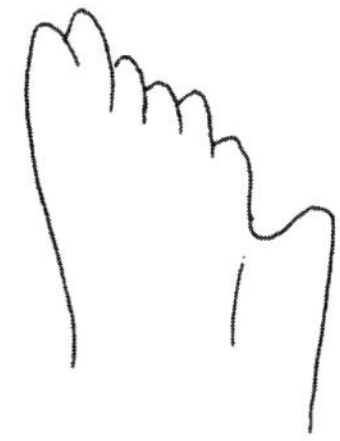

412

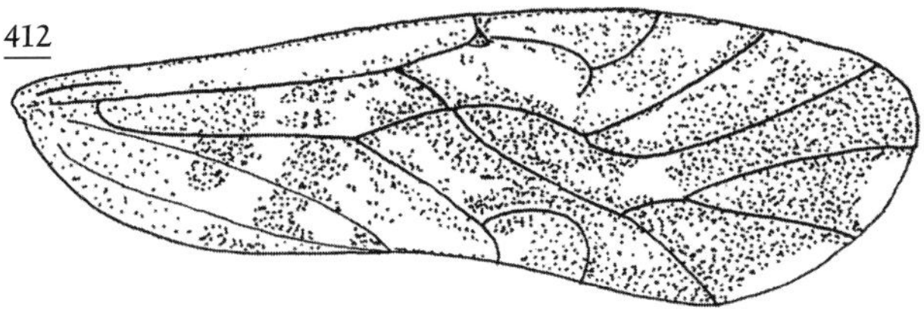

414

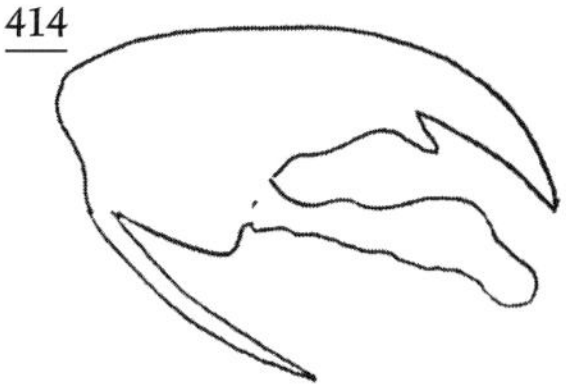

415

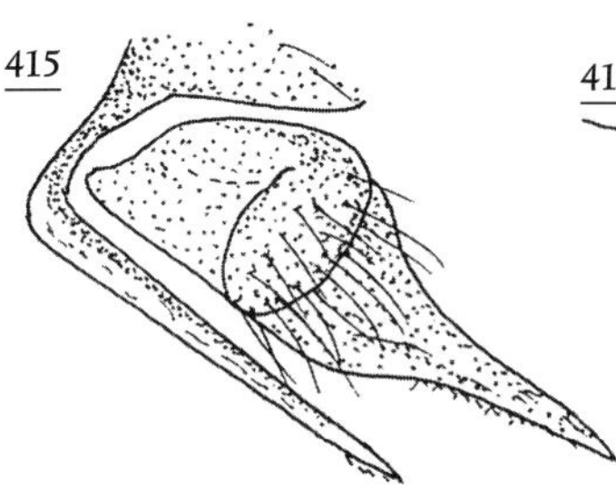

416

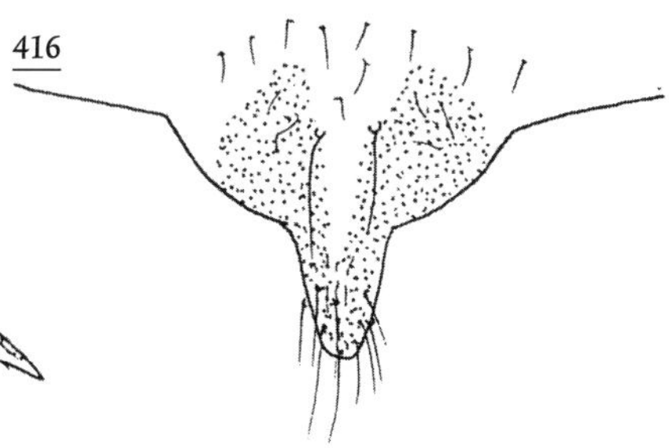

417

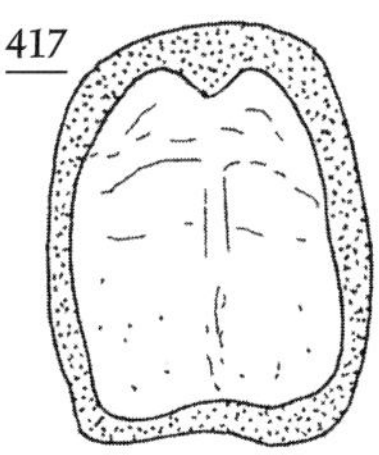

FIGURES 418 – 430

Psocidae, Amphigerontiinae. Fore wing venation of **418** *Amphigerontia* – **419** *Anomaloblaste* – **420** *Blaste* – **421** *Blastopsocus*; gonapophyses of **422** *Anomaloblaste* – **423** *Blastopsocus* – **424** *Blaste*; subgenital plate of **425** *Anomaloblaste* – **426** *Blaste* – **427** *Blastopsocus*; phallosome/parameres of **428** *Amphigerontia* – **429** *Blaste* – **430** *Blastopsocus*.

418

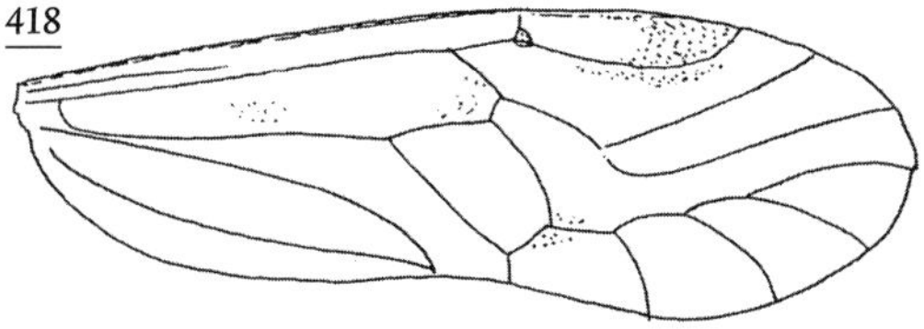

419

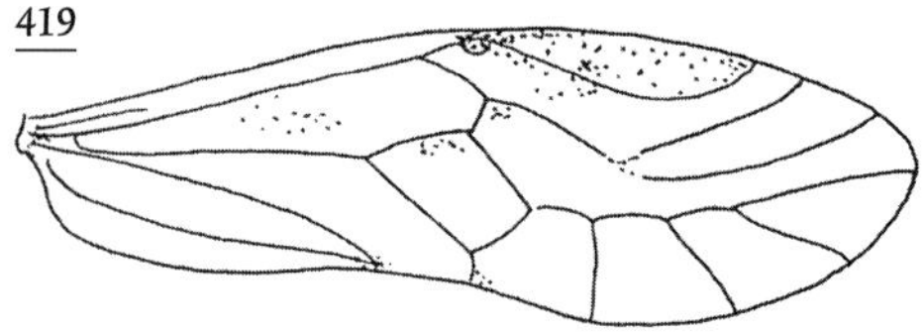

420

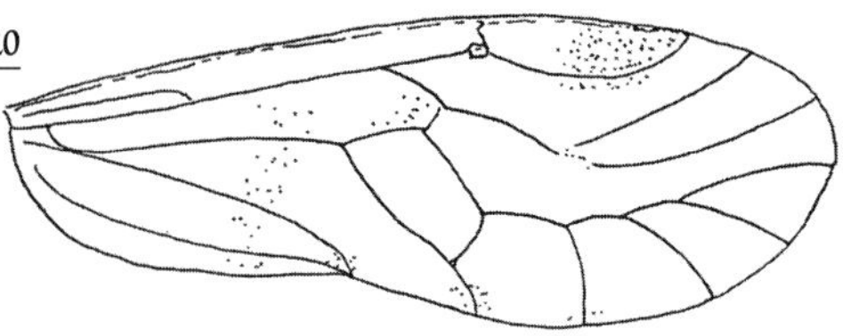

428

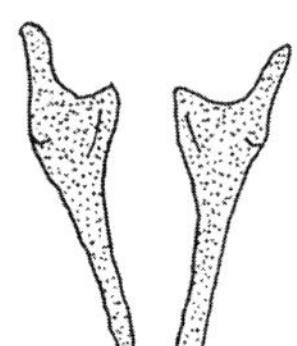

430

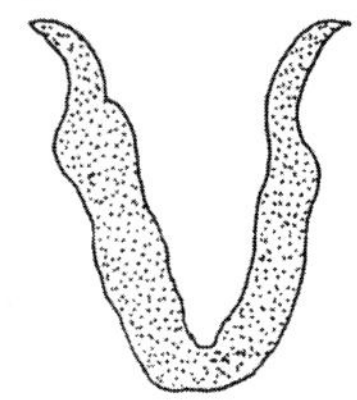

421

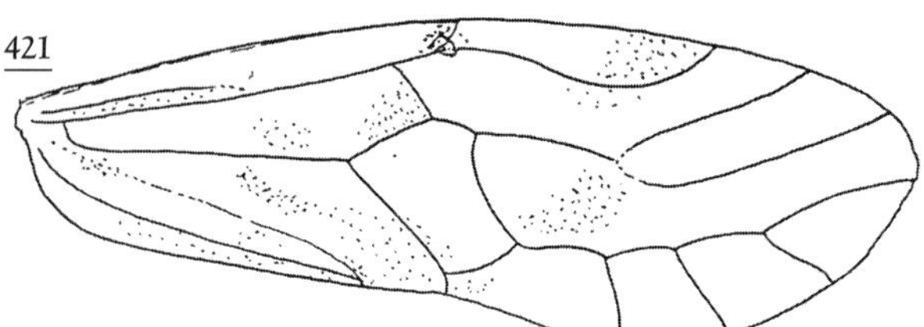

429

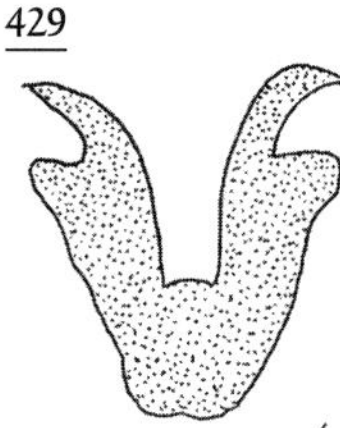

426

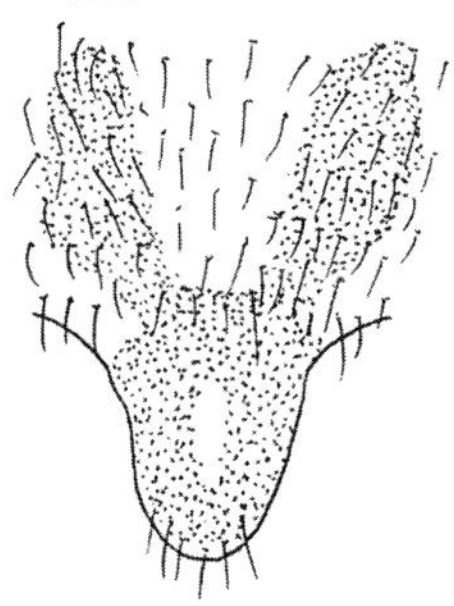

422

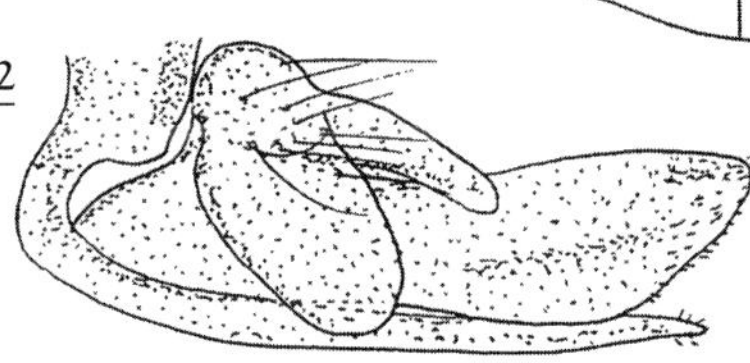

425

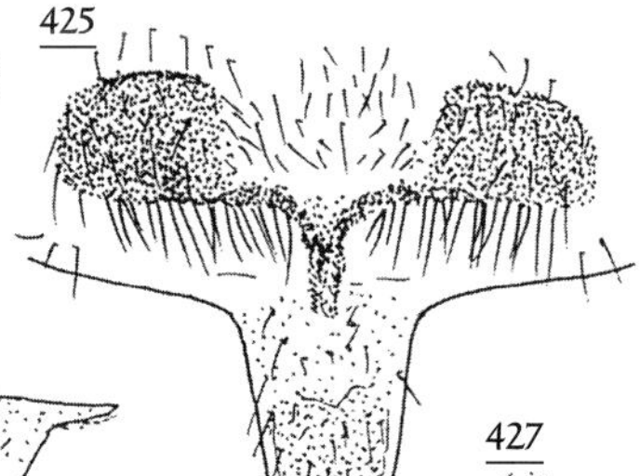

423

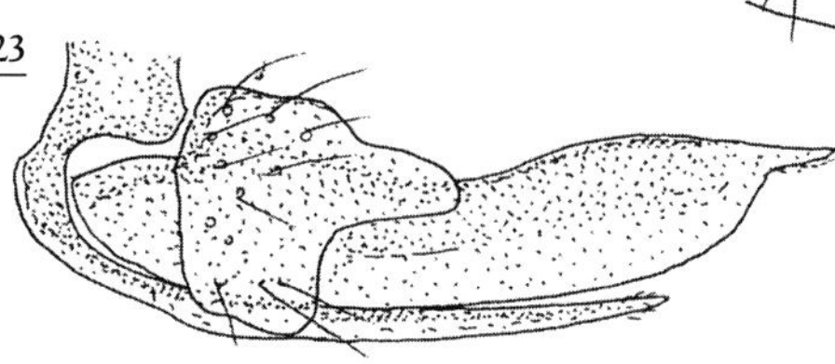

427

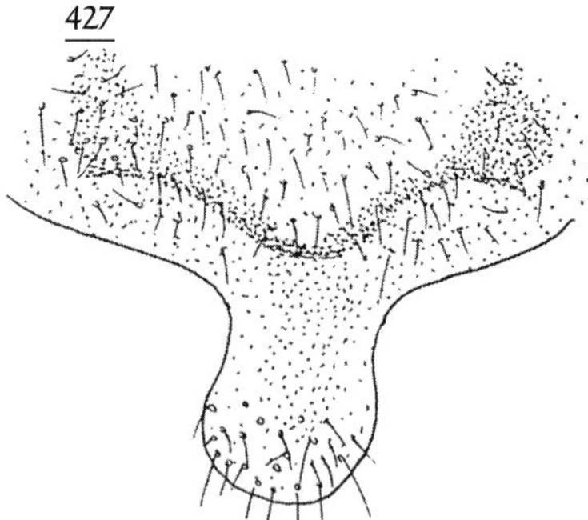

424

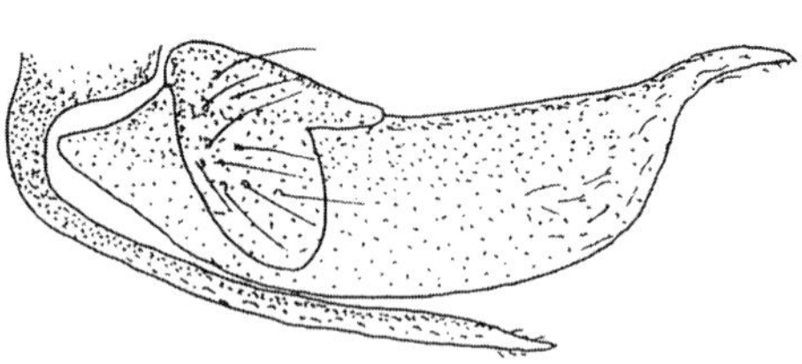

FIGURES 431 – 445 [FACING PAGE]

Psocidae, Amphigerontiinae. Wing venation of **431** *Euclismioides* – **432, 433** *Javablaste* – **434, 435** *Indoblaste* – **436, 437** *Neoblaste* (431, 432, 434, 436 fore wings; 433, 435, 437 hind wings); gonapophyses of **438** *Euclismioides* – **439** *Indoblaste* – **440** *Neoblaste* – **441** *Javablaste*; parameres of **442** *Euclismioides* – **443** *Indoblaste* – **444** *Neoblaste* – **445** *Javablaste*.

431

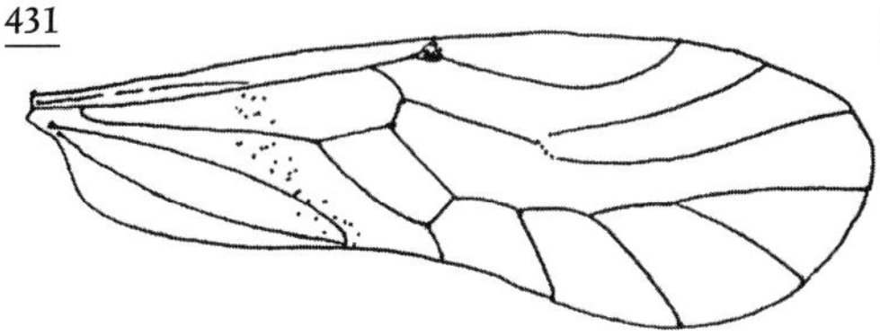

432

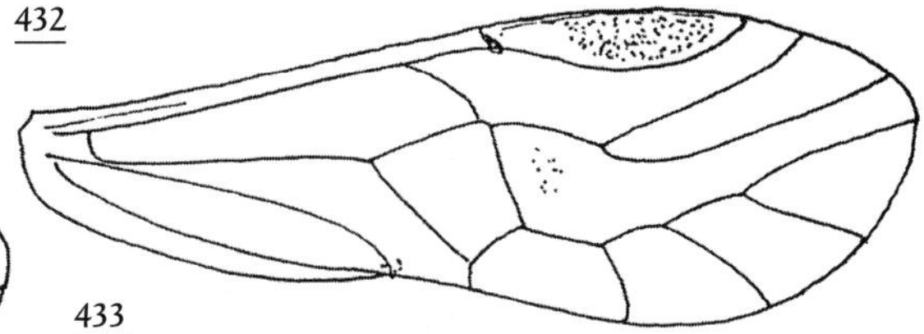

433

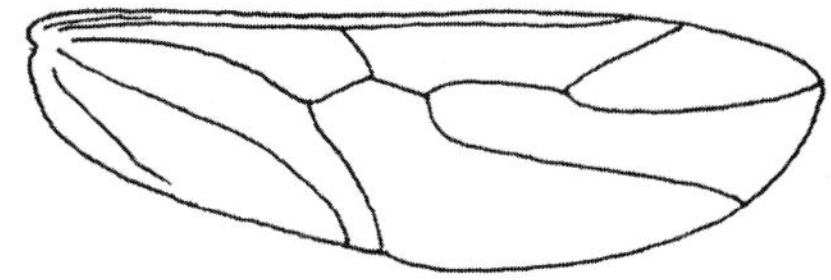

434

436

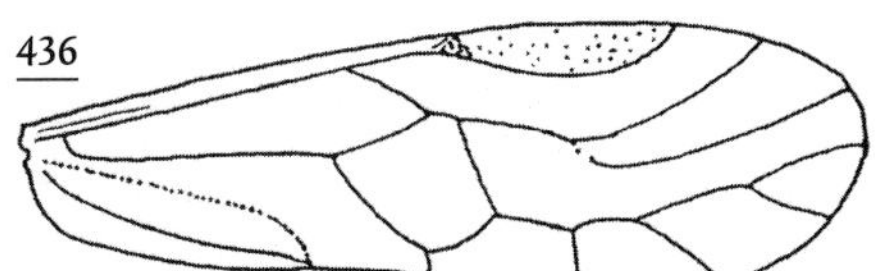

435

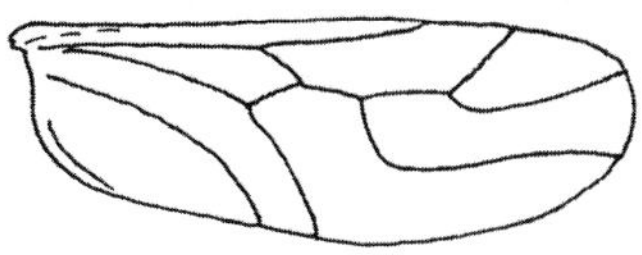

437

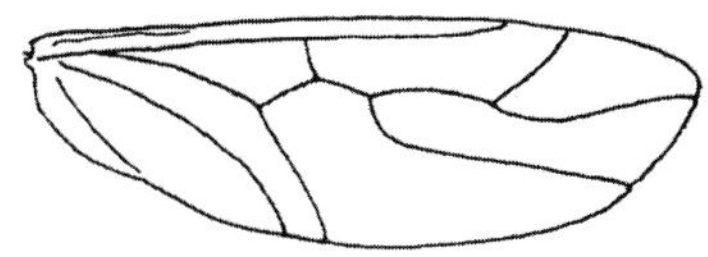

438

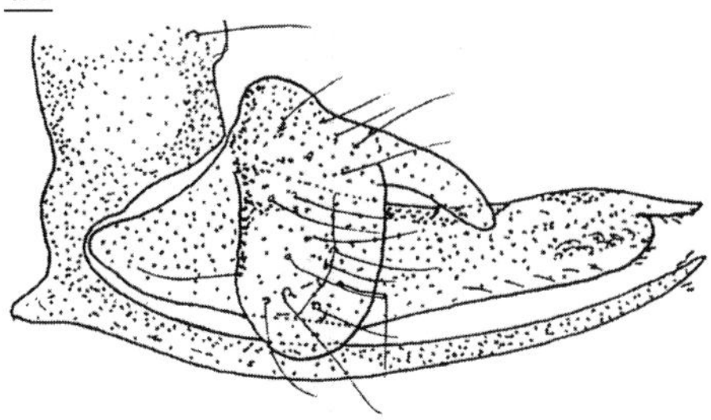

442

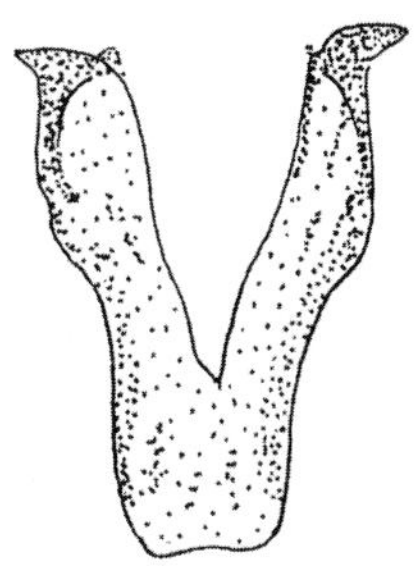

443

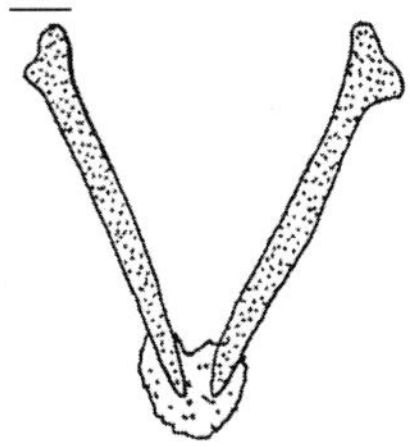

439

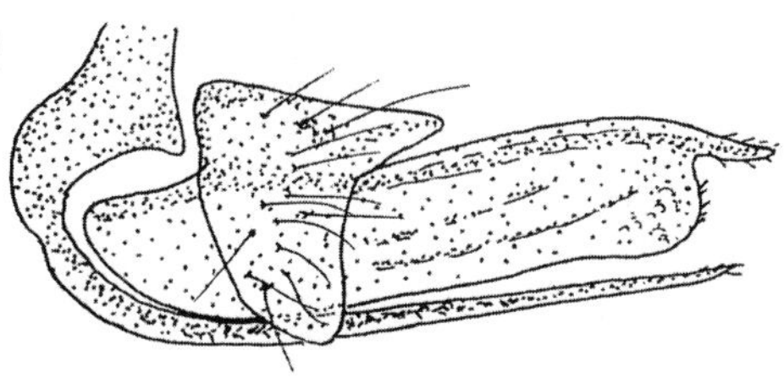

444

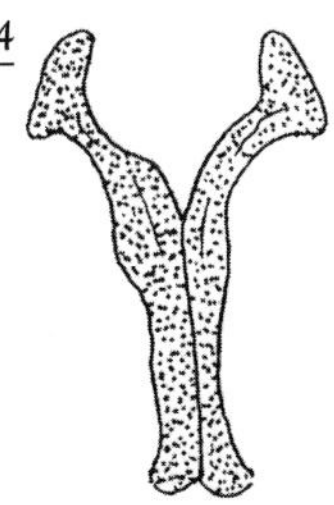

445

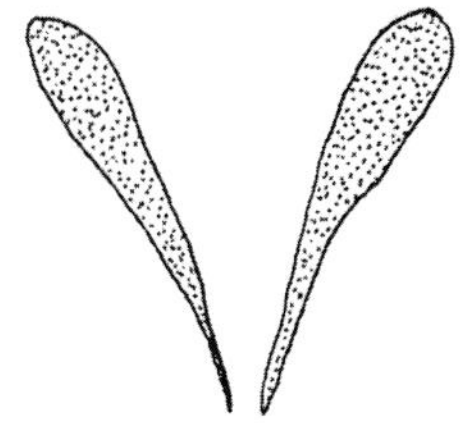

440

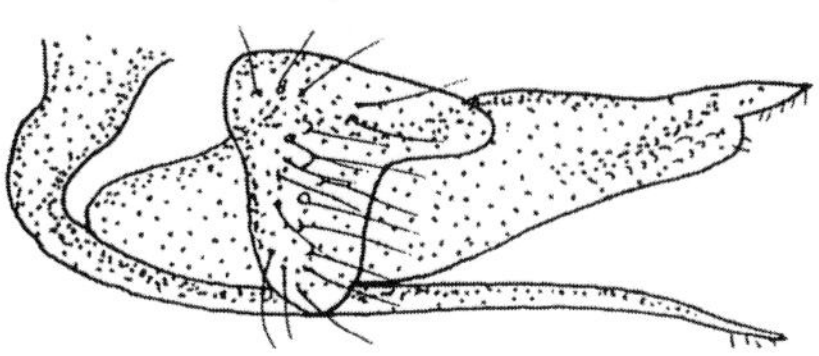

441

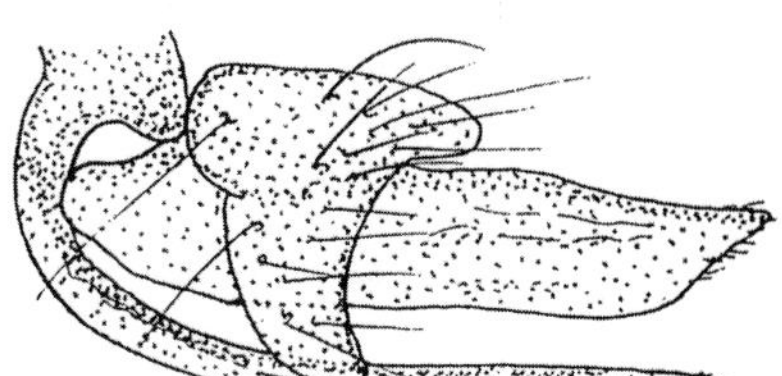

FIGURES 446 – 458

Psocidae. Fore wing venation of **446** *Clematoscenea* – **447** *Sigmatoneura* male (the figures on this plate are of a species formerly placed in *Podopterocus*; cf. figs. 473 on) – **448** *Sigmatoneura*, male hind leg to show swollen tibia; gonapophyses of **449** *Clematoscenea* – **450** *Sigmatoneura* – **451** *Clematoscenea*: male clunium expanded over base of epiproct; phallosome of **452** *Clematoscenea* – **453** *Sigmatoneura*; subgenital plate of **454** *Clematoscenea* – **455** *Sigmatoneura* – **456** *Neoblaste* – **457** *Indoblaste* – **458** *Javablaste*.

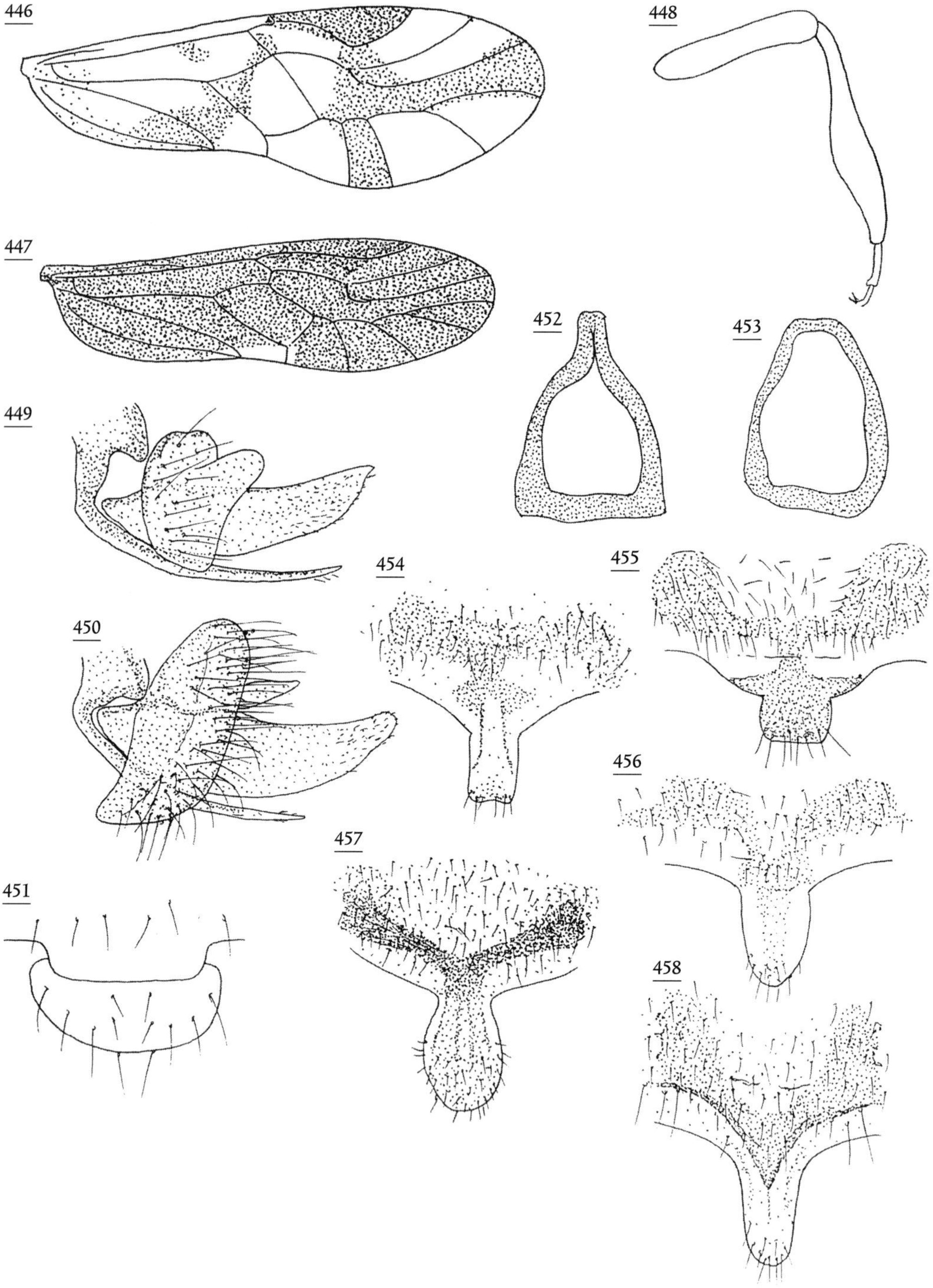

FIGURES 459 – 472

Psocidae. Fore wing venation of **459** *Psococerastis* (with **460** hind wing) – **461** *Cycetes* – **462** *Setopsocus* – **463** *Sundapsocus*; gonapophyses of **464** *Psococerastis* – **465** *Cycetes* – **466** *Sundapsocus*; subgenital plate of **467** *Sundapsocus* – **468** *Cycetes* – **469** *Psococerastis* – **470** *Setopsocus*; phallosome of **471** *Psococerastis* – **472** *Sundapsocus*.

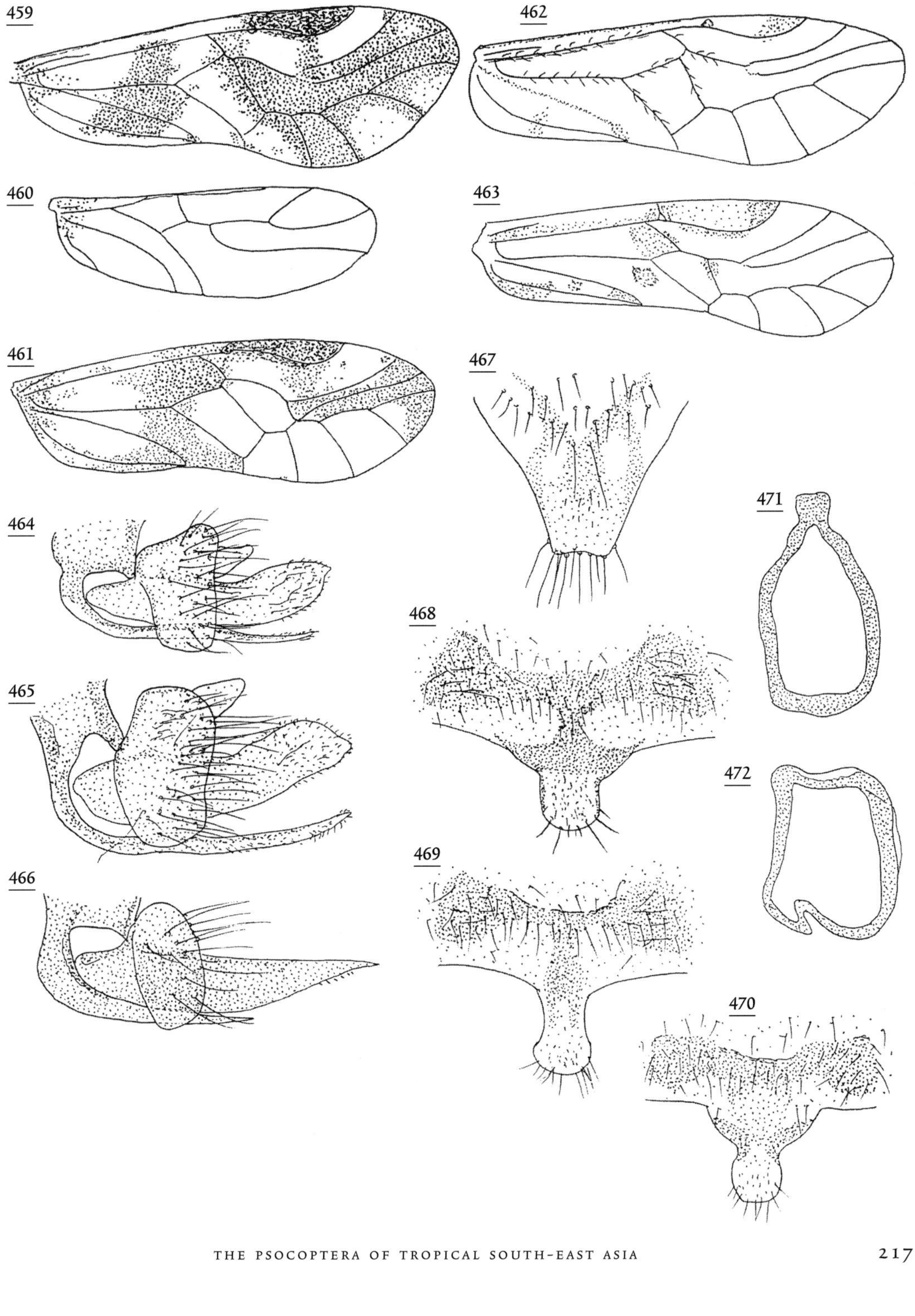

473

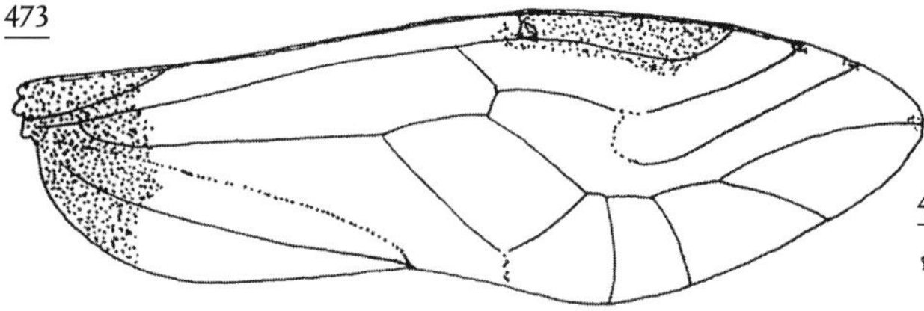

478

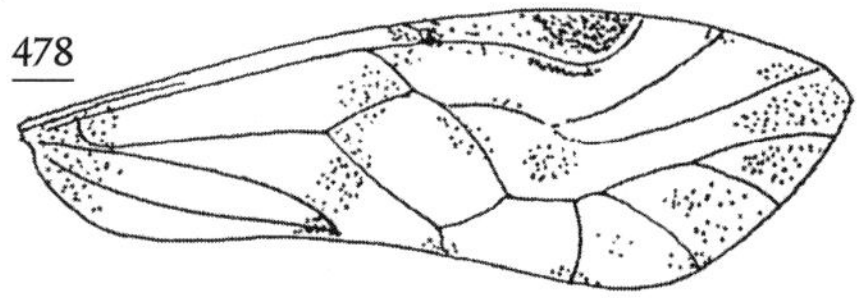

474

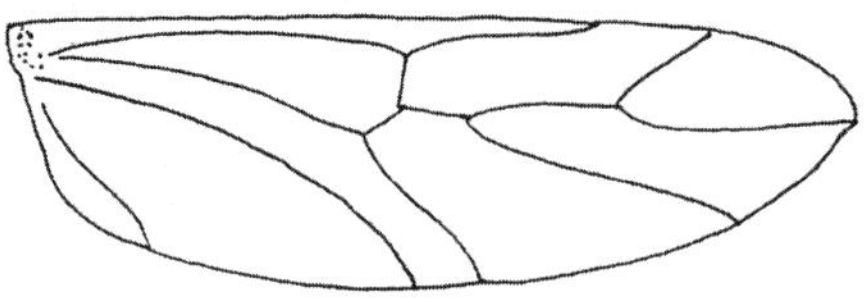

479

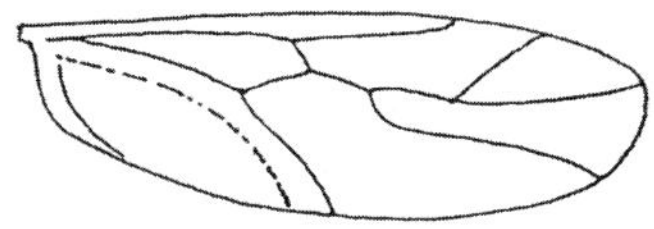

476

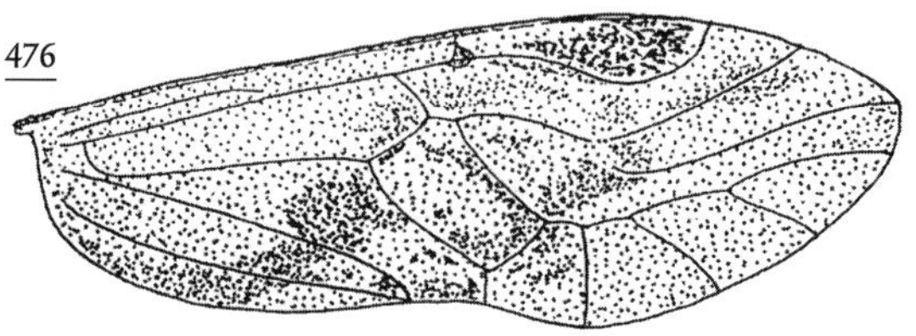

475

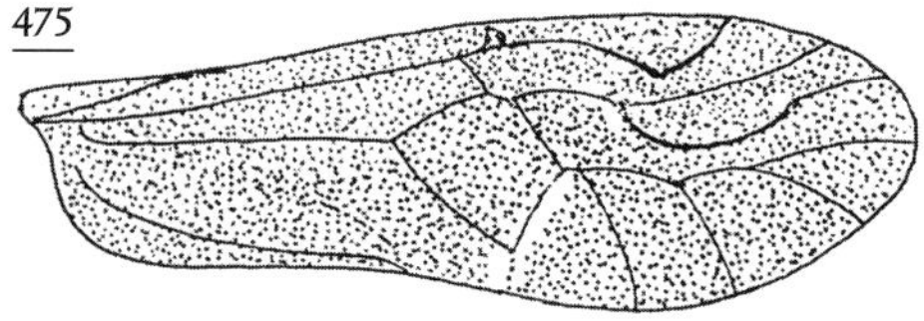

477

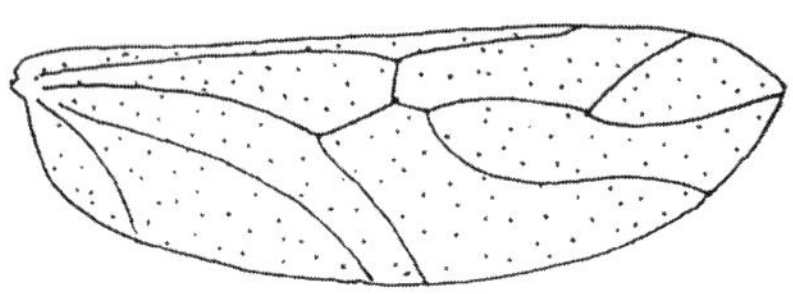

484

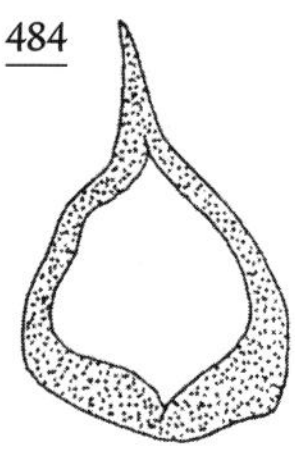

485

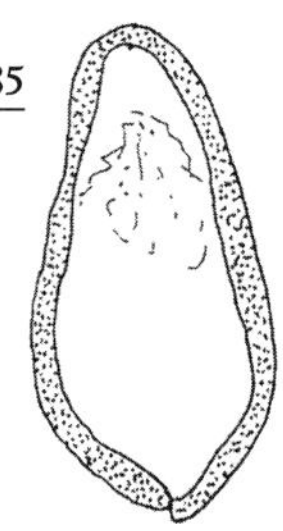

480

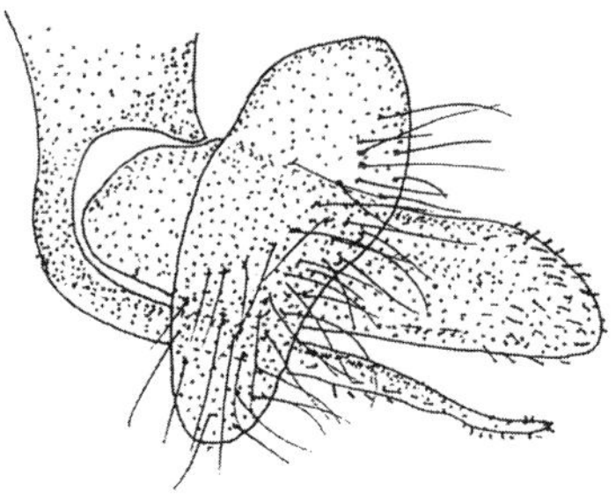

482

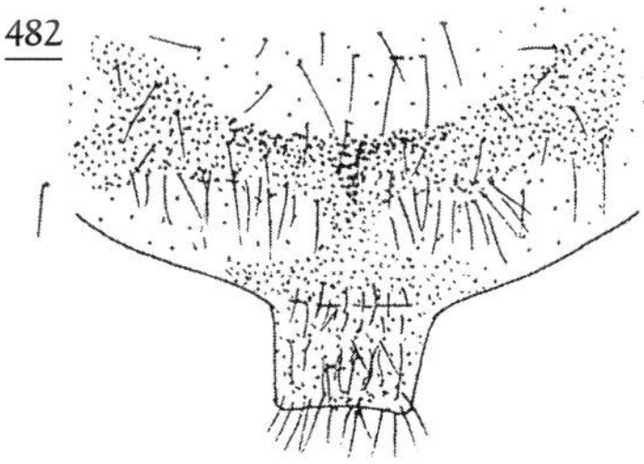

481

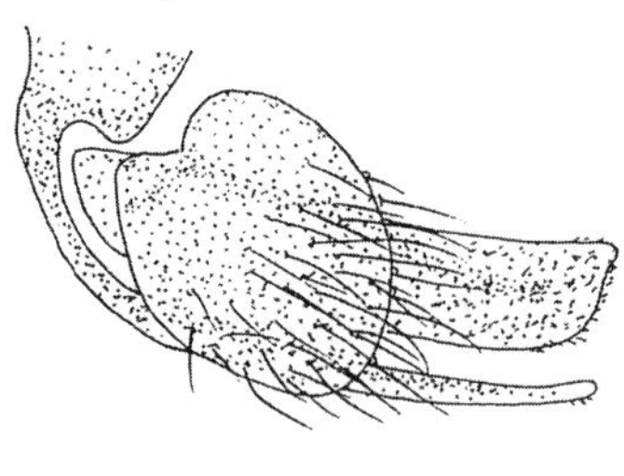

483

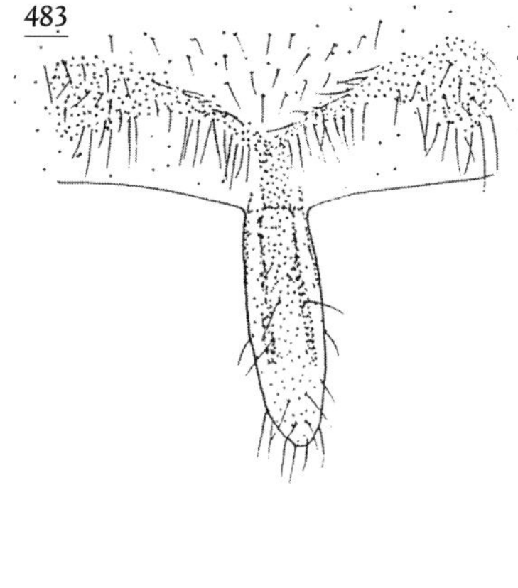

FIGURES 473 – 485 [FACING PAGE]
Psocidae. Wing venation of *Sigmatoneura*: **473, 474** male – **475** female fore wing to indicate strong sexual colour dimorphism in the genus – **476, 477** *Metylophorus* – **478, 479** *Atrichadenoctecnum*; gonapophyses of **480** *Sigmatoneura* – **481** *Metylophorus*; subgenital plate of **482** *Sigmatoneura* – **483** *Metylophorus*; phallosome of **484** *Metylophorus* – **485** *Sigmatoneura*.

FIGURES 486 – 493
Psocidae. Fore wing venation of **486** *Psocidus* s.str. – **487** *Clematostigma* – **488** *Ptycta* – **489** *Copostigma* – **490, 491** *Elytropsocus*, fore and hind wings; phallosome of **492** *Psocidus* s.str. – **493** *Ptycta*.

486
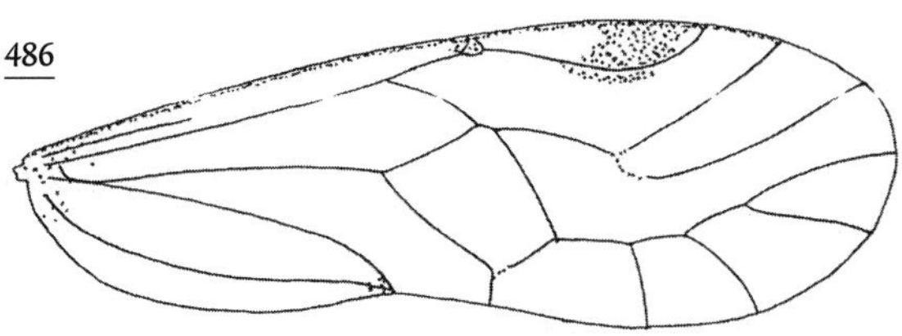

492
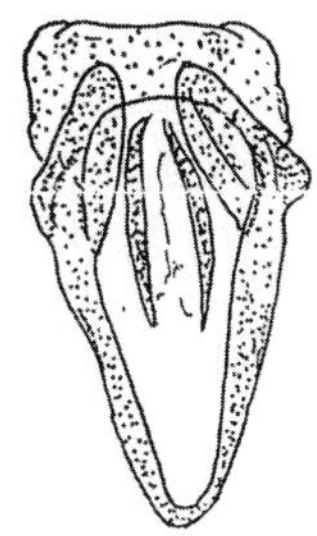

493
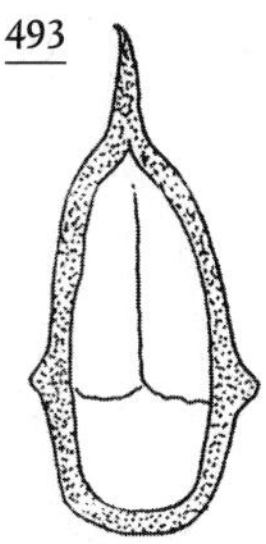

487
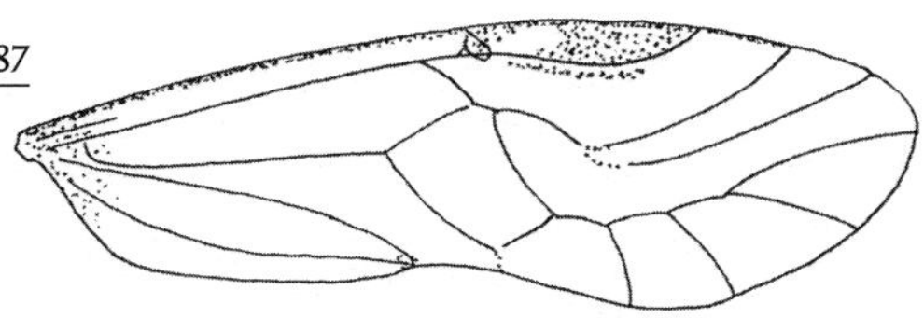

488
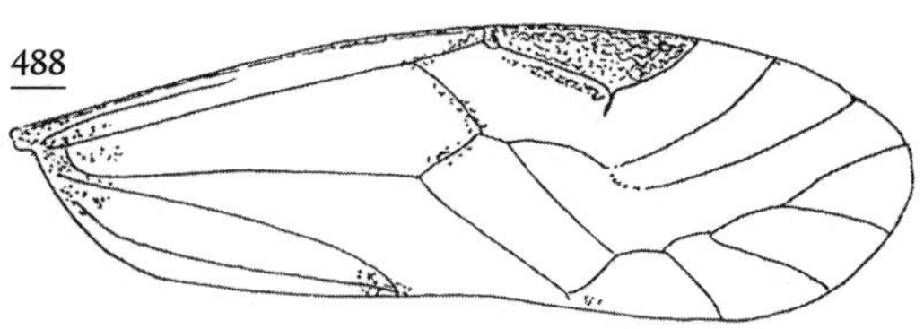

490
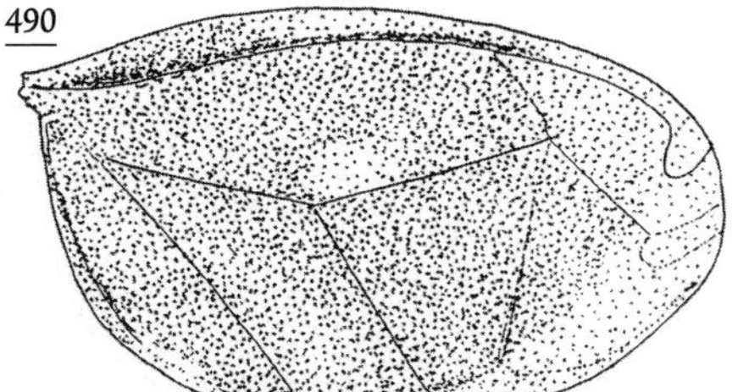

491
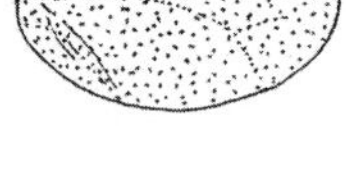

489
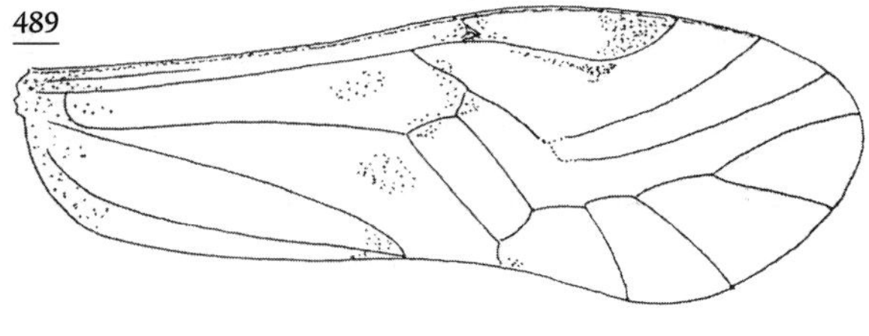

FIGURES 494 – 508 [FACING PAGE]
Psocidae. Fore wing venation of **494** *Javapsocus* – **495** *Kaindipsocus* – **496, 497** *Trichadenotecnum*, showing two basic patterns; gonapophyses of **498** *Javapsocus* – **499** *Trichadenotecnum* of *T. spiniserrulum*-group, short ventral valve – **500** *Trichadenotecnum*, of *T. medium*-group; subgenital plate of **501** *Kaindipsocus* – **502** *Javapsocus* – **503** *Trichadenotecnum*, of *T. spiniserrulum*-group – **504** *Trichadenotecnum*, of *T. medium*-group; phallosome of **505** *Javapsocus* – **506** *Kaindipsocus* – **507, 508** *Trichadenotecnum*, various forms.

494
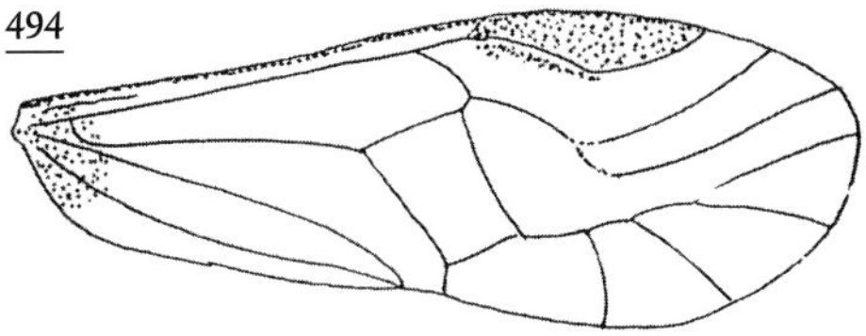

495
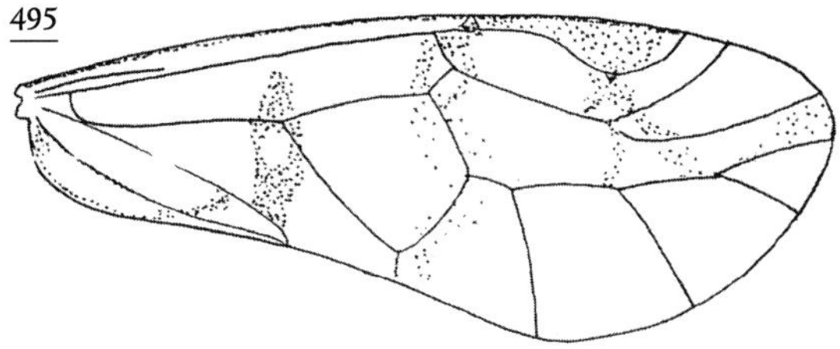

496
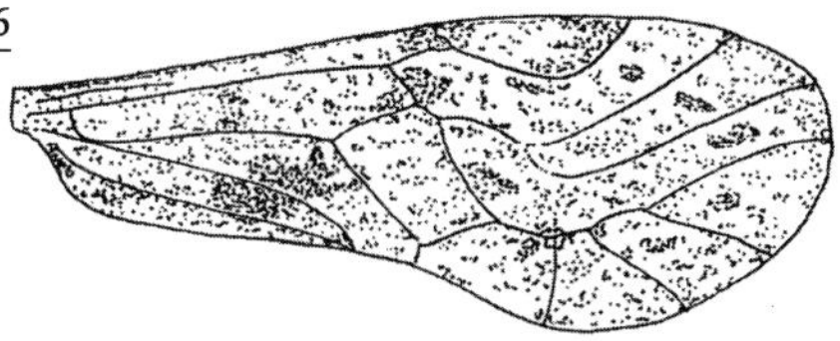

505
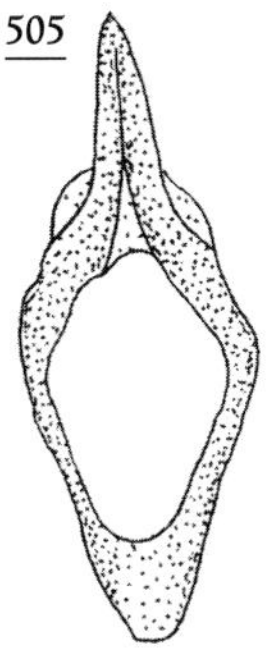

506
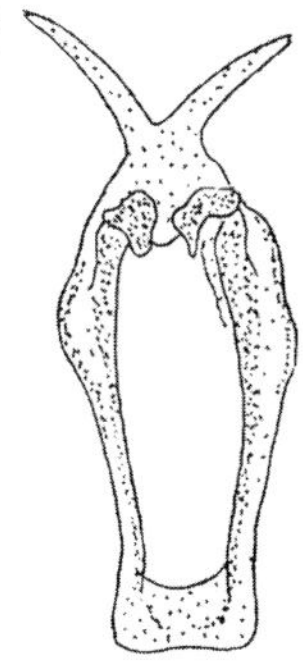

498
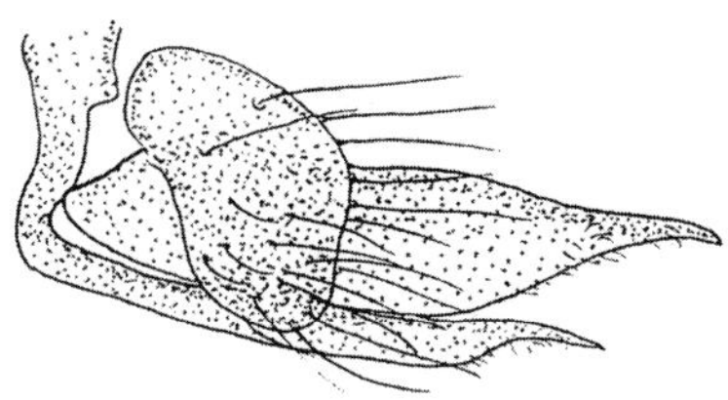

501
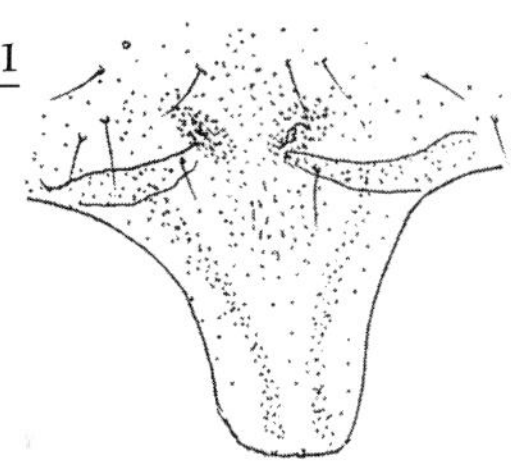

497
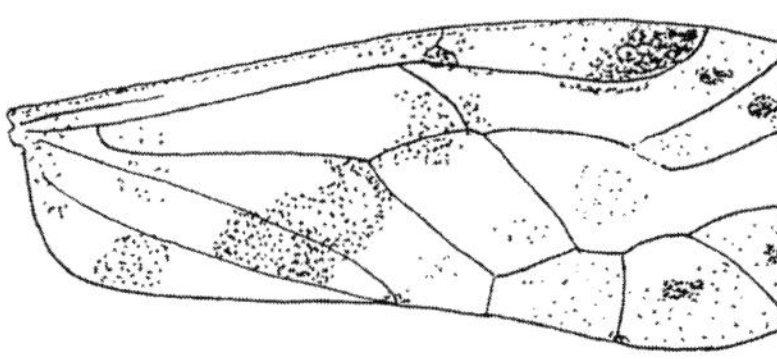

502
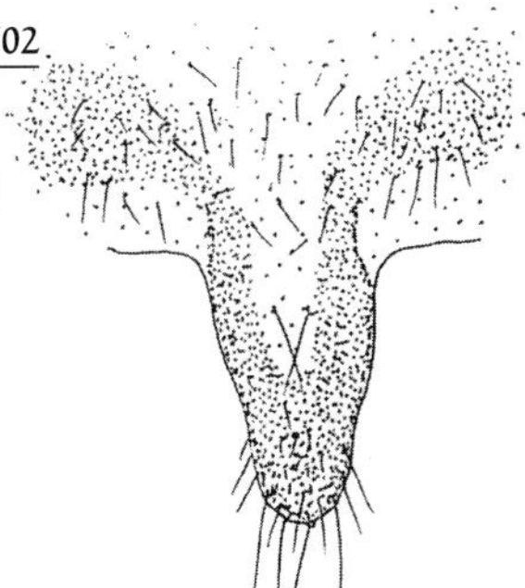

503
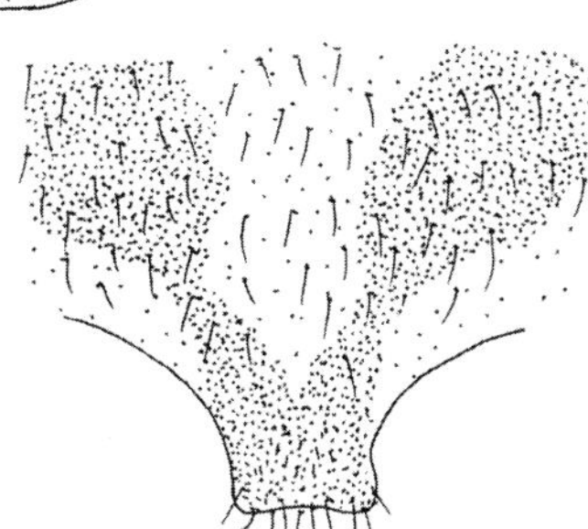

499
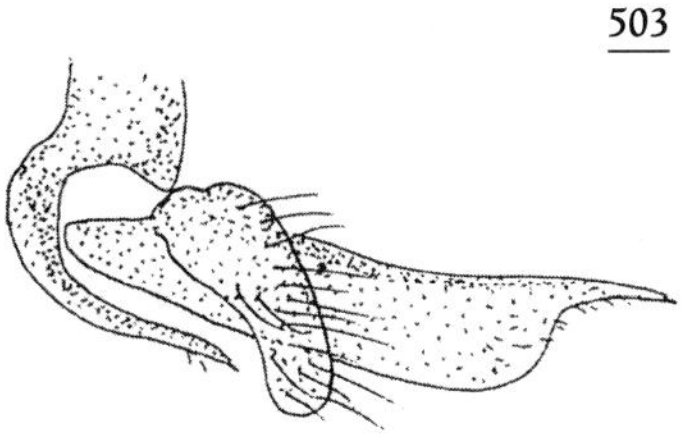

500
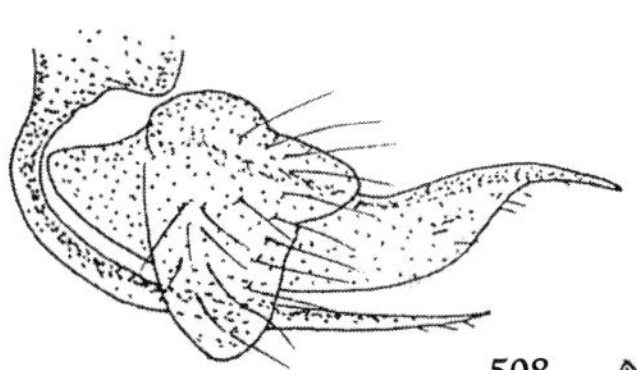

504
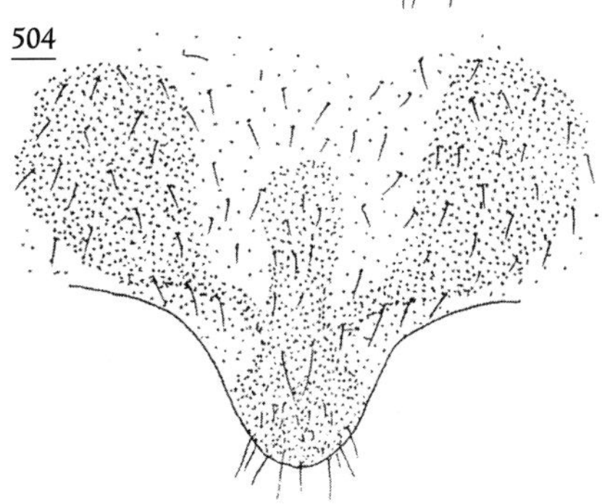

508
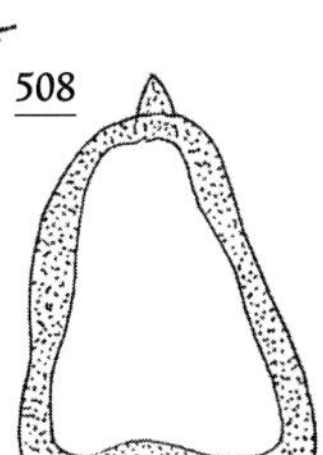

507
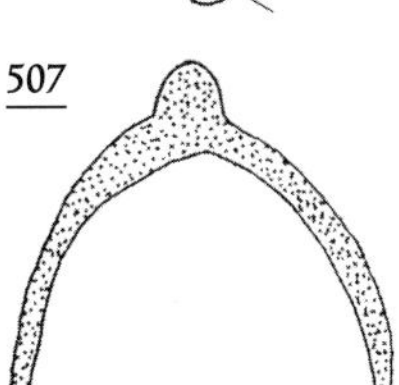

FIGURES 509 – 516

Psocidae. Representative hypandrium forms, postero-ventral aspect: **509** *Sigmatoneura* (simple, convex) – **510** *Neoblaste* – **511** *Kaindipsocus* – **512** *Ptycta* – **513** *Javapsocus* – **514** *Psocidus* s.str. – **515** *Psococerastis* – **516** *Trichadenotecnum*.

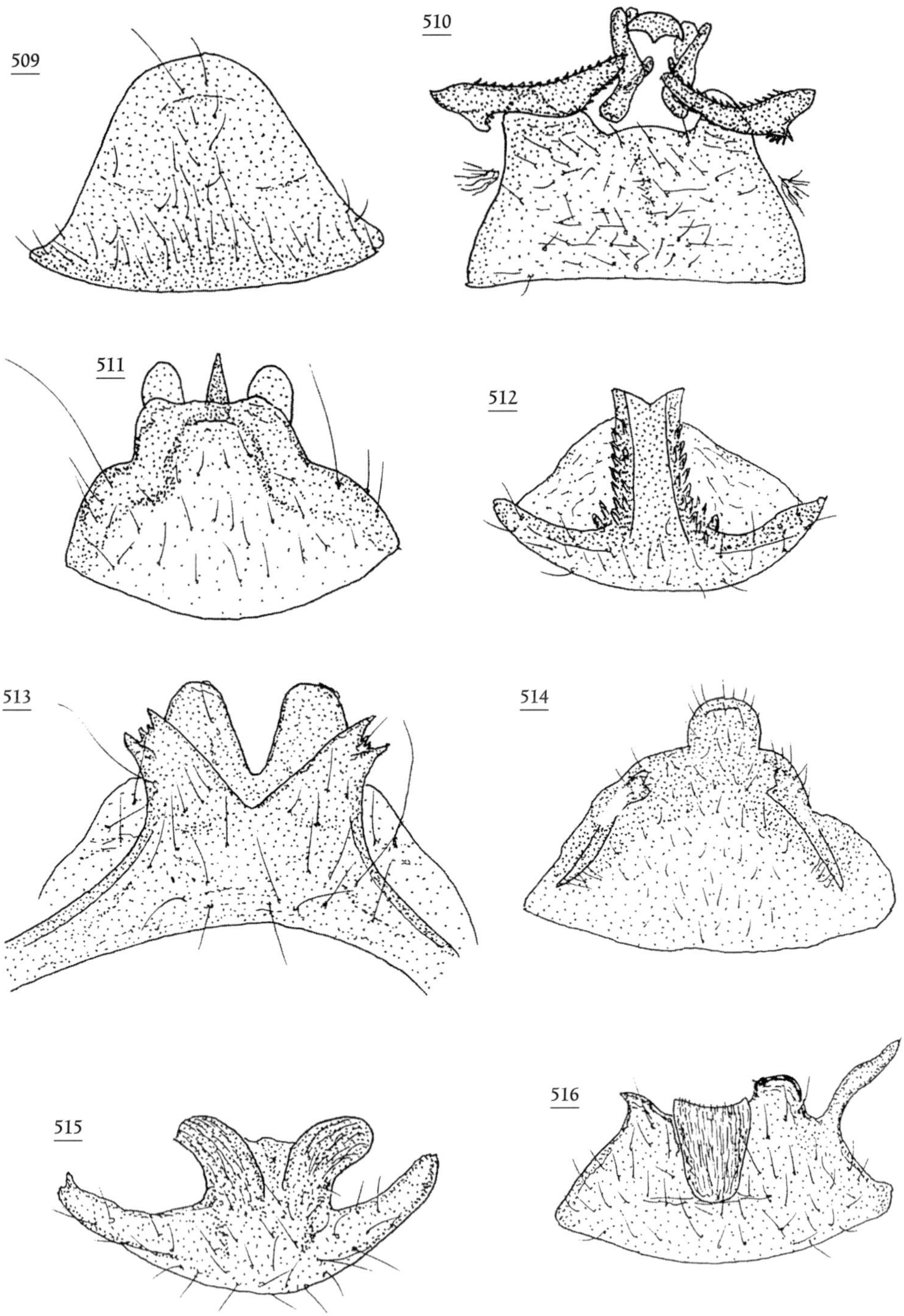

FIGURES 517 – 518

Psocidae. *Thyrsopsocopsis*: **517** fore wing and **518** hind wing (after Mockford, 2004).

517

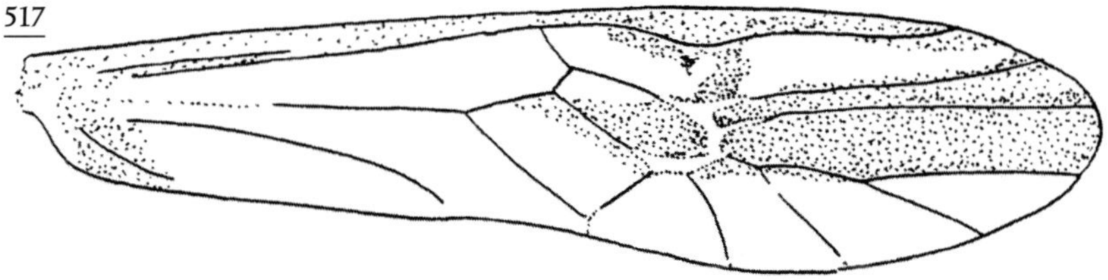

518

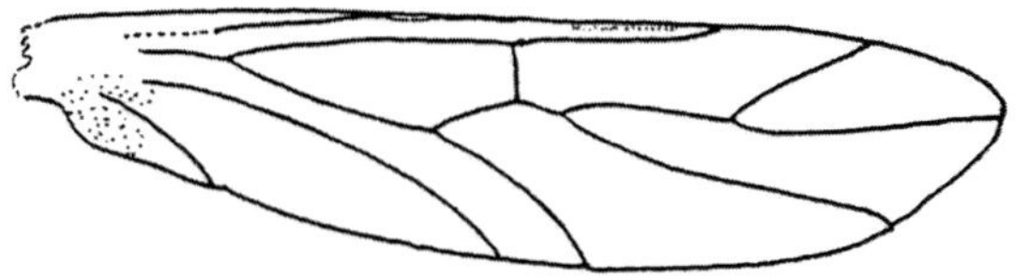

Chapter 7

Checklist

by C. Lienhard

This checklist is based on Lienhard & Smithers (2002) and Lienhard (2003a, 2004c, 2005, 2006). It contains all species for which published records exist up to the end of the year 2005. The family arrangement corresponds to chapter 5 (see also Table 2). Genera and species are listed alphabetically within each family and genus respectively. For synonymies see Lienhard & Smithers (2002). Countries are listed alphabetically. Only distribution in tropical South-east Asia is mentioned in this checklist, but for the species which have a wider distribution Lienhard & Smithers (2002) provide references to all published records for each species. Two Oriental representatives of Dolabellopsocidae and Elipsocidae, two families not yet recorded from tropical South-east Asia, are also mentioned in this list. For each species the bibliographical reference to primary taxonomic literature is cited, including the page number where the original description of the species begins. In the few cases where the valid name is a replacement name, only the primary reference to this nomenclatural act is cited; the latter automatically refers to the original description of the species. Published records of material merely identified down to genus level (e.g. *Perientomum* spec., *Mesopsocus* spec.) are only mentioned if the genus concerned is not represented in the region or in the particular country by at least one species level record. However, many recognized but still undescribed species have never been mentioned in the literature. These are ignored in this checklist. One of the consequences is that the distribution of some genera provides an incomplete picture.

List of countries from where psocids have been mentioned in the literature

- Brunei,
- Cambodia,
- Christmas Island (Australia),
- Indonesia (for Papua (Irian Jaya), see New Guinea),
- Laos,
- Malaysia,
- Myanmar (Burma),
- New Guinea (including Papua New Guinea, Papua (Irian Jaya), Admiralty Islands, Bismarck Archipelago and Bougainville),
- Philippines,
- Singapore,
- Thailand,
- Vietnam.

TROGIOMORPHA
ATROPETAE
LEPIDOPSOCIDAE

Cyptophania

marginata Thornton, Lee & Chui – Thornton et al., 1972: 63 ▸ Christmas Island

Echinopsocus

erinaceus Enderlein – Enderlein, 1903a: 332 ▸ New Guinea

Echmepteryx

frontalis Vaughan, Thornton & New – Vaughan et al., 1991: 105 ▸ Indonesia
lacinipennis Enderlein – Enderlein, 1926b: 68 ▸ Indonesia
lawrencei Smithers – Smithers, 1995d: 531 ▸ Christmas Island
lineata Vaughan, Thornton & New – Vaughan et al., 1991: 106 ▸ Indonesia
lunulata Thornton, Lee & Chui – Thornton et al., 1972: 64 ▸ Christmas Island, Indonesia
madagascariensis (Kolbe) – Kolbe, 1885a: 184 ▸ Indonesia
malayensis New – New, 1975d: 179 ▸ Indonesia, Singapore
pallida Smithers – Smithers, 1965a: 75 ▸ Christmas Island, Indonesia

Lepidopsocus

marmoratus (Banks) – Banks, 1931a: 439 ▸ Christmas Island, Indonesia
nepticulides Enderlein – Enderlein, 1903a: 330 ▸ Singapore
pretiosus (Banks) – Banks, 1942: 28 ▸ Christmas Island, Indonesia
pseudomaculatus Thornton – Thornton, 1981a: 28 ▸ Indonesia
sudarmani Vaughan, Thornton & New – Vaughan et al., 1989: 6 ▸ Indonesia

Nepticulomima

biroiana (Enderlein) – Enderlein, 1903a: 327 ▸ Indonesia, New Guinea
essigkeana Enderlein – Enderlein, 1906e: 97 ▸ Malaysia
hoesemanni (Enderlein) – Enderlein, 1903c: 3 ▸ Christmas Island
jacobsoni Enderlein – Enderlein, 1926b: 66 ▸ Indonesia
latisquama Enderlein – Enderlein, 1926b: 67 ▸ Indonesia
lusiae Thornton – Thornton, 1981a: 55 ▸ Christmas Island, Indonesia
orientalis New – New, 1975d: 177 ▸ Singapore
penicillata Enderlein – Enderlein, 1926b: 67 ▸ Indonesia
sakuntala Enderlein – Enderlein, 1906e: 96 ▸ Indonesia
sumatrensis Vaughan, Thornton & New – Vaughan et al., 1991: 108 ▸ Indonesia
uniformis Vaughan, Thornton & New – Vaughan et al., 1989: 11 ▸ Indonesia

Papuapsocus

caudatus Mockford – Mockford, 2005: 204 ▸ New Guinea

Parasoa

haploneura Thornton – Thornton, 1962b: 452 ▸ Christmas Island, Indonesia, Malaysia, Singapore

Perientomum

corticola Vaughan, Thornton & New – Vaughan et al., 1989: 8 ▸ Indonesia
[spec., see Deharveng & Leclerc, 1989 and Deharveng & Bedos, 2000] ▸ Thailand

Soa

dahliana Enderlein – Enderlein, 1904a: 110 ▸ New Guinea
flaviterminata Enderlein – Enderlein, 1906e: 79 ▸ Christmas Island, Indonesia

PSOQUILLIDAE

Eosilla

jacobsoni Ribaga – Ribaga, 1908a: 22 ▸ Indonesia

Psoquilla

marginepunctata Hagen – Hagen, 1865b: 123 ▸ Malaysia

Rhyopsocus

pandanicola Thornton, Lee & Chui – Thornton et al., 1972: 74 ▸ Christmas Island

TROGIIDAE

Cerobasis

guestfalica (Kolbe) – Kolbe, 1880c: 132 ▸ Indonesia

PSOCATHROPETAE

PRIONOGLARIDIDAE

Siamoglaris

zebrina Lienhard – Lienhard: 2004a: 866 ▸ Thailand

PSYLLIPSOCIDAE

Psocathropos

lachlani Ribaga – Ribaga, 1899: 157 ▸ Indonesia, New Guinea, Thailand

Psyllipsocus

batuensis Thornton – Thornton, 1962b: 442 ▸ Malaysia
hirsutus Thornton – Thornton, 1962b: 447 ▸ Malaysia

yongi New & Lee – New & Lee, 1992d: 391 ▸ Malaysia
[spec., see Deharveng & Leclerc, 1989] ▸ Thailand

TROCTOMORPHA
AMPHIENTOMETAE
AMPHIENTOMIDAE

Cymatopsocus
emarginatus New – New, 1975d: 187 ▸ Singapore
enderleini New – New, 1975d: 185 ▸ Singapore
enigmaticus New – New, 1975d: 189 ▸ Singapore
maculatus Lee – Lee, 1974: 110 ▸ Malaysia, Singapore
opalinus Enderlein – Enderlein, 1903a: 315 ▸ Malaysia, Singapore

Paramphientomum
malayense New – New, 1971d: 194 ▸ Singapore
nietneri Enderlein – Enderlein, 1906e: 63 ▸ Indonesia
sinuosum New – New, 1975d: 193 ▸ Singapore
striatum Thornton – Thornton, 1984b: 87 ▸ Indonesia
tristrigatum (Enderlein) – Enderlein, 1926b: 65 ▸ Indonesia

Pseudoseopsis
elegans New – New, 1973f: 227 ▸ Thailand

Seopsis
luzonica Banks – Banks, 1937a: 134 ▸ Philippines
tricolor Banks – Banks, 1937a: 133 ▸ Philippines
[spec., see Thornton et al., 1988] ▸ Indonesia

Stigmatopathus
horvathi Enderlein – Enderlein, 1903a: 313 ▸ Indonesia, Malaysia

Stimulopalpus
brunneus (New) – New, 1975d: 191 ▸ Christmas Island, Indonesia, Singapore
distinctus Smithers – Smithers, 1995d: 535 ▸ Christmas Island

Syllysis
samarangana Enderlein – Enderlein, 1926b: 64 ▸ Indonesia

Tineomorpha
jacobsoniana Enderlein – Enderlein, 1926b: 63 ▸ Indonesia
murphyi New – New, 1975d: 181 ▸ Singapore
timahensis New – New, 1975d: 182 ▸ Singapore

TROCTOPSOCIDAE

Coleotroctellus

burckhardti Lienhard – Lienhard, 1988b: 581 ▸ Thailand
loebli Lienhard – Lienhard, 1988b: 584 ▸ Thailand

Selenopsocus

schwendingeri Lienhard & Mockford – Lienhard & Mockford, 1997: 376 ▸ Malaysia

Thaipsocus

bau Lienhard & Mockford – Lienhard & Mockford, 1997: 371 ▸ Malaysia
borneensis Lienhard & Mockford – Lienhard & Mockford, 1997: 369 ▸ Malaysia
doisuthep Lienhard & Mockford – Lienhard & Mockford, 1997: 367 ▸ Thailand
orientalis (Lienhard) – Lienhard, 1990b: 340 ▸ Indonesia
sarawakensis Lienhard & Mockford – Lienhard & Mockford, 1997: 371 ▸ Malaysia
siamensis Lienhard & Mockford – Lienhard & Mockford, 1997: 363 ▸ Thailand
thamluang Lienhard & Mockford – Lienhard & Mockford, 1997: 367 ▸ Thailand

NANOPSOCETAE
LIPOSCELIDIDAE

Belaphopsocus

murphyi Lienhard – Lienhard, 1991: 80 ▸ Singapore

Belaphotroctes

ghesquierei Badonnel – Badonnel, 1949c: 20 ▸ Thailand

Embidopsocus

trichurensis Menon – Menon, 1942: 27 ▸ Philippines

Liposcelis

bostrychophila Badonnel – Badonnel, 1931a: 250 ▸ Indonesia, Singapore, Thailand, Vietnam
entomophila (Enderlein) – Enderlein, 1907d: 34 ▸ Indonesia, Philippines, Singapore, Thailand, Vietnam
paeta Pearman – Pearman, 1942: 289 ▸ Indonesia

PACHYTROCTIDAE

Nanopsocus

[spec., see Mockford, 1993a] ▸ Malaysia

Pachytroctes

dichromoscelis Badonnel – Badonnel, 1949a: 32 ▸ Christmas Island
robustus Vaughan, Thornton & New – Vaughan et al., 1991: 110 ▸ Indonesia
sinuosus Smithers – Smithers, 1995d: 537 ▸ Christmas Island
[spec., see Deharveng & Leclerc, 1989 and Deharveng & Bedos, 2000] ▸ Thailand

Psylloneura

simbangana Enderlein – Enderlein, 1903a: 318 ▸ New Guinea

Tapinella

baliensis Thornton – Thornton, 1984b: 89 ▸ Indonesia
clypeola Thornton – Thornton, 1984b: 89 ▸ Indonesia
curvatoides Smithers – Smithers, 1995d: 539 ▸ Christmas Island
nebulosa Vaughan, Thornton & New – Vaughan et al., 1991: 111 ▸ Indonesia
pictipenna Thornton, Lee & Chui – Thornton et al., 1972: 81 ▸ Indonesia, Philippines
spinosa Thornton – Thornton, 1984b: 91 ▸ Indonesia

PSOCOMORPHA
EPIPSOCETAE
EPIPSOCIDAE

Dicropsocus

complexus Smithers & Thornton – Smithers & Thornton, 1977: 72 ▸ New Guinea
montanus Smithers & Thornton – Smithers & Thornton, 1977: 67 ▸ New Guinea
rugosus Smithers & Thornton – Smithers & Thornton, 1977: 72 ▸ New Guinea

Epipsocopsis

apicalis New & Thornton – New & Thornton, 1976: 381 ▸ Malaysia
basalis New & Thornton – New & Thornton, 1976: 389 ▸ Malaysia
cameronensis New & Lee – New & Lee, 1991c: 121 ▸ Malaysia
completa (Banks) – Banks, 1916: 200 ▸ Philippines
costalis (Banks) – Banks, 1914a: 352 ▸ Indonesia, Malaysia
dubia (Karny) – Karny, 1925: 67 ▸ Malaysia
fasciata Smithers & Thornton – Smithers & Thornton, 1977: 63 ▸ New Guinea
fumipennis (Banks) – Banks, 1920: 310 ▸ Malaysia, Philippines
hyalina (Banks) – Banks, 1920: 310 ▸ Singapore
longiceps (Enderlein) – Enderlein, 1926b: 57 ▸ Indonesia
maclurei New & Thornton – New & Thornton, 1976: 384 ▸ Malaysia
macrostigma Smithers & Thornton – Smithers & Thornton, 1977: 65 ▸ New Guinea
maculata New & Thornton – New & Thornton, 1976: 385 ▸ Malaysia
magna (New & Thornton) – New & Thornton, 1976: 377 ▸ Malaysia

mouldsi Smithers – Smithers, 1976a: 32 ▸ New Guinea
murca (Enderlein) – Enderlein, 1903a: 257 ▸ Malaysia
murphyi New & Thornton – New & Thornton, 1976: 385 ▸ Singapore
nebulifera Smithers & Thornton – Smithers & Thornton, 1977: 63 ▸ New Guinea
nubilipennis (Karny) – Karny, 1925: 66 ▸ Malaysia
paraselena Vaughan, Thornton & New – Vaughan et al., 1989: 21 ▸ Indonesia
prominens (Banks) – Banks, 1937a: 132 ▸ Philippines
punctata Smithers & Thornton – Smithers & Thornton, 1977: 61 ▸ New Guinea
sclerota New & Thornton – New & Thornton, 1976: 389 ▸ Malaysia, Singapore
selena New & Thornton – New & Thornton, 1976: 383 ▸ Malaysia
setosa New & Thornton – New & Thornton, 1976: 391 ▸ Malaysia
singaporensis New & Thornton – New & Thornton, 1976: 387 ▸ Malaysia, Singapore
thailandensis New – New, 1973f: 229 ▸ Thailand

Epipsocus

marginatus Enderlein – Enderlein, 1903a: 255 ▸ New Guinea
pinnatus (Enderlein) – Enderlein, 1926b: 58 ▸ Indonesia
viiv (Enderlein) – Enderlein, 1926b: 57 ▸ Indonesia

Heteroepipsocus

inornatus (Banks) – Banks, 1916: 199 ▸ Indonesia, Philippines

[DOLABELLOPSOCIDAE]

Auroropsocus

orientis Eertmoed – Eertmoed, 1973: 388 ▸ China, India

CAECILIUSETAE
CAECILIUSIDAE

Species names in *Caecilius* described from South-east Asia, regarded as *incertae sedis* by Mockford (2000) and assigned by him to *Valenzuela* pending more detailed appraisal, are marked with an asterisk (*).

Asiocaecilius

narmadus (Thornton) – Thornton, 1984b: 101 ▸ Christmas Island, Indonesia
okanus (Thornton) – Thornton, 1984b: 102 ▸ Indonesia
singaporensis (Enderlein) – Enderlein, 1903a: 281 ▸ Singapore

Caecilius

albonigrus Thornton – Thornton 1984b: 95 ▸ Indonesia
fuscopterus (Latreille) – Latreille, 1799: 10 ▸ Vietnam

hornei Vaughan, Thornton & New – Vaughan et al., 1989: 31 ▸ Indonesia
marcidus Banks – Banks, 1937a: 129 ▸ Philippines
parvulus (Banks) – Banks, 1920: 313 ▸ Indonesia, Malaysia, Singapore
plagosus Banks – Banks, 1937a: 131 ▸ Philippines
reductus Banks – Banks, 1920: 310 ▸ Philippines

Coryphosmila
bushi (Vaughan, Thornton & New) – Vaughan et al., 1989: 33 ▸ Indonesia
dolabrata (Hagen) – Hagen, 1858a: 475 ▸ Indonesia, Singapore
jacobsoni (Enderlein) – Enderlein, 1926b: 55 ▸ Indonesia
soehardjani (Smithers) – Smithers, 1965c: 136 ▸ Indonesia

Dypsocus
apicatus Banks – Banks, 1916: 203 ▸ Philippines
coleoptratus (Hagen) – Hagen, 1858a: 474 ▸ Indonesia
corporaali Navás – Navás, 1924c: 139 ▸ Indonesia

Enderleinella
calta (Thornton) – Thornton, 1984b: 97 ▸ Indonesia
unicolor (Enderlein) – Enderlein, 1903a: 271 ▸ Indonesia, Singapore

Isophanopsis
[spec., see Mockford, 2000: 344] ▸ Indonesia

Maoripsocopsis
sumatrensis Mockford – Mockford, 2000: 346 ▸ Indonesia

Paracaecilius
cardaleae Smithers – Smithers, 1995d: 541 ▸ Christmas Island
hallae Vaughan, Thornton & New – Vaughan et al., 1989: 43 ▸ Indonesia
macrops (Enderlein) – Enderlein, 1903a: 273 ▸ Indonesia, Singapore
novoguineensis (Enderlein) – Enderlein, 1903a: 276 ▸ New Guinea
parviareola (Enderlein) – Enderlein, 1926b: 59 ▸ Indonesia
pygmaeus (Enderlein) – Enderlein, 1903a: 276 ▸ New Guinea

Pericaecilius
vaughani Mockford – Mockford, 2000: 339, 340 ▸ Indonesia
werneri Mockford – Mockford, 2000: 339 ▸ Philippines

Protodypsocus
fissiceps Enderlein – Enderlein, 1903a: 251 ▸ New Guinea

Stenocaecilius

analis (Banks) – Banks, 1931a: 437 ▸ Malaysia
casarum (Badonnel) – Badonnel, 1931a: 234 ▸ Indonesia, New Guinea

Thorntoniella

lalea (Thornton) – Thornton, 1984b: 99 ▸ Indonesia

Valenzuela

albipes (Thornton) – Thornton, 1984b: 93 ▸ Indonesia
**alticola* (Vaughan, Thornton & New) – Vaughan et al., 1991: 115 ▸ Indonesia
**andinus* (Vaughan, Thornton & New) – Vaughan et al., 1989: 23 ▸ Indonesia
**annulicornis* (Enderlein) – Enderlein, 1903a: 270 ▸ New Guinea
aridus (Hagen) – Hagen, 1858a: 474 ▸ Philippines
badiostigma (Okamoto) – Okamoto, 1910: 206 ▸ Indonesia
baliensis (Thornton) – Thornton, 1984b: 95 ▸ Indonesia
bambusae (Soehardjan & Hamann) – Soehardjan & Hamann, 1959: 6 ▸ Indonesia
bataviensis (Enderlein) – Enderlein, 1926b: 60 ▸ Indonesia
**borneensis* (Karny) – Karny, 1925: 69 ▸ Malaysia
**castellus* (Banks) – Banks, 1916: 202 ▸ Philippines
**cincticornis* (Banks) – Banks, 1920: 311 ▸ Philippines
**clarivenus* (Vaughan, Thornton & New) – Vaughan et al., 1991: 116 ▸ Indonesia
**conspicuus* (Banks) – Banks, 1937a: 129 ▸ Philippines
deceptor (Banks) – Banks, 1920: 311 ▸ Philippines
**farrelli* (Vaughan, Thornton & New) – Vaughan et al., 1989: 35 ▸ Indonesia
**hivesi* (Vaughan, Thornton & New) – Vaughan et al., 1989: 25 ▸ Indonesia
**javanus* (Enderlein) – Enderlein, 1907a: 116 ▸ Indonesia
lemniscellus (Enderlein) – Enderlein, 1907a: 118 ▸ Indonesia
**luridus* (Enderlein) – Enderlein, 1903a: 274 ▸ New Guinea
**maculistigma* (Enderlein) – Enderlein, 1903a: 265 ▸ Indonesia
mclareni (Vaughan, Thornton & New) – Vaughan et al., 1989: 40 ▸ Indonesia
medialunatus (Vaughan, Thornton & New) – Vaughan et al., 1991: 120 ▸ Indonesia
**melanocnemis* (Enderlein) – Enderlein, 1907a: 119 ▸ Indonesia
**mjoebergi* (Karny) – Karny, 1925: 70 ▸ Malaysia
mueggenburgi (Enderlein) – Enderlein, 1903a: 269 ▸ Indonesia, Philippines, Singapore
**nebulomaculatus* Mockford – Mockford, 2000: 382 ▸ Indonesia
**pelmus* (New & Thornton) – New & Thornton, 1976: 391 ▸ Indonesia, Malaysia
**rubinii* (Vaughan, Thornton & New) – Vaughan et al., 1991: 118 ▸ Indonesia
**rubiventer* (Thornton) – Thornton, 1984b: 104 ▸ Indonesia
serpentinus (Enderlein) – Enderlein, 1926b: 55 ▸ Indonesia
**singularis* (Smithers) – Smithers, 1995d: 543 ▸ Christmas Island
traceus (Thornton) – Thornton, 1984b: 107 ▸ Indonesia

**trigonostigma* (Enderlein) – Enderlein, 1907a: 117 ▸ Indonesia
velectus (Thornton) – Thornton, 1984b: 108 ▸ Indonesia
**vittidorsum* (Enderlein) – Enderlein, 1907a: 119 ▸ Indonesia

AMPHIPSOCIDAE

Amphipsocus

connexus Banks – Banks, 1916: 200 ▸ Philippines
heterothrix Thornton & Wong – Thornton & Wong, 1966: 13 ▸ Indonesia
javensis Vaughan, Thornton & New – Vaughan et al., 1989: 47 ▸ Indonesia
minor Thornton – Thornton, 1984b: 113 ▸ Indonesia
pilosus McLachlan – McLachlan, 1872b: 77 ▸ Indonesia, Malaysia
unitoides New & Thornton – New & Thornton, 1976: 395 ▸ Singapore
unitus Banks – Banks, 1916: 201 ▸ Philippines

Calocaecilius

apertus Thornton – Thornton, 1984b: 114 ▸ Indonesia
decipiens Mockford – Mockford, 1974d: 156 ▸ Malaysia, Philippines

Dasypsocus

solox (Enderlein) – Enderlein, 1903a: 278 ▸ Singapore

Kodamaius

ikomai (Kuwayama) – Kuwayama, 1961: 203 ▸ Thailand

Kolbia

bakeri Banks – Banks, 1916: 201 ▸ Philippines
maculipennis Karny – Karny, 1925: 71 ▸ Malaysia

Taeniostigma

baiyerense Smithers & Thornton – Smithers & Thornton, 1975b: 156 ▸ New Guinea
bimaculatum Banks – Banks, 1913c: 171 ▸ Philippines
elongatum (Hagen) – Hagen, 1858a: 474 ▸ Indonesia, Malaysia, New Guinea, Singapore, Thailand
ingens Enderlein – Enderlein, 1903a: 238 ▸ Singapore, Vietnam
irianum Smithers & Thornton – Smithers & Thornton, 1975b: 163 ▸ New Guinea
malayanum (McLachlan) – McLachlan, 1872b: 75 ▸ Indonesia
nigricorne Smithers & Thornton – Smithers & Thornton, 1975c: 159 ▸ New Guinea
perkinsi Banks – Banks, 1918: 5 ▸ New Guinea

phaeostigma Smithers & Thornton – Smithers & Thornton, 1975b: 162 ▸ New Guinea
sibilense Smithers & Thornton – Smithers & Thornton, 1975b: 160 ▸ New Guinea
tibiale Navás – Navás, 1924g: 215 ▸ Vietnam

Tagalopsocus
hyalinus Banks – Banks, 1931b: 58 ▸ Philippines
luzonensis Banks – Banks, 1916: 201 ▸ Philippines

STENOPSOCIDAE

Graphopsocus
infirmus Banks – Banks, 1920: 308 ▸ Philippines
subaequalis Banks – Banks, 1920: 308 ▸ Singapore

Malostenopsocus
cubitalis (Thornton) – Thornton, 1984b: 109 ▸ Indonesia

Stenopsocus
adisoemartoi Vaughan, Thornton & New – Vaughan et al., 1989: 44 ▸ Indonesia
albipileus Smithers – Smithers, 1974d: 21 ▸ Indonesia, New Guinea
dissimilis Banks – Banks, 1937a: 127 ▸ Philippines
jocosus Banks – Banks, 1939a: 135 ▸ Malaysia, Philippines
pilosus Enderlein – Enderlein, 1926b: 54 ▸ Indonesia
rubellus Thornton – Thornton, 1984b: 110 ▸ Indonesia
tonkinensis Enderlein – Enderlein, 1903a: 240 ▸ Malaysia, Vietnam
uniformis (Hagen) – Hagen, 1859: 200 ▸ Indonesia, Philippines

HOMILOPSOCIDEA
ARCHIPSOCIDAE

Archipsocopsis
fernandi (Pearman) – Pearman, 1934a: 112 ▸ Indonesia

Archipsocus
alternatus Vaughan, Thornton & New – Vaughan et al., 1989: 75 ▸ Indonesia
recens Enderlein – Enderlein, 1903a: 286 ▸ Indonesia, Singapore
sanurensis Thornton – Thornton, 1984b: 137 ▸ Indonesia
[spec., see New & Thornton, 1976] ▸ Malaysia

CALOPSOCIDAE

Callistoptera

anna Enderlein – Enderlein, 1903a: 248 ▸ New Guinea

Calopsocus

apicatus Thornton & Smithers – Thornton & Smithers, 1984: 190 ▸ New Guinea
furcatus (New) – New, 1978e: 313 ▸ Malaysia
gorokensis Thornton & Smithers – Thornton & Smithers, 1984: 188 ▸ New Guinea
guttatus Smithers – Smithers, 1967b: 61 ▸ New Guinea, Philippines
infelix (Hagen) – Hagen, 1858a: 475 ▸ Indonesia, Malaysia, New Guinea
iridescens Banks – Banks, 1920: 314 ▸ Malaysia, Thailand
javanus (Navás) – Navás, 1924b: 138 ▸ Indonesia, Malaysia
koridensis New – New, 1978e: 313 ▸ New Guinea
maculatus Thornton & Smithers – Thornton & Smithers, 1984: 201 ▸ New Guinea
marginatus Thornton & Smithers – Thornton & Smithers, 1984: 192 ▸ New Guinea
ovatus Thornton & Smithers – Thornton & Smithers, 1984: 209 ▸ Philippines
pallistigma Thornton & Smithers – Thornton & Smithers, 1984: 207 ▸ Philippines
reticulatus Thornton & Smithers – Thornton & Smithers, 1984: 211 ▸ New Guinea
reticuloides Thornton & Smithers – Thornton & Smithers, 1984: 213 ▸ New Guinea
rizali Banks – Banks, 1916: 198 ▸ Philippines
solomonensis Thornton & Smithers – Thornton & Smithers, 1984: 199 ▸ New Guinea
yatesi Smithers – Smithers, 1995d: 546 ▸ Christmas Island

Calosema

browningae Thornton & Smithers – Thornton & Smithers, 1984: 217 ▸ New Guinea

Cyclopsocus

gressitti Thornton & Smithers – Thornton & Smithers, 1984: 234 ▸ New Guinea
hyalinus Thornton & Smithers – Thornton & Smithers, 1984: 228 ▸ New Guinea
hyaloides Thornton & Smithers – Thornton & Smithers, 1984: 231 ▸ New Guinea
navasi (New) – New, 1978e: 315 ▸ New Guinea
straatmani Thornton & Smithers – Thornton & Smithers, 1984: 231 ▸ New Guinea

Dendropsocus

swartensis Thornton & Smithers – Thornton & Smithers, 1984: 234 ▸ New Guinea

Nemupsocus

lobatus New – New, 1978e: 317 ▸ New Guinea

Neurosema

alobum Thornton & Smithers – Thornton & Smithers, 1984: 226 ▸ New Guinea
apicale McLachlan – McLachlan, 1866b: 347 ▸ New Guinea
dimorphum Thornton & Smithers – Thornton & Smithers, 1984: 221 ▸ New Guinea
pulchroides Thornton & Smithers – Thornton & Smithers, 1984: 225 ▸ New Guinea
pulchrum New – New, 1978b: 53 ▸ New Guinea
[spec., see Thornton & Smithers, 1984: 219] ▸ Philippines

Torrepsocus

novoguinensis Thornton & Smithers – Thornton & Smithers, 1984: 236 ▸ New Guinea
pulleni (Smithers) – Smithers, 1981d: 24 ▸ New Guinea

ECTOPSOCIDAE

Ectopsocopsis

cognata Thornton & Wong – Thornton & Wong, 1968: 44 ▸ Malaysia
decorata (Thornton & Wong) – Thornton & Wong, 1968: 18 ▸ Indonesia

Ectopsocus

adelphos Thornton & Wong – Thornton & Wong, 1968: 81 ▸ New Guinea
aethiops (Hagen) – Hagen, 1859: 204 ▸ Philippines
amblyura Thornton & Wong – Thornton & Wong, 1968: 63 ▸ Philippines
amphithrix Thornton & Wong – Thornton & Wong, 1968: 47 ▸ Malaysia
argus Thornton & Wong – Thornton & Wong, 1968: 64 ▸ Philippines
baliosus Thornton & Wong – Thornton & Wong, 1968: 47 ▸ Christmas Island, Indonesia, Malaysia
barlowi New & Lee – New & Lee, 1992c: 171 ▸ Malaysia
basalis (Banks) – Banks, 1937a: 133 ▸ Malaysia, Philippines
bifurcatus Vaughan, Thornton & New – Vaughan et al., 1991: 125 ▸ Indonesia
briggsi McLachlan – McLachlan, 1899d: 277 ▸ Indonesia
brinchangensis New & Lee – New & Lee, 1992c: 174 ▸ Malaysia
cinctus Thornton – Thornton, 1962c: 305 ▸ Indonesia, Malaysia, Vietnam
cirratus Thornton & Wong – Thornton & Wong, 1968: 50 ▸ Malaysia

cluniatus Vaughan, Thornton & New – Vaughan et al., 1989: 53 ▸ Indonesia
coccophilus Ball – Ball, 1943: 22 ▸ Indonesia
comitus Thornton & Wong – Thornton & Wong, 1968: 30 ▸ Indonesia
crinitus Thornton & Wong – Thornton & Wong, 1968: 50 ▸ Malaysia
cristatus Thornton & Wong – Thornton & Wong, 1968: 81 ▸ New Guinea
denervus Thornton & Wong – Thornton & Wong, 1968: 95 ▸ Christmas Island, Philippines
denotatus Thornton & Wong – Thornton & Wong, 1968: 57 ▸ Vietnam
dicroglossus Thornton & Wong – Thornton & Wong, 1968: 85 ▸ New Guinea
drepanus Thornton – Thornton, 1984b: 117 ▸ Indonesia
erosus (Enderlein) – Enderlein, 1903a: 297 ▸ New Guinea
ewarti Vaughan, Thornton & New – Vaughan et al., 1989: 51 ▸ Indonesia
fumidus Thornton & Wong – Thornton & Wong, 1968: 68 ▸ Philippines
furcatus Thornton & Wong – Thornton & Wong, 1968: 115 ▸ Malaysia
gentingensis New & Lee – New & Lee, 1992c: 172 ▸ Malaysia
heurni (Navás) – Navás, 1924c: 140 ▸ Indonesia
hirsutus Thornton – Thornton, 1962c: 303 ▸ Indonesia
hypandrus Thornton – Thornton, 1984b: 118 ▸ Indonesia, Malaysia
innotatus Thornton & Wong – Thornton & Wong, 1968: 54 ▸ Malaysia
intersitus Thornton & Wong – Thornton & Wong, 1968: 68 ▸ Philippines
kepongensis New & Lee – New & Lee, 1992c: 173 ▸ Malaysia
maindroni Badonnel – Badonnel, 1935d: 81 ▸ Indonesia, Malaysia
meridionalis Ribaga – Ribaga, 1904: 296 ▸ Malaysia
myrmecophilus (Enderlein) – Enderlein, 1903a: 298 ▸ New Guinea
nidicolus Thornton & Wong – Thornton & Wong, 1968: 85 ▸ New Guinea
ornatus Thornton – Thornton, 1962c: 308 ▸ Christmas Island, Indonesia
paraplesius Thornton & Wong – Thornton & Wong, 1968: 106 ▸ Indonesia
pilosus Badonnel – Badonnel, 1967a: 162 ▸ Cambodia, Malaysia
predanus Vaughan, Thornton & New – Vaughan et al., 1989: 49 ▸ Indonesia, Malaysia
proctus New & Lee – New & Lee, 1992c: 171 ▸ Malaysia
ridderi Vaughan, Thornton & New – Vaughan et al., 1989: 55 ▸ Indonesia
salpinx Thornton & Wong – Thornton & Wong, 1968: 70 ▸ Indonesia, Malaysia, Philippines, Singapore
speciosus Thornton & Wong – Thornton & Wong, 1968: 87 ▸ New Guinea, Philippines
stictus Thornton & Wong – Thornton & Wong, 1968: 73 ▸ Malaysia, Philippines
sumatrensis Vaughan, Thornton, & New – Vaughan et al., 1991: 127 ▸ Indonesia
tenellus Thornton & Wong – Thornton & Wong, 1968: 54 ▸ Malaysia
thoi New & Lee – New & Lee, 1992c: 172 ▸ Malaysia
thysanus Thornton & Wong – Thornton & Wong, 1968: 110 ▸ Indonesia
tinctus Navás – Navás, 1924c: 140 ▸ Indonesia
titschacki Jentsch – Jentsch, 1939b: 120 ▸ Indonesia, Philippines

triangulus Thornton & Wong – Thornton & Wong, 1968: 89 ▸ Malaysia, New Guinea
vacuus New & Lee – New & Lee, 1992c: 166 ▸ Malaysia
vannus Thornton & Wong – Thornton & Wong, 1968: 55 ▸ Malaysia
venosus Lienhard – Lienhard, 2002: 391 ▸ Malaysia
vilhenai Badonnel – Badonnel, 1955: 189 ▸ Christmas Island
waterstradti (Enderlein) – Enderlein, 1901: 547 ▸ Indonesia, Malaysia, New Guinea
yongi New & Lee – New & Lee, 1992c: 167 ▸ Malaysia

[ELIPSOCIDAE]

Elipsocus

oligotrichus Thornton – Thornton, 1959b: 345 ▸ Hong Kong

LACHESILLIDAE

Lachesilla

brinchangensis New & Lee – New & Lee, 1991d: 128 ▸ Malaysia
hirsuta Lienhard – Lienhard, 2002: 388 ▸ Thailand
pedicularia (Linnaeus) – Linnaeus, 1758: 551 ▸ Malaysia

MESOPSOCIDAE

Mesopsocus

[spec., see New & Thornton, 1976] ▸ Singapore

PERIPSOCIDAE

Diplopsocus

anoplus (Thornton & Wong) – Thornton & Wong, 1968: 33 ▸ Malaysia
circularis (New & Thornton) – New & Thornton, 1976: 405 ▸ Singapore
fasciatus (Thornton) – Thornton, 1959a: 45 ▸ Indonesia, Malaysia
selene (Thornton & Wong) – Thornton & Wong, 1968: 40 ▸ Malaysia, Philippines

Peripsocus

anura Vaughan, Thornton & New – Vaughan et al., 1991: 129 ▸ Christmas Island, Indonesia
apicatus Vaughan, Thornton & New – Vaughan et al., 1989: 62 ▸ Indonesia
barunus Vaughan, Thornton & New – Vaughan et al., 1991: 130 ▸ Indonesia
bifidus Thornton – Thornton, 1984b: 121 ▸ Christmas Island, Indonesia, Malaysia

brachyura Thornton & Wong – Thornton & Wong, 1968: 59 ▸ Indonesia, Philippines
brinchangensis New & Lee – New & Lee, 1991e: 189 ▸ Malaysia
circinus Thornton & Wong – Thornton & Wong, 1968: 34 ▸ Indonesia, Malaysia
constrictus Thornton & Wong – Thornton & Wong, 1968: 35 ▸ Malaysia
crassicosta New & Lee – New & Lee, 1991e: 185 ▸ Malaysia
crenulatus Thornton & Wong – Thornton & Wong, 1968: 76 ▸ New Guinea
denticulatus Thornton & Wong – Thornton & Wong, 1968: 79 ▸ New Guinea
elongatus New & Thornton – New & Thornton, 1976: 405 ▸ Indonesia, Malaysia
ferrugineus Thornton & Wong – Thornton & Wong, 1968: 91 ▸ Indonesia
frimensis New & Lee – New & Lee, 1991e: 187 ▸ Malaysia
fulvescens Navás – Navás, 1920c: 93 ▸ Vietnam
hongkongensis Thornton & Wong – Thornton & Wong, 1968: 24 ▸ Indonesia
ignis Okamoto – Okamoto, 1910: 189 ▸ Malaysia
limi New & Thornton – New & Thornton, 1976: 407 ▸ Malaysia
nanus Navás – Navás, 1922d: 60 ▸ Vietnam
oculatus Enderlein – Enderlein, 1926b: 62 ▸ Indonesia
orientalis Garcia Aldrete – Garcia Aldrete, 1999b: 243 ▸ Malaysia
papillatus Vaughan, Thornton & New – Vaughan et al., 1989: 67 ▸ Indonesia
parareicherti Vaughan, Thornton & New – Vaughan et al., 1989: 60 ▸ Indonesia
paraspinosus New & Lee – New & Lee, 1991e: 187 ▸ Malaysia
pauliani Badonnel – Badonnel, 1949a: 42 ▸ Malaysia, Philippines
quercicola Enderlein – Enderlein, 1906c: 316 ▸ Malaysia, Singapore
reicherti Enderlein – Enderlein, 1903a: 290 ▸ Indonesia, Malaysia, Singapore
similis Enderlein – Enderlein, 1903a: 290 ▸ Indonesia, Singapore
stigmatus Thornton & Wong – Thornton & Wong, 1968: 40 ▸ Malaysia
suffitus Enderlein – Enderlein, 1903a: 293 ▸ New Guinea
taipingensis New & Lee – New & Lee, 1991e: 191 ▸ Malaysia
tinctus Vaughan, Thornton, & New – Vaughan et al., 1989: 64 ▸ Indonesia
uniformis New & Thornton – New & Thornton, 1976: 408 ▸ Malaysia
valvulus Thornton & Wong – Thornton & Wong, 1968: 42 ▸ Malaysia
variatus Soehardjan & Hamann – Soehardjan & Hamann, 1959: 4 ▸ Indonesia
yongi New & Lee – New & Lee, 1991e: 186 ▸ Malaysia

Periterminalis

hiatus (Thornton & Wong) – Thornton & Wong, 1968: 38 ▸ Malaysia

Turriperipsocus

chiangensis (New) – New, 1973f: 230 ▸ Thailand

PHILOTARSIDAE

Aaroniella

antennata Thornton & Smithers – Thornton & Smithers, 1977: 423 ▸ New Guinea
bakeri Thornton – Thornton, 1981g: 429 ▸ Philippines
gressitti Thornton, Lee & Chui – Thornton et al., 1972: 92 ▸ New Guinea
guttulata (Banks) – Banks, 1916: 202 ▸ Philippines
kepongensis New & Lee – New & Lee, 1992b: 159 ▸ Malaysia
lobata Vaughan, Thornton & New – Vaughan et al., 1989: 80 ▸ Christmas Island, Indonesia
lombokensis Thornton – Thornton, 1981g: 429 ▸ Indonesia
nebulosa Vaughan, Thornton & New – Vaughan et al., 1991: 147 ▸ Indonesia
pedunculata Thornton & Smithers – Thornton & Smithers, 1977: 427 ▸ New Guinea
spenceri Thornton – Thornton, 1981g: 428 ▸ Vietnam
sudarmani Vaughan, Thornton & New – Vaughan et al., 1991: 146 ▸ Indonesia

Haplophallus

comptoni Vaughan, Thornton & New – Vaughan et al., 1989: 78 ▸ Indonesia
cubitalis New & Thornton – New & Thornton, 1976: 403 ▸ Malaysia, Singapore
oblongatus Thornton & Smithers – Thornton & Smithers, 1977: 434 ▸ New Guinea
orientalis Thornton – Thornton, 1959b: 336 ▸ Christmas Island, Indonesia, Malaysia
separatus Thornton & Smithers – Thornton & Smithers, 1977: 436 ▸ New Guinea
talilus Thornton & New – Thornton & New, 1977a: 454 ▸ New Guinea

Philotarsopsis

alticola (Thornton & Smithers) – Thornton & Smithers, 1977: 423 ▸ New Guinea
elongata (Thornton & Smithers) – Thornton & Smithers, 1977: 433 ▸ New Guinea
kaindiensis (Thornton & Smithers) – Thornton & Smithers, 1977: 434 ▸ New Guinea
manus (Thornton & New) – Thornton & New, 1977a: 454 ▸ New Guinea
pseudosigma (Thornton & Smithers) – Thornton & Smithers, 1977: 429 ▸ New Guinea
setosa (Thornton & Smithers) – Thornton & Smithers, 1977: 438 ▸ New Guinea
sigma (Thornton & Smithers) – Thornton & Smithers, 1977: 429 ▸ New Guinea
sigmoides (Thornton & Smithers) – Thornton & Smithers, 1977: 432 ▸ New Guinea
stigmata (Thornton & Smithers) – Thornton & Smithers, 1977: 440 ▸ New Guinea

PSEUDOCAECILIIDAE

Allocaecilius

elongatus Lee & Thornton – Lee & Thornton, 1967: 33 ▸	Malaysia
fusciceps Lee & Thornton – Lee & Thornton, 1967: 33 ▸	Malaysia
gombakensis Lee & New – Lee & New, 1992: 1089 ▸	Malaysia
maculatus Lee & Thornton – Lee & Thornton, 1967: 31 ▸	Malaysia
selangorensis New & Thornton – New & Thornton, 1975: 397 ▸	Malaysia
smithersi Lee & New – Lee & New, 1992: 1089 ▸	Malaysia

Allopsocus

marginalis Banks – Banks, 1920: 312 ▸	Malaysia, Philippines

Austropsocus

novoguinensis (Thornton & Smithers) – Thornton & Smithers, 1977: 441 ▸	New Guinea
perforatus Thornton & Smithers – Thornton & Smithers, 1977: 445 ▸	New Guinea

Heterocaecilius

analis Vaughan, Thornton & New – Vaughan et al., 1991: 132 ▸	Indonesia
anareolatus New – New, 1993: 339 ▸	Malaysia
brevifurca Thornton – Thornton, 1984b: 125 ▸	Indonesia
challengerius Lee & New – Lee & New, 1992: 1110 ▸	Malaysia
fasciatus Lee & Thornton – Lee & Thornton, 1967: 40 ▸	Philippines
flaviformis Lee & New – Lee & New, 1992: 1103 ▸	Malaysia
fumosus Thornton – Thornton, 1984b: 125 ▸	Indonesia
fuscipalpus Lee & Thornton – Lee & Thornton, 1967: 15 ▸	Malaysia
gressitti Lee & Thornton – Lee & Thornton, 1967: 50 ▸	New Guinea
hoplites Lee & New – Lee & New, 1992: 1101 ▸	Malaysia
kobus Lee & Thornton – Lee & Thornton, 1967: 47 ▸	Malaysia, New Guinea
leei New – New, 1993: 338 ▸	Malaysia
longipenna Lee & Thornton – Lee & Thornton, 1967: 38 ▸	Malaysia
longispinus Lee & Thornton – Lee & Thornton, 1967: 40 ▸	Philippines
maculosus Lee & New – Lee & New, 1992: 1108 ▸	Malaysia
marginatus Lee & New – Lee & New, 1992: 1101 ▸	Malaysia
orientalis Lee & New – Lee & New, 1992: 1107 ▸	Malaysia
ornatus (Enderlein) – Enderlein, 1903a: 262 ▸	Indonesia, Singapore
papillosus Vaughan, Thornton & New – Vaughan et al., 1991: 134 ▸	Indonesia
reticulatus Lee & New – Lee & New, 1992: 1105 ▸	Malaysia
rhadnus Thornton – Thornton, 1984b: 128 ▸	Indonesia
stiliger Lee & Thornton – Lee & Thornton, 1967: 53 ▸	New Guinea
taipingensis Garcia Aldrete – Garcia Aldrete, 1999b: 243 ▸	Malaysia

Levucaecilius

hilli Smithers – Smithers, 1995d: 549 ▸ Christmas Island

Lobocaecilius

arcuatus Vaughan, Thornton & New – Vaughan et al., 1989: 69 ▸ Indonesia, Malaysia

drepanus New & Thornton – New & Thornton, 1976: 401 ▸ Singapore

ligur Lee & New – Lee & New, 1992: 1086 ▸ Malaysia

longifurca Lee & Thornton – Lee & Thornton, 1967: 28 ▸ Malaysia, Singapore

pahangensis Lee & New – Lee & New, 1992: 1087 ▸ Malaysia

pilosus Lee & New – Lee & New, 1992: 1082 ▸ Malaysia

thorntoni Lee & New – Lee & New, 1992: 1079 ▸ Malaysia

viridis Lee & New – Lee & New, 1992: 1084 ▸ Malaysia

Mepleres

cornutus (Lee & New) – Lee & New, 1992: 1122 ▸ Malaysia

indicus (Lee & New) – Lee & New, 1992: 1118 ▸ Malaysia

nigroguttatus (Karny) – Karny, 1925: 73 ▸ Indonesia, Malaysia

pisiformis (Lee & New) – Lee & New, 1992: 1117 ▸ Malaysia

radialis (Thornton) – Thornton, 1984b: 134 ▸ Indonesia

raphalis (Lee & New) – Lee & New, 1992: 1124 ▸ Malaysia

similis (Lee & New) – Lee & New, 1992: 1125 ▸ Malaysia

solenocerus (Lee & New) – Lee & New, 1992: 1120 ▸ Malaysia

suella (Thornton) – Thornton, 1984b: 134 ▸ Indonesia, Malaysia

transversus (Banks) – Banks, 1937b: 265 ▸ Thailand

urbanus (Vaughan, Thornton & New) – Vaughan et al., 1991: 144 ▸ Indonesia, Malaysia

vilus (Lee & New) – Lee & New, 1992: 1122 ▸ Malaysia

Neocaecilius

granulatus (Lee & New) – Lee & New, 1992: 1099 ▸ Malaysia

mutatus (Lee & New) – Lee & New, 1992: 1114 ▸ Malaysia

Novopsocus

stenopterus (Thornton & Smithers) – Thornton & Smithers, 1977: 443 ▸ New Guinea

Ophiodopelma

glyptocephalum Lienhard – Lienhard, 1985c: 313 ▸ Malaysia

hieroglyphicum (Enderlein) – Enderlein, 1903a: 279 ▸ Indonesia, Malaysia

multipunctatum (Hagen) – Hagen, 1859: 204 ▸ Philippines

permaculatum Lee & Thornton – Lee & Thornton, 1967: 55 ▸ Indonesia, Malaysia, New Guinea

pictipenna Lee & Thornton – Lee & Thornton, 1967: 43 ▸ Indonesia, Philippines
semiceps Lee & Thornton – Lee & Thornton, 1967: 45
▸ Indonesia, Malaysia, Philippines

Phallocaecilius

hirsutus (Thornton) – Thornton, 1961b: 148 ▸ Christmas Island, Indonesia

Pseudocaecilius

angustus (Enderlein) – Enderlein, 1903a: 273 ▸ New Guinea
citricola (Ashmead) – Ashmead, 1879: 228
▸ Indonesia, Malaysia, Philippines, Singapore, Thailand
claggi (Banks) – Banks, 1937a: 130 ▸ Philippines
clunialis Vaughan, Thornton & New – Vaughan et al., 1991: 136 ▸ Indonesia
cornutus Lee & Thornton – Lee & Thornton, 1967: 53 ▸ New Guinea
danaus Lee & New – Lee & New, 1992: 1093 ▸ Malaysia
helicoides Lee & Thornton – Lee & Thornton, 1967: 26 ▸ Malaysia
inaequalis (Banks) – Banks, 1916: 202 ▸ Philippines
innotatus Banks – Banks, 1937a: 132 ▸ Philippines
nitoris (Banks) – Banks, 1937a: 130 ▸ Philippines
otiosus (Banks) – Banks, 1937a: 131 ▸ Philippines
productus New & Thornton – New & Thornton, 1976: 401 ▸ Singapore
pusillus (Banks) – Banks, 1931b: 56 ▸ Philippines
ranus Thornton – Thornton, 1984b: 132 ▸ Indonesia
serratus Lee & New – Lee & New, 1992: 1098 ▸ Malaysia
setifer Vaughan, Thornton & New – Vaughan et al., 1991: 138 ▸ Indonesia
tenellus Enderlein – Enderlein, 1926b: 59 ▸ Indonesia
testaceus Enderlein – Enderlein, 1903a: 262 ▸ Indonesia, New Guinea
wellsae Lee & New – Lee & New, 1992: 1094 ▸ Malaysia

PSOCETAE
HEMIPSOCIDAE

Hemipsocus

africanus Enderlein – Enderlein, 1907g: 33 ▸ Indonesia
chloroticus (Hagen) – Hagen, 1858a: 474
▸ Indonesia, Malaysia, Philippines, Singapore, Vietnam
luridus Enderlein – Enderlein, 1903a: 235
▸ Christmas Island, Indonesia, Malaysia, New Guinea, Singapore
maculatus New – New, 1973f: 231 ▸ Indonesia, Thailand
roseus (Hagen) – Hagen, 1859: 203
▸ Indonesia, New Guinea, Philippines, Singapore, Thailand, Vietnam

MYOPSOCIDAE

Lichenomima

ampla (Smithers & Thornton) – Smithers & Thornton, 1974b: 95 ▸ Indonesia, New Guinea

cubitalis Lienhard – Lienhard, 2004b: 198 ▸ Malaysia

fenestrata Enderlein – Enderlein, 1926b: 62 ▸ Indonesia

merapi Thornton – Thornton, 1992, in: Thornton & Browning, 1992: 348 ▸ Indonesia

sinuosa Lienhard – Lienhard, 2004b: 195 ▸ Indonesia

sumatrana (Enderlein) – Enderlein, 1906g: 87 ▸ Indonesia, Thailand

Lophopterygella

camelina Enderlein – Enderlein, 1907a: 122 ▸ Indonesia, Laos, Philippines, Singapore, Thailand, Vietnam

ridderi Vaughan, Thornton & New – Vaughan et al., 1991: 161 ▸ Indonesia

spilota Smithers & Thornton – Smithers & Thornton, 1979: 516 ▸ New Guinea

triangularis Lienhard – Lienhard, 2004b: 193 ▸ Philippines

Mouldsia

inocellata (Smithers & Thornton) – Smithers & Thornton, 1974b: 115 ▸ New Guinea

marmorata Smithers & Thornton – Smithers & Thornton, 1979: 520 ▸ New Guinea

Myopsocus

angulatus Vaughan, Thornton & New – Vaughan et al., 1991: 159 ▸ Indonesia

apicalis (Thornton) – Thornton, 1984b: 162 ▸ Indonesia

bakeri Banks – Banks, 1916: 199 ▸ Philippines

bomasa (Smithers & Thornton) – Smithers & Thornton, 1974b: 105 ▸ New Guinea

brunneigena (Smithers & Thornton) – Smithers & Thornton, 1979: 522 ▸ New Guinea

brunneipes Smithers – Smithers, 1995d: 554 ▸ Christmas Island

corticola Vaughan, Thornton & New – Vaughan et al., 1991: 155 ▸ Indonesia

dentatus (Smithers & Thornton) – Smithers & Thornton, 1974b: 114 ▸ New Guinea

enderleini Banks – Banks, 1913a: 205 ▸ Philippines

fenestratus (Smithers & Thornton) – Smithers & Thornton, 1974b: 110 ▸ New Guinea

gressitti (Smithers & Thornton) – Smithers & Thornton, 1974b: 115 ▸ New Guinea

horripilis Lienhard – Lienhard, 2004b: 200 ▸ Malaysia, Thailand

javensis Vaughan, Thornton & New – Vaughan et al., 1991: 157 ▸ Indonesia
kolbei Enderlein – Enderlein, 1903a: 302 ▸ New Guinea
lineatus (Smithers & Thornton) – Smithers & Thornton, 1979: 524 ▸ New Guinea
lombokensis (Thornton) – Thornton, 1984b: 164 ▸ Indonesia
longigena (Enderlein) – Enderlein, 1926b: 63 ▸ Indonesia
loriai Ribaga – Ribaga, 1908b: 107 ▸ New Guinea
luteolus (Banks) – Banks, 1920: 301 ▸ Malaysia, Singapore
maculatus (Smithers & Thornton) – Smithers & Thornton, 1974b: 97 ▸ New Guinea
marginatus (Smithers & Thornton) – Smithers & Thornton, 1974b: 112 ▸ New Guinea
mjoebergi (Karny) – Karny, 1925: 63 ▸ Malaysia
peltatus (Smithers & Thornton) – Smithers & Thornton, 1974b: 114 ▸ New Guinea
pilipes (Smithers & Thornton) – Smithers & Thornton, 1974b: 99 ▸ New Guinea
platyvalvula (Smithers & Thornton) – Smithers & Thornton, 1979: 537 ▸ New Guinea
preclarus (Smithers & Thornton) – Smithers & Thornton, 1974b: 106 ▸ New Guinea
pulchellus (Banks) – Banks, 1920: 300 ▸ Singapore
rimosus (Smithers & Thornton) – Smithers & Thornton, 1974b: 108 ▸ New Guinea
scabiosus (Smithers & Thornton) – Smithers & Thornton, 1974b: 112 ▸ New Guinea
spatulatus Thornton – Thornton, 1992, in: Thornton & Browning, 1992: 354 ▸ Indonesia
thecatus (New & Thornton) – New & Thornton. 1976: 409 ▸ Malaysia
toxeres (Smithers & Thornton) – Smithers & Thornton, 1974b: 101 ▸ New Guinea
unduosus (Hagen) – Hagen, 1859: 201 ▸ Malaysia
venustus (Smithers & Thornton) – Smithers & Thornton, 1974b: 109 ▸ New Guinea

Smithersia

angusta (Vaughan, Thornton & New) – Vaughan et al., 1991: 159 ▸ Indonesia
hermosa (Banks) – Banks, 1920: 300 ▸ Singapore
hermosoides (Vaughan, Thornton & New) – Vaughan et al., 1991: 158 ▸ Indonesia
newi Thornton – Thornton, 1992, in: Thornton & Browning, 1992: 359 ▸ Indonesia, Singapore

Thorntonodes

lobatus (New & Thornton) – New & Thornton, 1976: 411 ▸ Singapore, Thailand

quadrituberculatus Lienhard – Lienhard, 2004b: 177 ▸ Thailand

PSILOPSOCIDAE

Psilopsocus

malayanus New & Lee – New & Lee, 1991a: 63 ▸ Malaysia

manus Smithers & Thornton – Smithers & Thornton, 1979: 514 ▸ New Guinea

marmoratus Smithers & Thornton – Smithers & Thornton, 1973: 99 ▸ New Guinea

nebulosus Mockford – Mockford, 1961a: 40 ▸ Philippines

nigricornis Enderlein – Enderlein, 1903a: 306. 74 ▸ New Guinea

pulchripennis Smithers & Thornton – Smithers & Thornton, 1973: 98 ▸ New Guinea

PSOCIDAE

Amphigerontia

feai Ribaga – Ribaga, 1908b: 103 ▸ Myanmar

incerta Ribaga – Ribaga, 1908b: 109 ▸ Myanmar

lata Enderlein – Enderlein, 1926b: 52 ▸ Indonesia

namiana Navás – Navás, 1920b: 27 ▸ Vietnam

Anomaloblaste

spinivalva Endang, Thornton & New – Endang et al., 2002: 120 ▸ Indonesia

treubia Endang, Thornton & New – Endang et al., 2002: 120 ▸ Indonesia

Atrichadenotecnum

nebulosum (Vaughan, Thornton & New) – Vaughan et al., 1991: 154 ▸ Indonesia

umbratum (New & Thornton) – New & Thornton, 1976: 415 ▸ Malaysia

yoshizawai Endang, Thornton & New – Endang et al., 2002: 142 ▸ Indonesia

Blaste

binotata (Enderlein) – Enderlein, 1926b: 57 ▸ Indonesia

longivalva New – New, 1975e: 245 ▸ Singapore

nana New – New, 1975e: 243 ▸ Singapore

serrata New – New, 1973f: 233 ▸ Thailand

[spec., see New & Thornton, 1976] ▸ Malaysia

Blastopsocus

angustus Smithers & Thornton – Smithers & Thornton, 1981a: 923 ▸ New Guinea

Clematoscenea

apta Smithers & Thornton – Smithers & Thornton, 1981a: 937 ▸ New Guinea

biloba Smithers & Thornton – Smithers & Thornton, 1981a: 941 ▸ New Guinea

biroi (Enderlein) – Enderlein, 1903a: 222 ▸ New Guinea

dimorpha Smithers & Thornton – Smithers & Thornton, 1990: 432 ▸ New Guinea

goilala Smithers & Thornton – Smithers & Thornton, 1981a: 942 ▸ New Guinea

gressitti Smithers & Thornton – Smithers & Thornton, 1981a: 942 ▸ New Guinea

lemniscata (Enderlein) – Enderlein, 1903a: 218 ▸ Indonesia, Malaysia

longitarsus Smithers & Thornton – Smithers & Thornton, 1981a: 939 ▸ New Guinea

macalpinei Smithers & Thornton – Smithers & Thornton, 1981a: 930 ▸ New Guinea

mosomensis Smithers & Thornton – Smithers & Thornton, 1981a: 933 ▸ New Guinea

newi Smithers & Thornton – Smithers & Thornton, 1981a: 929 ▸ Malaysia

pulchra Smithers & Thornton – Smithers & Thornton, 1981a: 935 ▸ New Guinea

Clematostigma

brevistylus (Enderlein) – Enderlein, 1903a: 233 ▸ New Guinea

fumatum (Enderlein) – Enderlein, 1903a: 231 ▸ New Guinea

sasmoko Endang, Thornton & New – Endang et al., 2002: 144 ▸ Indonesia

Copostigma

bilineatum (Smithers & Thornton) – Smithers & Thornton, 1981a: 965 ▸ New Guinea

dorsopunctatum Enderlein – Enderlein, 1903a: 30 ▸ New Guinea

hyalinum (Smithers & Thornton) – Smithers & Thornton, 1981a: 964 ▸ New Guinea

majus (Smithers & Thornton) – Smithers & Thornton, 1981a: 962 ▸ New Guinea

montanum (Smithers & Thornton) – Smithers & Thornton, 1981a: 960 ▸ New Guinea

trimaculatum (Hagen) – Hagen, 1858a: 473 ▸ Malaysia

Cycetes

thyrsophoroides Enderlein – Enderlein, 1907a: 109 ▸ Indonesia

Elytropsocus

coleoptratus Smithers & Thornton – Smithers & Thornton, 1981a: 953 ▸ New Guinea

Euclismioides

inocellata Smithers & Thornton – Smithers & Thornton, 1981a: 926 ▸ New Guinea

Indoblaste

discoloba Endang, Thornton & New – Endang et al., 2002: 112 ▸ Indonesia
lawuensis Endang, Thornton & New – Endang et al., 2002: 111 ▸ Indonesia
lienhardi Endang & Thornton – Endang & Thornton, 1992: 328 ▸ Indonesia
sastrawani Endang & Thornton – Endang & Thornton, 1992: 332 ▸ Indonesia

Javablaste

aidab Endang, Thornton & New – Endang et al., 2002: 122 ▸ Indonesia
merapiensis Endang, Thornton & New – Endang et al., 2002: 122 ▸ Indonesia

Javapsocus

intimadentatus Endang, Thornton & New – Endang et al., 2002: 168 ▸ Indonesia
smithersi Endang, Thornton & New – Endang et al., 2002: 171 ▸ Indonesia
soegiri Endang, Thornton & New – Endang et al., 2002: 169 ▸ Indonesia
ungulatus Endang, Thornton & New – Endang et al., 2002: 170 ▸ Indonesia

Kaindipsocus

mixtus Smithers & Thornton – Smithers & Thornton, 1981a: 956 ▸ New Guinea

Loensia

fuscimacula Enderlein – Enderlein, 1926b: 51 ▸ Indonesia
glabridorsum Enderlein – Enderlein, 1926b: 51 ▸ Indonesia

Metylophorus

elongatus Endang, Thornton & New – Endang et al., 2002: 133 ▸ Indonesia
javensis Endang, Thornton & New – Endang et al., 2002: 136 ▸ Indonesia
latespinosus Endang, Thornton & New – Endang et al., 2002: 135 ▸ Indonesia
lisae Thornton – Thornton, 1984b: 140 ▸ Indonesia

Neoblaste

alticola Thornton – Thornton, 1984b: 143 ▸ Indonesia
brunnea Endang & Thornton – Endang & Thornton, 1992: 325 ▸ Indonesia
complexa Endang, Thornton & New – Endang et al., 2002: 118 ▸ Indonesia
cubitalis (Enderlein) – Enderlein, 1919: 36 ▸ Indonesia, Singapore
javensis Endang, Thornton & New – Endang et al., 2002: 114 ▸ Indonesia

parasetosa Endang, Thornton & New – Endang et al., 2002: 117 ▸ Indonesia
soehardjani Endang, Thornton & New – Endang et al., 2002: 116 ▸ Indonesia
timorensis Endang, Thornton & New – Endang et al., 2002: 118 ▸ Indonesia

Psocidus

albovarius (Banks) – Banks, 1920: 299 ▸ Malaysia
apertus Thornton – Thornton, 1984b: 146 ▸ Indonesia
arenosus (Navás) – Navás, 1922e: 198 ▸ Vietnam
borneensis (Banks) – Banks, 1920: 302 ▸ Malaysia
divisus (Navás) – Navás, 1922d: 58 ▸ Vietnam
gloriosus (Banks) – Banks, 1920: 301 ▸ Malaysia
jacobsoni (Enderlein) – Enderlein, 1907a: 112 ▸ Indonesia
javanicus (Enderlein) – Enderlein, 1907a: 114 ▸ Indonesia
lacroixi (Navás) – Navás, 1920c: 91 ▸ Vietnam
lestagei (Navás) – Navás, 1922d: 57 ▸ Vietnam
lobatus Endang & Thornton – Endang & Thornton, 1992: 353 ▸ Indonesia
masinus (Navás) – Navás, 1923b: 22 ▸ Philippines
murphyi New – New, 1975e: 251 ▸ Indonesia, Singapore
peregrans Vaughan, Thornton & New – Vaughan et al., 1989: 82 ▸ Indonesia
proctus Vaughan, Thornton & New – Vaughan et al., 1989: 84 ▸ Indonesia
reidi Thornton – Thornton, 1984b: 149 ▸ Indonesia
relativus (Banks) – Banks, 1920: 302 ▸ Singapore
segmentatus Endang & Thornton – Endang & Thornton, 1992: 350 ▸ Indonesia
similaris (Banks) – Banks, 1920: 299 ▸ Singapore
tergatus Endang & Thornton – Endang & Thornton, 1992: 349 ▸ Indonesia
validus Thornton – Thornton, 1960: 252 ▸ Indonesia
viscayanus (Banks) – Banks, 1920: 302 ▸ Philippines
vitalisi (Navás) – Navás, 1922d: 58 ▸ Vietnam

Psococerastis

annae Thornton – Thornton, 1984b: 151 ▸ Indonesia
aspinosa Endang, Thornton & New – Endang et al., 2002: 127 ▸ Indonesia
bakeri (Banks) – Banks, 1913c: 171 ▸ Philippines
bengalensis (Kolbe) – Kolbe, 1883b: 153 ▸ Malaysia
bispinosa Endang, Thornton & New – Endang et al., 2002: 128 ▸ Indonesia
collessi Smithers & Thornton – Smithers & Thornton, 1981a: 948 ▸ New Guinea
cosmoptera (McLachlan) – McLachlan, 1866b: 350
▸ Indonesia, Malaysia, Myanmar, Singapore, Thailand, Vietnam
discalis (Navás) – Navás, 1920c: 90 ▸ Vietnam
flavistigma (Kolbe) – Kolbe, 1885c: 329 ▸ Indonesia
joannisi (Navás) – Navás, 1934c: 7 ▸ Vietnam
lifashengi Endang, Thornton & New – Endang et al., 2002: 126 ▸ Indonesia
lombokensis (Navás) – Navás, 1927b: 40 ▸ Indonesia

luroris (Soehardjan & Hamann) – Soehardjan & Hamann, 1959: 1 ▸ Indonesia
luzonensis (Banks) – Banks, 1916: 198 ▸ Philippines
murudensis (Karny) – Karny, 1925: 64 ▸ Malaysia
parasinensis New – New, 1975e: 247 ▸ Singapore
taprobanes (Hagen) – Hagen, 1858a: 473 ▸ Indonesia, Malaysia, Vietnam

Psocus

alticolus Banks – Banks, 1937a: 126 ▸ Philippines
dolorosus Banks – Banks, 1937a: 125 ▸ Philippines
illotus Banks – Banks, 1939a: 134 ▸ Philippines
incomptus Banks – Banks, 1937a: 126 ▸ Philippines
omissus Banks – Banks, 1939a: 134 ▸ Philippines
rizali Banks – Banks, 1939a: 134 ▸ Philippines

Ptycta

afasciata Endang, Thornton & New – Endang et al., 2002: 152 ▸ Indonesia
breuschi Smithers & Thornton – Smithers & Thornton, 1990: 440 ▸ New Guinea
distincta Smithers & Thornton – Smithers & Thornton, 1990: 442 ▸ New Guinea
floresensis Endang, Thornton & New – Endang et al., 2002: 149 ▸ Indonesia
frontalis Thornton – Thornton, 1984b: 154 ▸ Indonesia
incerta Smithers & Thornton – Smithers & Thornton, 1990: 436 ▸ New Guinea
incurvata Thornton – Thornton, 1960: 245 ▸ Indonesia
krakatau Vaughan, Thornton & New – Vaughan et al., 1989: 86 ▸ Indonesia
lugosensis Smithers & Thornton – Smithers & Thornton, 1990: 434 ▸ New Guinea
merapiensis Endang, Thornton & New – Endang et al., 2002: 151 ▸ Indonesia
novohibernica Smithers & Thornton – Smithers & Thornton, 1990: 438 ▸ New Guinea
pitallo Endang, Thornton & New – Endang et al., 2002: 153 ▸ Indonesia
precincta Thornton – Thornton, 1984b: 155 ▸ Indonesia
sardjani Endang & Thornton – Endang & Thornton, 1992: 359 ▸ Indonesia
schillei (Enderlein) – Enderlein, 1906c: 310 ▸ Indonesia
serrata Endang, Thornton & New – Endang et al., 2002: 147 ▸ Indonesia
tapensis Endang, Thornton & New – Endang et al., 2002: 148 ▸ Indonesia
timorensis Endang, Thornton & New – Endang et al., 2002: 150 ▸ Indonesia
tsutsumii New & Thornton – New & Thornton, 1976: 417 ▸ Indonesia, Malaysia
verticalis Vaughan, Thornton & New – Vaughan et al., 1991: 152 ▸ Indonesia

Setopsocus

hirsutus (New) – New, 1975e: 255 ▸ Indonesia, Singapore
magnus Smithers & Thornton – Smithers & Thornton, 1981a: 945 ▸ New Guinea

Sigmatoneura

basalis Endang & Thornton – Endang & Thornton, 1992: 341 ▸ Indonesia
kakisayap Yoshizawa & Lienhard – Yoshizawa & Lienhard, 2005, in: Yoshizawa et al., 2005: 833 ▸ Indonesia, Malaysia
kolbei (Enderlein) – Enderlein, 1906a: 246 ▸ Indonesia, Singapore
lemahsayap Yoshizawa & Lienhard – Yoshizawa & Lienhard, 2005, in: Yoshizawa et al., 2005: 837 ▸ Indonesia
longicornis (Banks) – Banks, 1920: 308 ▸ Brunei, Malaysia, Philippines, Singapore
orientalis (New) – New, 1975e: 250 ▸ Singapore
scitula Smithers & Thornton – Smithers & Thornton, 1981a: 950 ▸ Indonesia, New Guinea
semicolorata (Banks) – Banks, 1920: 307 ▸ Philippines
subcostalis (Enderlein) – Enderlein, 1903a: 215 ▸ Indonesia, Malaysia, Singapore

Sundapsocus

robertsoni Smithers – Smithers, 1995d: 553 ▸ Christmas Island

Thyrsopsocopsis

thorntoni Mockford – Mockford, 2004: 153 ▸ Vietnam

Trichadenotecnum

adika Endang, Thornton & New – Endang et al., 2002: 166 ▸ Indonesia
alinguum Endang, Thornton & New – Endang et al., 2002: 163 ▸ Indonesia
alobum Endang & New – Endang & New, 2005: 30 ▸ Indonesia
arciforme Thornton – Thornton, 1961a: 11 ▸ Indonesia, Malaysia
bidens Thornton – Thornton, 1961a: 6 ▸ Indonesia
bidentatum Thornton – Thornton, 1984b: 158 ▸ Indonesia
bromoense Endang, Thornton & New – Endang et al., 2002: 157 ▸ Indonesia
cheahae Endang, Thornton & New – Endang et al., 2002: 166 ▸ Indonesia
cinnamonum Endang & New – Endang & New, 2005: 35 ▸ Indonesia
circulare (Hagen) – Hagen, 1859: 201 ▸ Vietnam
circularoides Badonnel – Badonnel, 1955: 229 ▸ Malaysia
cornutum Endang & New – Endang & New, 2005: 28 ▸ Indonesia
galihi Endang, Thornton & New – Endang et al., 2002: 159 ▸ Indonesia
godavarense New – New, 1971c: 197 ▸ Indonesia
gombakense New & Lee – New & Lee, 1992a: 155 ▸ Indonesia, Malaysia
hammani Endang, Thornton & New – Endang et al., 2002: 161 ▸ Indonesia
imrum New & Thornton – New & Thornton, 1976: 414 ▸ Malaysia
jambiense Endang & New – Endang & New, 2005: 34 ▸ Indonesia
kalibiruense Endang & New – Endang & New, 2005: 27 ▸ Indonesia
kerinciense Endang & New – Endang & New, 2005: 26 ▸ Indonesia
krucilense Endang, Thornton & New – Endang et al., 2002: 165 ▸ Indonesia
laticornutum Endang, Thornton & New – Endang et al., 2002: 158 ▸ Indonesia

malayense New – New, 1975e: 257 ▸ Indonesia, Malaysia
marginatum New & Thornton – New & Thornton, 1976: 413 ▸ Malaysia
mclachlani New – New, 1973f: 235 ▸ Thailand
minutum Enderlein – Enderlein, 1926b: 50 ▸ Indonesia
muaraense Endang & New – Endang & New, 2005: 31 ▸ Indonesia
paradika Endang & New – Endang & New, 2005: 32 ▸ Indonesia
pardus Badonnel – Badonnel, 1955: 231 ▸ Indonesia, New Guinea, Singapore
proctum Endang & New – Endang & New, 2005: 33 ▸ Indonesia
quadrispinosum Endang, Thornton & New – Endang et al., 2002: 167 ▸ Indonesia
rachimi Endang & Thornton – Endang & Thornton, 1992: 363 ▸ Indonesia
santosai Endang & Thornton – Endang & Thornton, 1992: 364 ▸ Indonesia
sibolangitense Endang & New – Endang & New, 2005: 27 ▸ Indonesia
soekarmanni Endang, Thornton & New – Endang et al., 2002: 160 ▸ Indonesia
soenarti Endang, Thornton & New – Endang et al., 2002: 162 ▸ Indonesia
sumatrense Endang & New – Endang & New, 2005: 29 ▸ Indonesia
vaughani Endang, Thornton & New – Endang et al., 2002: 164 ▸ Indonesia
waykambasense Endang & New – Endang & New, 2005: 34 ▸ Indonesia
waykananense Endang & New – Endang & New, 2005: 25 ▸ Indonesia

Glossary

This glossary explains some the morphological terms used in keys and descriptions of taxa of Psocoptera. Some terms are very general, while others reflect the specialised terminology developed expecially for Psocoptera.

a *amphitoky* – a form of parthenogenetic reproduction by which offspring of both sexes are produced.

annulation(s) – see secondary annulations.

anogenital complex – see terminalia.

apterous – without wings.

areolate – in the form of a 'network' of enclosed areas; a form of integumental sculpturing with such a pattern and sometimes of diagnostic value (*Liposcelis*).

areola postica – a marginal cell on the hind border of the fore wing of many psocids, formed by forking of vein Cu1.

b *body length* – the greatest length from the front of the head (without antennae) in normal orientation to the tip of the abdomen.

boss – a small rounded prominence or dome on the body.

brachia – literally 'arms'; here, the lateral processes extending from the median blade of the oviruptor, the structure by which psocid pronymphs leave the egg.

brachypterous – short winged, wing venation present to varying extents.

c *chorion* – the 'egg shell'.

cibarial – relating to the extraoral food cavity above the hypopharynx in the psocopteran mouthparts.

clunium – the margin of the last apparent abdominal tergite abutting the epiproct and which may be variously ornamented in some psocids.

coriaceous (wings) – 'leathery', refers to toughened texture of fore wing of some psocids.

corticolous (cortical) – living on bark of trees.

ctenidiobothria – long sensory bristles, usually on the tibia or tarsus and each sited in an enlarged protruding socket.

d *diapause* – a state of physiological arrest (as a form of dormancy) in the life cycle, whereby insects pass inclement parts of the year.

e *edaphic* – 'ground-dwelling': refers to insects living on or near the ground surface, as in leaf-litter.

elytriform – refers to hardened fore wing of some psocids, so as to resemble a beetle elytron.

endophallus – complex of rugose areas or more clearly-defined sclerites within the phallosome (q.v.)

epiphytic – pertaining to the leaf surface; applied to psocids that are 'foliage frequenters'.

epiproct – the small median dorsal posterior abdominal sclerite above the anus and between the paraprocts.

exopterygote – having an incomplete or gradual metamorphosis through which the wings develop externally as progressively enlarging wing buds.

f *filiform* (antenna) – slender or threadlike (see also moniliform).

flagellar segment – one unit of the flagellum (q.v.).

flagellomere – a flagellar 'segment'.

flagellum – the antenna beyond the two basal segments (scape and pedicel, respectively).

g *galea* (*galeae*) – a distal part of the maxilla(e) in the mouthparts.

glabrous (wings) – without macrotrichia (setae) on the wing membrane or veins.

gonapophyses – the constituent parts (valves) of the ovipositor of a female insect; three pairs are the basic pattern, which may become modified in many ways to provide diagnostic characters for psocids.

h *homology* – applied to structures having the same morphological origin.

hypandrium (*hypandria*) – the enlarged ventral plate subtending the phallosome in male psocids.

hypopharynx – a structure in the oral cavity forming the floor of the functional mouth of an insect.

i IO/D – the ratio of interocular distance (IO) to eye diameter (D), used under head measurements. There are two methods, see p. 35 for explanation.

l *lacinia* – the rodlike 'pick' as a modified part of the maxilla in psocid mouthparts.

laterocervical sclerite – the small lateral sclerites in the 'neck' (the membranous area between the head and thorax) of psocids.

m *macropterous* – fully winged.

micropyle – a small pore in the chorion of the egg, through which sperm pass during fertilisation.

micropterous – wings strongly reduced, sometimes to tiny lobes, and usually with little or no venation.

moniliform (antenna) – 'beadlike', referring to presence of distinct constrictions between flagellar segments (see also filiform).

n *neopterous* – belonging to the infraclass Neoptera, having the ability to flex the wings back over the abdomen.

neoteny – the expression of sexual maturity with persistence of nymphal features in the adult form.

nodulus – in the psocopteran fore wing: the position on the hind margin where veins Cu2 and 1A meet and where a wing-coupling structure may be formed.

o *ommatidia* – the individual facets of the insect compound eye.

p *paraproct* – a lateral lobe of the last apparent abdominal tergite, bordering the anus, often with a field of sensory trichobothria (q.v.).

parthenogenesis – reproduction without fertilisation. See also amphitoky, thelytoky.

phallosome – the male genital organ, the aedeagus.

phragmotic – adapted to blocking a hole. Applied to the few psocids known to live in tunnels in wood and in which the apex of the abdomen may be modified to seal the entrance to the cavity.

placoid sensillum (sensilla) – a kind of sense organ, usually circular in outline and with an enlarged membrane covering an enlarged pore canal.

PO/D – the ratio of shortest eye axis (PO) to greatest eye axis (D) as seen in dorsal aspect, thus indicating the extent of prominence of the eye. See p. 36 for explanation.

postclypeus – the posterior region of the clypeus, in many psocids characteristically expanded and 'globular'.

pronymph – the immature psocid at the stage of hatching from the egg and before moulting to reveal the true free-living first instar nymph.

pulvillus (pulvilli) – small ventral thread-like or broadened structures below the pretarsal claws.

r *rugose* – 'wrinkled', applied to the texture or appearance of insect integument.

s *sclerite* – any distinct hardened portion of the insect integument.

secondary annulations – refers to a characteristic appearance of the flagellar segment of some psocids, in which these are ornamented with encircling lines or 'annulations'.

sensillum (sensilla) – simple sense organ(s).

setiform – appearing like a bristle or seta.

setose – covered with setae.

sitophore – part of the hypopharynx; the dorsal surface of the hypopharynx constituting part of the channel through which food passes.

spermapore – the ventral external opening of the spermathecal duct.

spermatheca – the receptacle for sperm in the female psocid.

spur vein – a short vein extending from another vein to end free in the wing membrane. In psocids used most commonly for a short vein produced from the hind angle of the pterostigma in the fore wing.

sternite – the ventral structure of an insect body segment.

stigmasac – a small ventral projection at the base of the fore wing pterostigma, which retains the costa of the hind wing when the psocid is at rest.

subcortical – living or found under bark.

t *tagma* – a major body region; in insects the three tagmata are head, thorax and abdomen.

telson – epiproct and paraprocts together.

tergite – a dorsal structure comprising the upper side of an insect body segment.

terminalia – the collective term for the apical abdominal structures comprising the telson (q.v.) and genital structures.

thelytoky – a form of parthenogenetic reproduction in which all offspring are female.

trichobothria – long slender sensory bristles, each sited in a depressed pit socket.

trichoid – hairlike.

tympanic organ – an auditory organ, most commonly as a membranous area.

u *univoltinism* – the state of having one generation each year.

v *vertex* – the dorsalmost part of the head.

vesicle – a small extensible membranous bladder. Here applied to eversible structures found ventrally on the abdomen of some foliage-frequenting psocids.

voltinism – the number of generations passed each year.

References

The following list includes taxonomic papers relevant to chapter 7, as well as those noted in earlier chapters. Notations 'a', 'b', et cetera refer to the sequence given in the World Catalogue and Bibliography (Lienhard & Smithers, 2002) to facilitate cross-referencing.

Anonymous, 1996. Opinion 1830. Caeciliidae Kolbe, 1880 (Insecta, Psocoptera): spelling emended to Caeciliusidae, so removing the homonymy with Caeciliidae Rafinesque, 1814 (Amphibia, Gymnophiona).– Bulletin of Zoological Nomenclature 53: 68-69.

Ashmead, W.H., 1879. On a new *Psocus*.– Canadian Entomologist 11: 228-229.

Badonnel, A., 1931a. Contribution à l'étude de la faune du Mozambique. Voyage de M. P. Lesne (1928-1929). 4e note. Copéognathes.– Annales des Sciences naturelles, Zoologie (10)14 (16): 229-260.

Badonnel, A., 1935d. Psocoptères nouveaux d'Afrique et d'Arabie.– Revue française d'Entomologie 2: 76-82.

Badonnel, A., 1943a. Psocoptères.– Faune de France 42: 1-164. Paris.

Badonnel, A., 1949a. Psocoptères de la Côte d'Ivoire. Mission Paulian-Delamare (1945).– Revue française d'Entomologie 16: 20-46.

Badonnel, A., 1949c. Psocoptères du Congo Belge (3e note).– Bulletin de l'Institut Royal des Sciences naturelles de Belgique 25: 1-64.

Badonnel, A., 1951. Ordre des Psocoptères.– In: Grassé, P.P. (ed.). Traité de zoologie, anatomie, systématique, biologie. Traité de Zoologie 10(2): 1301-1340. Paris, Masson.

Badonnel, A., 1955. Psocoptères de l'Angola.– Publicacoes culturais da Companhia de Diamantes de Angola 26: 1-267.

Badonnel, A., 1962a. Psocoptères.– Biologie de l'Amerique australe 1: 185-229.

Badonnel, A., 1963a. Psocoptères terricoles, lapidicoles et corticicoles du Chili.– Biologie de l'Amerique australe 2: 291-338.

Badonnel, A., 1966a. Sur le genre *Archipsocus* Hagen (Psocoptera, Archipsocidae).– Bulletin du Museum national d'Histoire naturelle, Paris, 2e serie 38: 409-415.

Badonnel, A., 1967a. Insectes Psocoptères.– Faune de Madagascar XXIII. ORSTOM, Paris: 1-238.

Badonnel, A., 1967b. Psocopteres edaphiques du Chili (2e note).– Biologie de l'Amerique australe 3: 541-585.

Badonnel, A., 1969. Psocoptères de l'Angola et de pays voisins avec revision de types africains d'Enderlein (1902) et de Ribaga (1911).– Publicacoes culturais da Companhia de Diamantes de Angola 79: 1-152.

Badonnel, A., 1972. Psocoptères édaphiques du Chile (3e note).– Bulletin du Museum national d'Histoire naturelle, Paris, Zoologie 1: 1-38.

Badonnel, A., 1973a. Espèces bresiliennes de la sous-famille des Embidopsocinae (Psocoptera: Liposcelidae).– Bulletin du Museum national d'Histoire naturelle, Paris, 3e serie, no. 87, Zoologie 66: 1097-1139.

Badonnel, A. & Lienhard, C., 1988. Revision de la famille des Mesopsocidae (Insecta, Psocoptera).– Bulletin du Museum national d'Histoire naturelle, Paris (4) 10A: 375-412.

Badonnel, A., Mockford, E.L. & García-Aldrete, A.N., 1984. Pararchipsocinae, nouvelle subdivision des Archipsocidae (Insecta: Psocoptera), avec description de *Pararchipsocus* n.g. et de onze espèces inédites de cette sous-familie.– Bulletin du Museum national d'Histoire naturelle. Paris (4) 6A: 741-768.

Ball, A., 1943. Contribution B l'étude des Psocoptères. III. *Ectopsocus* du Congo belge, avec une remarque sur le rapport IO/D.– Bulletin du Musée royal d'Histoire naturelle de Belgique 19(38): 1-28.

Banks, N., 1913a. Synopses and descriptions of exotic Neuroptera.– Transactions of the American Entomological Society 39: 201-242.
Banks, N., 1913c. On a collection of Neuropteroid insects from the Philippine Islands.– Proceedings of the Entomological Society of Washington 15: 170-180.
Banks, N., 1914a. Zoological Results of the Abor Expedition, 1911-1912. XXVI. Neuropteroid insects.– Records of the Indian Museum 8: 351-356.
Banks, N., 1916. Neuropteroid insects of the Philippine islands.– Philippine Journal of Science (D) 11: 195-217.
Banks, N., 1918. New Neuropteroid insects.– Bulletin of the Museum of Comparative Zoology, Harvard College 62: 1-22.
Banks, N., 1920. New neuropteroid insects.– Bulletin of the Museum of Comparative Zoology, Harvard 64: 299-362.
Banks, N., 1931a. On some Psocidae from the Hawaiian Islands.– Proceedings of the Hawaiian Entomological Society 7: 437-441.
Banks, N., 1931b. Some Oriental Neuropteroid insects.– Psyche 38: 56-70.
Banks, N., 1934. Supplementary Neuropteroid insects from the Malay Peninsula and from Mt. Kinabalu, Borneo.– Journal of the Federated Malay States Museum 17: 567-578.
Banks, N., 1937a. Philippine neuropteroid insects.– Philippine Journal of Science 63: 125-174.
Banks, N., 1937b. Neuropteroid insects from Formosa.– Philippine Journal of Science 62: 255-291.
Banks, N., 1939a. Neuropteroid insects from the Philippines.– Philippine Journal of Science 69: 133-145.
Barlow, H. S., 1988. Forest Lepidoptera.– In: Cranbrook, G. G.-H., Earl of (ed.). Key environments. Malaysia. Pergamon Press, Oxford: 212-224.
Betz, B. W., 1983c. Systematics of the *Trichadenotecnum alexanderae* species complex (Psocoptera, Psocidae) based on an investigation of modes of reproduction and morphology.– Canadian Entomologist 115: 1329-1354.
Broadhead, E., 1947d. The life-history of *Embidopsocus enderleini* (Ribaga) (Corrodentia, Liposcelidae).– Entomologist's Monthly Magazine 83: 200-203.
Broadhead, E., 1950a. A revision of the genus *Liposcelis* Motschulsky with notes on the position of this genus in the order Corrodentia and on the variability of ten *Liposcelis* species.– Transactions of the Royal Entomological Society of London 101: 335-388.
Broadhead, E., 1954a. The infestation of warehouses and ships' holds by psocids in Britain.– Entomologist's Monthly Magazine 90: 103-105.
Broadhead, E., 1954b. A new parthenogenetic psocid from stored products, with observations on parthenogenesis in other psocids.– Entomologist's Monthly Magazine 90: 10-16.
Broadhead, E., 1958a. The psocid fauna of larch trees in northern England – an ecological study of mixed species populations exploiting a common resource.– Journal of Animal Ecology 27: 217-263.
Broadhead, E., 1961. The biology of *Psoquilla marginepunctata* (Hagen) (Corrodentia, Trogiidae).– Transactions of the Society for British Entomology 14: 223-236.
Broadhead, E., 1983. The assessment of faunal diversity and guild size in tropical forests with particular reference to the Psocoptera.– In: Sutton, S. L., Whitmore, T. C. & Chadwick, A. C. (eds). Tropical rain forest: ecology and management. British Ecological Society Special Publication 2: 107-119.
Broadhead, E. & Cheke, R. A., 1975. Host spatial pattern, parasitoid interference and the modelling of the dynamics of *Alaptus fusculus* (Hym.: Mymaridae), a parasitoid of two *Mesopsocus* species (Psocoptera).– Journal of Animal Ecology 44: 767-793.

Broadhead, E. & Wapshere, A. J., 1966b. *Mesopsocus* populations on larch in England – the distribution and dynamics of two closely-related coexisting species of Psocoptera sharing the same food resource.– Ecological Monographs 36: 327-388.

Broadhead, E. & Wolda, H., 1985. The diversity of Psocoptera in two tropical forests in Panama.– Journal of Animal Ecology 54: 739-754.

Burmeister, H., 1839. Handbuch der Entomologie. Vol. 2, Berlin.

Carpenter, F. M., 1992. Superclass Hexapoda. Treatise on Invertebrate Palaeontology.– Part R. Arthropoda 4, Volumes 3, 4. Geological Society of America, Boulder, Colorado and University of Kansas, Lawrence, Kansas.

Curtis, J., 1837. A guide to an arrangement of British insects: being a catalogue of all the named species hitherto discovered in Great Britain and Ireland. London, Pigot.VI + 294 pp.

Dallai, R. & Afzelius, B. A., 1991. Sperm flagellum of insects belonging to orders Psocoptera, Mallophaga and Anoplura. Ultrastructural and phylogenetic aspects.– Bolletino di Zoologia 58: 211-216.

Danks, H. V., 1970. Biology of some stem-nesting aculeate Hymenoptera.– Transactions of the Royal Entomological Society of London 122: 323-395.

Deharveng, L. & Bedos, A., 2000. The cave fauna of Southeast Asia. Origin, evolution and ecology.– In: Wilkens, H., Culver, D. C. & Humphreys, W. F. (eds). Subterranean ecosystems. Ecosystems of the World, Vol. 30. Elsevier, Amsterdam: 603-632.

Deharveng, L. & Leclerc, P., 1989. Recherches sur les faunes cavernicoles d'Asie du Sud-Est.– Mémoires de Biospeleologie 16: 91-110.

Dickersen, R. E., Merrill, E. D., McGregor, R. C., Schultze, W., Taylor, E. H. & Herre, A. W. C. T., 1928. Distribution of life in the Philippines. Manila.– Bureau of Sciences Monograph 21: 1-322. (Reprint 1975, Bookmark, Manila).

Eertmoed, G. E., 1973. The phenetic relationships of the Epipsocetae (Psocoptera): the higher taxa and the species of two new families.– Transactions of the American Entomological Society 99: 373-414.

Endang, S. K. & New, T. R. 2005. The genus *Trichadenotecnum* (Insecta: Psocoptera: Psocidae) in Sumatra, Indonesia, with description of thirteen new species.– Records of the Australian Museum 57: 15-38.

Endang, S. K. & Thornton, I. W. B., 1992. Psocidae (Insecta: Psocoptera) from the islands of Bali and Lombok, Indonesia.– Treubia 30: 319-379.

Endang, S. K., Thornton, I. W. B. & New, T. R., 2002. The Psocidae (Insecta: Psocoptera) of Java and the eastern islands of Indonesia.– Invertebrate Systematics 16: 107-176.

Enderlein, G., 1901. Neue deutsche und exotische Psociden sowie Bermerkungen zur Systematik.– Zoologischer Jahrbücher (Abteilung Systematik) 14: 537-548.

Enderlein, G. 1903a. Die Copeognathen des indo-australischen Faunengebietes.– Annales historico-naturales Musei nationalis Hungarici 1: 179-344.

Enderlein, G., 1903c. Neue Copeognathen aus Kamerun.– Zoologische Jahrbücher (Abteilung Systematik) 19: 1-8.

Enderlein, G., 1903d. Über die Morphologie, Gruppierung und systematische Stellung der Corrodentien.– Zoologischer Anzeiger 26: 423-437.

Enderlein, G., 1904a. Die von Herrn Prof. Dr. Friedr. Dahl im Bismarck-Archipel gesammelten Copeognathen, nebst Bemerkungen ueber die physiologische Bedeutung des Stigmasackes.– Zoologische Jahrbücher (Abteilung Systematik) 20: 105-112.

Enderlein, G., 1905c. Results of the Swedish Zoological Expedition to Egypt and the White Nile 1901.– No. 18. Morphologie, Systematik und Biologie der Atropiden und Troctiden sowie Zusammenstellung aller bisher bekannten recenten und fossilen Formen: 1-58. Uppsala.

Enderlein, G., 1906a. Die Copeognathen-Fauna Japans.– Zoologische Jahrbücher (Abteilung Systematik) 23: 243-256.
Enderlein, G., 1906c. Zehn neue aussereuropäische Copeognathen.– Stettiner Entomologische Zeitung 67: 306-316.
Enderlein, G., 1906e. The scaly-winged Copeognatha (monograph of the Amphientomidae, Lepidopsocidae and Lepidillidae in relation to their morphology and taxonomy).– Spolia Zeylandica 4: 39-122.
Enderlein, G., 1906g. Aussereuropäische Copeognathen aus dem Stettiner Museum.– Zoologische Jahrbücher (Abteilung Systematik) 24: 81-90.
Enderlein, G., 1907a. Die Copeognathen Javas.– Notes from the Leyden Museum 29: 107-126.
Enderlein, G., 1907d. *Troctes entomophilus*, ein neuer Insektenliebhaber aus Columbien.– Stettiner Entomologische Zeitung 68: 34-36.
Enderlein, G., 1907e. Neue Beiträge zur Kenntnis der Copeognathen Japans.– Stettiner Entomologische Zeitung 68: 90-106.
Enderlein, G., 1907g. Copeognatha.– Wissenschaftliche Ergebnisse der Schwedischen Zoologischen Expedition nach dem Kilimandjaro, dem Meru und den umgebenden Massaisteppen Deutsch-Ostafrikas, 1905-1906, unter Leitung von Prof. Dr. Yngve Sjöstedt 15(2): 29-39.
Enderlein, G., 1908d. Die Copeognathenfauna der Insel Formosa.– Zoologischer Anzeiger 33: 759-779.
Enderlein, G., 1910a. Eine Dekade neuer Copeognathengattungen.– Sitzungsberichte der Gesellschaft naturforschender Freunde zu Berlin 2: 63-77.
Enderlein, G., 1919. Copeognatha.– Collections Zoologiques du Baron Edm. de Selys Longchamps. Catalogue systématique et descriptif 3(2): 1-55.
Enderlein, G., 1925a. Beiträge zur Kenntnis der Copeognathen IX.– Konowia 4: 97-108.
Enderlein, G., 1926b. Die Copeognathenfauna Javas.–– Zoologische Mededeelingen, Leiden 9: 50-70.
García Aldrete, A. N., 1974a. A classification above species level of the genus *Lachesilla* Westwood (Psocoptera: Lachesillidae).– Folia Entomologica Mexicana 27: 3-88.
García Aldrete, A. N., 1988a. A new Oriental *Lachesilla* (Psocoptera: Lachesillidae).– Oriental Insects 22: 501-504.
García Aldrete, A. N., 1990a. Insecta: Psocoptera. In: Dindal, D. L. (ed.). Soil Biology Guide. John Wiley & Sons, New York: 1033-1052.
García Aldrete, A. N., 1999b. Replacement names for some Psocoptera (Insecta) because of homonymy.– Entomologist's Monthly Magazine 135: 243-244.
García Aldrete, A. N. & Mockford, E. L., 1996. A new Peruvian musapsocid genus and species (Psocoptera: Musapsocidae).– Entomological News 107: 88-92.
Ghani, M. A. & Sweetman, H. L., 1951. Ecological studies of the book louse, *Liposcelis divinatorius* (Müll.).– Ecology 32: 230-244.
Glick, P. A., 1939. The distribution of insects, spiders and mites in the air.– Technical Bulletin of the United States Department of Agriculture 673: 1-150.
Gressitt, J. L., 1959. Wallace's line and insect distribution.– In: Hewer, H. R. & Riley, N. D. (eds.). Proceedings of the XVth International Congress of Zoology, London, 16-23 July 1958: 66-68.
Günther, K. K., 1974a. Staubläuse, Psocoptera.– Die Tierwelt Deutschlands 61: 1-314. Jena.
Gurney, A. B., 1939. Nomenclatorial notes on Corrodentia, with descriptions of two new species of *Archipsocus*.– Journal of the Washington Academy of Sciences 29: 501-515.
Gurney, A. B., 1943. A synopsis of the psocids of the tribe Psyllipsocini, including the description of an unusual new genus from Arizona (Corrodentia, Empheriidae, Empheriinae).– Annals of the Entomological Society of America 36: 195-220.

Hagen, H., 1858a. Synopsis der Neuroptera Ceylons. Pars I.– Verhandlungen der Zoologisch-Botanischen Gesellschaft Wien 8: 471-488.

Hagen, H., 1859. Synopsis der Neuroptera Ceylons. Pars II.– Verhandlungen der Zoologisch-Botanischen Gesellschaft Wien 9: 199-212.

Hagen, H.A., 1861b. Synopsis of the Neuroptera of North America with a list of South American species.– Smithsonian miscellaneous Collections 4: 1-358.

Hagen, H., 1865b. Synopsis of the Psocina without ocelli.– Entomologist's Monthly Magazine 2: 121-124.

Hamilton, W.D., 1978. Evolution and diversity under bark.– Symposia of the Royal Entomological Society of London 9: 154-175.

Hardy, A.C. & Milne, P.S., 1938. Studies in the distribution of insects by aerial currents. Experiments in aerial tow-netting from kites.– Journal of Animal Ecology 7: 199-229.

Jacobson, E., 1924. Zur Kenntniss der Biologie javanischer Copeognathen.– Tijdschrift voor Entomologie 67 (suppl.): 65-71.

Jentsch, S., 1939b. Beiträge zur Kenntnis der Überordnung Psocoidea. 8. Die Gattung *Ectopsocus* (Psocoptera).– Zoologische Jahrbücher (Abteilung Systematik) 73: 111-128.

Johnson, K.P. & Mockford, E.L., 2003. Molecular systematics of Psocomorpha (Psocoptera).– Systematic Entomology 28: 409-416.

Johnson, K.P., Yoshizawa, K. & Smith, V.S., 2004. Multiple origins of parasitism in lice.– Proceedings of the Royal Society of London B 271: 1771-1776.

Karny, H.H., 1925. On the Copeognatha from Mt Murud and Mt Dulit, Sarawak.– Sarawak Museum Journal 3: 63-74.

Karny, H.H., 1930. Zur Systematik der Orthopteroiden Insekten. II.– Treubia 12: 431-461.

Kato, M., 1990. The fern flora of Seram.– In: Baas, P., Kalkman, K. & Geesink, R. (eds.) The plant diversity of Malesia. Kluwer Academic Publishers, Dordrecht: 225-234.

Kolbe, H.J., 1880c. Monographie der deutschen Psociden.– Jahresbericht des Westfälischen Provinzial-Vereins für Wissenschaft und Kunst. 1879-80: 73-142.

Kolbe, H.J., 1883b. Ueber die Racen des *Psocus Taprobanes* Hagen in Ostindien.– Entomologische Nachrichten 9: 152-154.

Kolbe, H.J., 1885a. Zur Kenntniss der Psociden-Fauna Madagaskars.– Berliner Entomologische Zeitschrift 29: 183-192.

Kolbe, H.J., 1885c. Die Borneo-Rasse des *Psocus taprobanes* Hag.– Entomologische Nachrichten 11: 328-330.

Kuwayama, S., 1961. Corrodentia of Thailand.– Nature and Life in Southeast Asia 1: 203-205.

Larsson, S.G., 1978. Baltic amber – a Palaeobiological study.– Entomonographs 1: 1-192.

Latreille, P.A., 1794. Extrait d'un mémoire pour servir de suite a l'histoire des *Termes*, ou fourmis blanches.– Bulletin des Sciences de la Societé philomathique de Paris 1: 84-85.

Latreille, P.A., 1799. Le genre *Psocus*.– In: Coquebert de Montbret, A.J. Illustratio Iconographica Insectorum, quae in Musaeis parisinis observavit et in lucem edidit Joh. Christ. Fabricius, praemissis eiusdem descriptionibus. Accedunt species plurimae, vel minus vel nondum cognitae. Paris. Part 1: 8-14.

Laumonier, Y., 1997. The vegetation and physiography of Sumatra.– Kluwer Academic Publishers, Dordrecht: ix + 222 pp.

Lee, S.S., 1965. The life history of *Psoquilla marginepunctata* (Hagen), a dimorphic psocopteran.– Malayan Nature Journal 19: 214-222.

Lee, S.S., 1974. A new species of *Cymatopsocus* from Malaysia and further descriptions of *Calopsocus iridescens* Banks (Psocoptera).– Entomologist's Monthly Magazine 110: 7-12.

Lee, S.S., 1987. Environmental factors inducing macroptery in the psocid *Psoquilla marginepunctata*.– Entomologia Experimentalis et Applicata 44: 89-95.

Lee, S. S. & New, T. R., 1992. Pseudocaeciliidae (Insecta: Psocoptera) from West Malaysia.– Invertebrate Taxonomy 6: 1077-1126.

Lee, S. S. & Thornton, I. W. B., 1967. The family Pseudocaeciliidae (Psocoptera) – a reappraisal based on the discovery of new Oriental and Pacific species.– Pacific Insects Monographs 16: 1-116.

Lienhard, C., 1982a. Les psoques de nos maisons ou les tours malicieux des appareils Berlese.– Musées de Geneve 222: 16-21.

Lienhard, C., 1985c. A new and unusual species of *Ophiodopelma* Enderlein (Insecta, Psocoptera, Pseudocaeciliidae) from Borneo (Sabah, East Malaysia).– Zoologica Scripta 14: 313-317.

Lienhard, C., 1988b. Three new extra-neotropical species of Troctopsocidae (Insecta: Psocoptera).– Journal of Natural History 22: 575-587.

Lienhard, C., 1990a. A new Oriental species of Troctopsocidae (Insecta: Psocoptera).– Revue suisse de Zoologie 97: 339-344.

Lienhard, C., 1990b. Revision of the western Palaearctic species of *Liposcelis* Motschulsky (Psocoptera: Liposcelididae).– Zoologische Jahrbücher (Abteilung Systematik) 117: 117-174.

Lienhard, C., 1991. New records and species of *Belaphopsocus* (Psocoptera: Liposcelididae).– Raffles Bulletin of Zoology 39: 75-85.

Lienhard, C., 1998a. Psocoptères euro-méditerranéens.– Faune de France 83: 1-517.

Lienhard, C., 2000b. A new genus of Prionoglarididae from a Namibian cave (Insecta: Psocoptera).– Revue suisse de Zoologie 107: 871-882.

Lienhard,C., 2002. Three extraordinary new species of Psocoptera (Insecta) from Colombia, Malaysia and Thailand (Epipsocidae, Lachesillidae, Ectopsocidae).– Revue suisse de Zoologie 109: 383-395.

Lienhard, C., 2003a. Additions and corrections (Part 1) to Lienhard & Smithers 2002: 'Psocoptera (Insecta) – World Catalogue and Bibliography' (pp. 1-24).– In: Yoshizawa, K. (ed.). Psocid News. The Psocidologists' Newsletter. No. 4 (February 17, 2003). Systematic Entomology, Faculty of Agriculture, Hokkaido University, Sapporo, Japan (see PsocoNet homepage, presently at: http://insect3.agr.hokudai.ac.jp/psoco-net/index.html).

Lienhard, C., 2003b. Additions and corrections (Part 2) to Lienhard & Smithers 2002: 'Psocoptera (Insecta). World Catalogue and Bibliography' (pp. 2-37).– In: Yoshizawa, K. (ed.). Psocid News. The Psocidologists' Newsletter. No. 5 (25 August 2003). Systematic Entomology, Faculty of Agriculture, Hokkaido University, Sapporo, Japan (see PsocoNet homepage, presently at: http://insect3.agr.hokudai.ac.jp/psoco-net/index.html).

Lienhard, C., 2003c. Nomenclatural amendments concerning Chinese Psocoptera (Insecta), with remarks on species richness.– Revue suisse de Zoologie 110: 695-721.

Lienhard, C., 2004a. *Siamoglaris zebrina* gen. n., sp. n., the first representative of Prionoglarididae from the Oriental Region (Insecta: Psocoptera).– Revue suisse de Zoologie 111: 865-875.

Lienhard, C., 2004b [2005]. New Myopsocidae (Insecta: Psocoptera) from Southeast Asia, with redescription of *Lophopterygella camelina* Enderlein. Thorntoniana. A commemorative volume for Ian W.B. Thornton (1926-2002).– Publicaciones Especiales del Instituto de Biologia, Universidad Nacional Autonoma de Mexico 20: 169-205.

Lienhard, C., 2004c. Additions and corrections (Part 3) to Lienhard & Smithers, 2002: 'Psocoptera (Insecta) – World Catalogue and Bibliography' (pp. 1-23).– In: Yoshizawa, K. (ed.). Psocid News. The Psocidologists' Newsletter. No. 6 (February 29, 2004). Systematic Entomology, Faculty of Agriculture, Hokkaido University, Sapporo, Japan. See also PsocoNet Homepage, presently at: http://insect3.agr.hokudai.ac.jp/psoco-net/index.html.

Lienhard, C., 2005. Additions and corrections (Part 4) to Lienhard & Smithers, 2002: 'Psocoptera (Insecta) – World Catalogue and Bibliography' (pp. 1-16).– In: Yoshizawa, K. (ed.). Psocid News. The Psocidologists' Newsletter. No. 7 (February 28, 2005). Systematic Entomology, Faculty of Agriculture, Hokkaido University, Sapporo, Japan. (See PsocoNet homepage, presently at: http://insect3.agr.hokudai.ac.jp/psoco-net/index.html).

Lienhard, C., 2006. Additions and corrections (Part 5) to Lienhard & Smithers, 2002: 'Psocoptera (Insecta) – World Catalogue and Bibliography' (pp. 1-18).– In: Yoshizawa, K. (ed.). Psocid News. The Psocidologists' Newsletter. No. 8 (February 28, 2006). Systematic Entomology, Faculty of Agriculture, Hokkaido University, Sapporo, Japan. (See PsocoNet Homepage, presently at: http://insect3.agr.hokudai.ac.jp/psoco-net/index.html).

Lienhard, C. & Ashmole, N. P., 1999. *Sphaeropsocopsis myrtleae* sp.n., a blind subterranean psocid from St Helena (Psocoptera: Sphaeropsocidae).– Revue suisse de Zoologie 106: 905-912.

Lienhard, C., Burckhardt, D. & Hauser, B., 1987. An improved beating tray and aspirator for collecting small arthropods.– Mitteilungen der Schweizerischen Entomologischen Gesellschaft 60: 107-112.

Lienhard, C. & Mockford, E. L., 1997. New taxa and records of Troctopsocidae (Psocoptera).– Mitteilungen Schweizerischen Entomologischen Gesellschaft 70: 361-385.

Lienhard, C. & Smithers, C. N., 2002. Psocoptera (Insecta). World Catalogue and Bibliography.– Instrumenta Biodiversitatis No 5. Muséum d'histoire naturelle, Geneva: xli+745 pp.

Li Fasheng, 1992d. Notes on Stenopsocidae and one new genus and seven new species (Psocoptera: Psocomorpha) from South China.– Entomotaxonomia 14: 245-256.

Li Fasheng, 1993a. [Psocoptera from National Chebaling Nature Reserve (Insecta: Psocoptera)].– [Collected papers for investigation in National Chebaling Nature Reserve]: 313-430. Science & Technology Publishing House of Guandong Province.

Li Fasheng, 1999. Psocoptera.– Fauna of insects, Fujian Province of China 3: 1-64.

Li Fasheng, 2000. A review of *Heterocaecilius* (s.l.) with the description of two new genera and new species (Psocoptera: Pseudocaeciliidae).– Contributions from the Tianjin Natural History Museum 17: 18-24.

Li Fasheng, 2002. [Psocoptera of China.] Science Press, Beijing: xlvi + 1976. (2 vols.)

Li Fasheng & Mockford, E. L. 1993. A description and notes on *Diplopsocus* gen. nov. and twenty-one new species from China (Psocoptera: Peripsocidae).– Oriental Insects 27: 55-91.

Linnaeus, C., 1758. Systema naturae. Tomus I. Regnum animale. 10 Edition Decima, Reformata Holmiae: vi + 824.

Lyal, C. H. C., 1985. Phylogeny and classification of the Psocodea with particular reference to the lice (Psocodea: Phthiraptera).– Systematic Entomology 10: 145-165.

MacKinnon, J. & MacKinnon, K., 1986. Review of the protected areas system in the Indo Malayan realm. IUCN, Gland: xii+ 284 pp.

McLachlan, R., 1866b. New genera and species of Psocidae.– Transactions of the Entomological Society of London (Third Series) 5: 345-352.

McLachlan, R., 1872b. Description of a new genus and five new species of exotic Psocidae.– Entomologist's Monthly Magazine 9: 74-78.

McLachlan, R., 1899d. *Ectopsocus briggsi*, a new genus and species of Psocidae found in England.– Entomologist's Monthly Magazine 35: 277-278.

Meinander, M., 1979a. Revision of the European Psocoptera recorded by Navás. – Miscellanea Zoologica 5: 59-67.

Meinander, M., Halkka, O. & Soderlund, V., 1974. Chromosomal evolution in the Psocoptera.– Notulae Entomologicae 54: 81-84.

Menon, R., 1938b. Coxal interlocking in the Lepidopsocidae (Copeognatha) and its probable taxonomic value.– Current Science 7: 66-67.

Menon, R., 1942. Studies on Indian Copeognatha (Psocoptera). II. Nanopsocetae and Psocatropetae.– Indian Journal of Entomology 4: 23-42.

Mey, W. & Tietze, F., 1979. Zur Indikation von Luftverunreinigungen mittels Psocoptera.– Hercynia, Neue Folge, Leipzig 16: 417-419.

Mockford, E. L., 1953b. Additional notes on Indiana Psocoptera.– Proceedings of the Indiana Academy of Science 62: 198-199.
Mockford, E. L., 1957a. Life history studies on some Florida insects of the genus *Archipsocus* (Psocoptera).– Bulletin of the Florida State Museum, Biological Sciences 1: 253-274.
Mockford, E. L., 1961a. The rediscovery and probable phylogenetic position of *Psilopsocus* (Psocoptera).– Psyche 68: 38-44.
Mockford, E. L., 1965b. The genus *Caecilius* (Psocoptera: Caeciliidae). Part I. Species groups and the North American species of the *flavidus* group.– Transactions of the American Entomological Society 91: 121-166.
Mockford, E. L., 1965c. A new genus of hump-backed psocids from Mexico and southwestern United States (Psocoptera: Psocidae).– Folia Entomologica Mexicana 11: 3-15.
Mockford, E. L., 1965e. Polymorphism in the Psocoptera: a review.– Proceedings of the North Central Branch of the Entomological Society of America. 20: 82-86.
Mockford, E. L., 1966a. The genus *Caecilius* (Psocoptera: Caeciliidae). Part II. Revision of the species groups and the North American species of the *fasciatus*, *confluens* and *africanus* groups.– Transactions of the American Entomological Society 92: 133-172.
Mockford, E. L., 1967a. Some Psocoptera from plumage of birds.– Proceedings of the Entomological Society of Washington 69: 307-309.
Mockford, E. L., 1967b. The Electrentomoid psocids (Psocoptera).– Psyche 74: 118-165.
Mockford, E. L., 1969a. The genus *Caecilius* (Psocoptera: Caeciliidae). Part III. The North American species of the *alcinus*, *caligonus* and *subflavus* groups.– Transactions of the American Entomological Society 95: 77-151.
Mockford, E. L., 1969b. Fossil insects of the order Psocoptera from Tertiary Amber of Chiapas, Mexico.– Journal of Paleontology 43: 1267-1273.
Mockford, E. L., 1974a. Records and descriptions of Cuban Psocoptera.– Entomologica Americana 48: 103-215.
Mockford, E. L., 1974d. New genus and species of Amphipsocidae from southeastern Asia (Psocoptera).– Entomological News 85: 155-158.
Mockford, E. L., 1975b. Genus *Eremopsocus* McLachlan: distinction from *Cerastipsocus* Kolbe and review of species (Psocoptera: Psocidae).– Psyche 82: 244-258.
Mockford, E. L., 1978a. A generic classification of family Amphipsocidae (Psocoptera: Caecilietae).– Transactions of the American Entomological Society 104: 139-190.
Mockford, E. L., 1981. Systematics of New World genera of Cerastipsocini and species of *Psococerastis* Pearman (Psocoptera: Psocidae: Cerastipsocinae).– Transactions of the American Entomological Society 107: 249-298.
Mockford, E. L., 1982. Redescription of the type species of *Myopsocus*, *M. unduosus* (Hagen), and resulting nomenclatural changes in genera and species of Myopsocidae (Psocoptera).– Psyche 89: 211-220.
Mockford, E. L., 1983a. Systematics of Asiopsocidae (Psocoptera) including *Pronotiopsocus amazonicus* n. gen. n. sp.– Florida Entomologist 66: 241-249.
Mockford, E. L., 1984d. Relationships among philotarsid and pseudocaeciliid genera and a proposed new family Bryopsocidae (Psocoptera).– Psyche 91: 309-318.
Mockford, E. L., 1987a. Order Psocoptera.– In: Stehr, F. (ed.). Immature Insects. Vol. 1. Kendall Hunt, Dubuque: 196-214.
Mockford, E. L., 1989a. *Xanthocaecilius* (Psocoptera: Caeciliidae), a new genus from the Western Hemisphere: 1. Description, species complexes, and species of the *quillayute* and *granulosus* complexes.– Transactions of the American Entomological Society 114: 265-294.
Mockford, E. L., 1991a. Psocids (Psocoptera).– In: United States Department of Agriculture and United States Department of Health and Human Services (eds.). Agriculture Handbook, no. 655: 371-402.

Mockford, E. L., 1993a. North American Psocoptera (Insecta).– Sandhill Crane Press, Gainesville, Florida: xviii + 455 pp.
Mockford, E. L., 1996a. New species and records of Psocoptera from northern Venezuela.– Tropical Zoology (Special Issue) 2: 1-98.
Mockford, E. l., 1997a. A new species of *Dicopomorpha* (Hymenoptera: Mymaridae) with diminutive, apterous males.– Annals of the Entomological Society of America 90: 115-120.
Mockford, E. L., 1998a. Generic definitions and species assignments in the family Epipsocidae (Psocoptera).– Insecta Mundi 12: 81-91.
Mockford, E. L., 1998b. Description of the male of *Calocaecilus decipiens* (Psocoptera: Amphipsocidae) with an assessment of the taxonomic position of the genus.– Oriental Insects 32: 79-81.
Mockford, E. L., 2000. A classification of the psocopteran family Caeciliusidae (Caeciliidae Auct.).– Transactions of the American Entomological Society 125 (1999): 325-417.
Mockford, E. L., 2001. Additions and corrections to 'A classification of the psocopteran family Caeciliusidae (Caeciliidae Auct.)'.– Transactions of the American Entomological Society 127: 79-84.
Mockford, E. L., 2004 [2005]. A new genus and species of thyrsophorine psocids (Insecta: Psocoptera: Psocidae: Thyrsophorinae) from Vietnam. Thorntoniana. A commemorative volume for Ian W. B. Thornton (1926-2002).– Publicaciones Especiales del Instituto de Biologia, Universidad Nacional Autonoma de Mexico 20: 151-158.
Mockford, E. L. 2005. A new genus of perientomine psocids (Psocoptera: Lepidopsocidae) with a review of the perientomine genera.– Transactions of the American Entomological Society 131: 201-215.
Mockford, E. L. & García Aldrete, A. N., 1976. A new species and notes on the taxonomic position of *Asiopsocus* Gunther (Psocoptera).– Southwestern Naturalist 21: 335-346.
Mockford, E. L. & Gurney, A. B., 1956. A review of the psocids, or book-lice and bark-lice, of Texas (Psocoptera).– Journal of the Washington Academy of Sciences 46: 353-368.
Mockford, E. L. & Sullivan, D. M., 1986. Systematics of the graphocaeciliine psocids, with a proposed higher classification of the family Lachesillidae (Psocoptera).– Transactions of the American Entomological Society 112: 1-80.
Mockford, E. L. & Wong, S. K., 1969. The genus *Kaestneriella* (Psocoptera: Peripsocidae).– Journal of the New York Entomological Society 77: 221-249.
Montgomery, S. L., 1982. Biogeography of the moth genus *Eupithecia* in Oceania and the evolution of ambush predation in Hawaiian caterpillars (Lepidoptera: Geometridae).– Entomologica Generalis 8: 27-34.
Navás, L., 1920b. Insecta nova. VII series.– Memorie della Pontifica Accademia Romana dei Nuovi Lincei, ser. 2, 5: 21-29.
Navás, L., 1920c. Socópteros nuevos.– Boletin de la Sociedad Entomologica de España 4: 90-94.
Navás, L., 1922d. Insectos exoticos.– Broteria, Serie Zoologica 20: 49-63.
Navás, L., 1922e. Estudias sobre Neuropteros (Insectes).– Arxius de l'Instituto de Ciencias de Barcelona 7: 179-203.
Navás, L., 1923b. Insecta nova. X series.– Memorie della Pontifica Accademia Romana dei Nuovi Lincei 6: 19-27.
Navás, L., 1924b. Insectos de la America Central.– Broteria, Serie Zoologica 21: 55-86.
Navás, L., 1924c. Mis excursiones entomológicas de verano de 1924.– Broteria, Serie Zoológica 21: 115-150.
Navás, L., 1924g. Insecta orientalia. II Serie.– Memorie della Pontifica Accademia delle Scienze Nuovi Lincei 7: 211-216.

Navás, L., 1927b. Comunicaciones entomológicas. 8. Socópteros del Museo de Hamburgo.– Revista de la Academia de Ciencias exactas, fisico-quimicas y naturales de Zaragoza 11: 37–52.
Navás, L., 1934c. Névroptères et insectes voisins. Chine et pays environnants. Septième série.– Notes d'Entomologie chinoise 2: 1–16.
New, T. R., 1969a. Observations on the biology of Psocoptera found in leaf litter in southern England.– Transactions of the Society for British Entomology 18: 169–180.
New, T. R., 1969b. The biology of some species of *Alaptus* (Mymaridae) parasitising eggs of Psocoptera.– Transactions of the Society for British Entomology 18: 181–193.
New, T. R., 1969c. The early stages and life histories of some British foliage-frequenting Psocoptera, with notes on the overwintering stages of British arboreal Psocoptera.– Transactions of the Royal Entomological Society of London 121: 59–77.
New, T. R., 1969d. Aerial dispersal of some British Psocoptera, as indicated by suction trap catches.– Proceedings of the Royal Entomological Society of London (A) 44: 49–61.
New, T. R., 1970b. The relative abundance of some British Psocoptera on different species of trees.– Journal of Animal Ecology 39: 521–540.
New, T. R., 1970c. Observations on reproduction in some British foliage-frequenting Psocoptera.– Entomologist's Gazette 21: 143–149.
New, T. R., 1971c. The Psocoptera of the Canadian Nepal Expedition.– Canadian Entomologist 103: 188–213.
New, T. R., 1971d. The Psocoptera of Curtis Island, Tasmania.– Journal of the Australian Entomological Society 10: 223–229.
New, T. R., 1971e. A new species of *Belaphapsocus* Badonnel from Brazil, with notes on its early stages and bionomics (Psocoptera).– Entomologist 104: 124–133.
New, T. R., 1971f. Species of *Lachesilla* (Psocoptera) associated with palm trees in central Brazil.– Zoological Journal of the Linnean Society 50: 431–447.
New, T. R., 1972a. Some Epipsocetae (Psocoptera) from central Brazil.– Transactions of the Royal Entomological Society of London 123: 455–497.
New, T. R., 1973c. The Archipsocidae of South America (Psocoptera).– Transactions of the Royal Entomological Society of London 125: 57–105.
New, T. R., 1973f. Some Psocoptera from Thailand.– Oriental Insects 7: 225–237.
New, T. R., 1973j. Local distribution of Psocoptera in the Mato Grosso, central Brazil.– Papeis Avulsos de Zoología, Sao Paulo 27: 115–144.
New, T. R., 1974e. Structural variation in psocopteran wing-coupling mechanisms.– International Journal of Insect Morphology and Embryology 3: 193–201.
New, T. R., 1974h. Psocoptera. [Handbooks for the identification of British insects.]– Royal Entomological Society of London: 102.
New, T. R., 1975a. Aerial dispersal of some Victorian Psocoptera, as indicated by suction trap catches.– Journal of the Australian Entomological Society 14: 179–184.
New, T. R., 1975d. Lepidopsocidae and Amphientomidae (Psocoptera) from Malaysia and Singapore.– Oriental Insects 9: 177–194.
New, T. R., 1975e. Psocidae (Psocoptera) from Malaysia and Singapore.– Oriental Insects 9: 243–259.
New, T. R., 1976a. Notes on two Indonesian Psocidae (Psocoptera) described by L. Navás.– Oriental Insects 10: 363–368.
New, T. R., 1977a. Notes on the identification of nymphs of the British Psocoptera.– Entomologist's Gazette 28: 61–71.
New, T. R., 1977h. Psocoptera of the Oriental Region: a review of published taxonomic information with keys to families and genera.– Oriental Insects, Supplement 6: 1–83.
New, T. R., 1977i. A reappraisal of the genus *Callistoptera* Enderlein (Psocoptera).– Australian Entomological Magazine 4: 52–54.

New, T.R., 1978a. An appraisal of the Thyrsophorinae stat. nov. (Psocoptera, Psocidae) and of its constituent genera from the Neotropics.– Systematic Entomology 3: 35-49.

New, T.R., 1978b. The genus *Neurosema* (Psocoptera, Calopsocidae) from New Guinea.– Systematic Entomology 3: 51-57.

New, T.R., 1978c. Notes on the Oriental species of *Trichadenotecnum* Enderlein s.l. (Psocoptera: Psocidae).– Oriental Insects 12: 33-41.

New, T.R., 1978d. Notes on some Oriental genera of Cerastipsocinae (Psocoptera: Psocidae).– Oriental Insects 12: 43-48.

New, T.R., 1978e. Notes on Calopsocidae (Psocoptera).– Oriental Insects 12: 305-318.

New, T.R., 1978f. Ergebnisse der Bhutan Expedition 1972 des Naturhistorischen Museums in Basel. Psocoptera.– Entomologica Basiliensia 3: 67-86.

New, T.R., 1978h. Notes on the complex of species allied to *Psococerastis taprobanes* (Hagen) (Psocoptera, Psocidae).– Entomologica Basiliensia 3: 235-248.

New, T.R., 1984. Trap catches of Psocoptera in the Reserva Ducke, Amazonas.– Acta Amazonica 13: 591-599.

New, T.R., 1985b. Communal oviposition and egg-brooding in a psocid, *Peripsocus nitens* (Insecta, Psocoptera) in Chile.– Journal of Natural History 19: 419-423.

New, T.R., 1987a. Biology of the Psocoptera.– Oriental Insects 21: 1-109.

New, T.R., 1993. Pseudocaeciliidae (Insecta: Psocoptera) from East Malaysia.– Oriental Insects 27: 335-345.

New, T.R., 2000. How to conserve the 'meek inheritors'.– Journal of Insect Conservation 4: 151-152.

New, T.R., 2002. The Neuroptera (Insecta) of Malesia.– Fauna Malesiana Handbook no 4. Brill, Leiden: viii + 204 pp.

New, T.R. & Collins, N.M., 1987. 'Herd-grazing' in tropical Psocoptera.– Entomologist's Monthly Magazine 123: 229-230.

New, T.R. & Lee, S.S., 1991a. A new species of *Psilopsocus* Enderlein (Psocoptera: Psilopsocidae) from Malaysia.– Journal of the Australian Entomological Society 30: 63-65.

New, T.R. & Lee, S.S., 1991b. An unusual new species of Psyllipsocidae (Psocoptera) from Malaysia.– Oriental Insects 25: 117-119.

New, T.R. & Lee, S.S., 1991c. Epipsocidae (Psocoptera) from West Malaysia.– Oriental Insects 25: 121-126.

New, T.R. & Lee, S.S., 1991d. *Lachesilla* Westwood (Psocoptera: Lachesillidae) in Malaysia.– Oriental Insects 25: 127-129.

New, T.R. & Lee, S.S., 1991e. Peripsocidae (Psocoptera) from West Malaysia.– Oriental Insects 25: 183-198.

New, T.R. & Lee. S.S., 1992a. *Trichadenotecnum* Enderlein (Psocoptera: Psocidae) from West Malaysia.– Oriental Insects 26: 153-158.

New, T.R. & Lee, S.S., 1992b. Philotarsidae (Psocoptera) from West Malaysia.– Oriental Insects 26: 159-164.

New, T.R. & Lee, S.S., 1992c. Ectopsocidae (Psocoptera) from West Malaysia.– Oriental Insects 26: 165-184.

New, T.R. & Lee, S.S., 1992d. Replacement name for *Psyllipsocus ornatus* New & Lee (Psocoptera: Psyllipsocidae).– Oriental Insects 26: 391.

New, T.R. & Thornton, I.W.B., 1975. Psocomorpha (Psocoptera) collected on recent expeditions to South America.– Journal of Entomology (B) 44: 27-80.

New, T.R. & Thornton, I.W.B., 1976. Psocomorpha (Psocoptera) from the Malayan Peninsula.– Oriental Insects 9 (1975): 375-418.

Okamoto, H. 1910. Die Caeciliiden Japans.– Annales historico-naturales Musei nationalis hungarici 8: 185-212.

Pearman, J.V., 1928b. On sound production in the Psocoptera and on a presumed stridulatory organ.– Entomologist's Monthly Magazine 64: 179-186.
Pearman, J.V., 1928c. Biological observations on British Psocoptera.– Entomologist's Monthly Magazine 64: 209-218, 239-243, 263-268.
Pearman, J.V., 1932a. Notes on the genus *Psocus*, with special reference to the British species.– Entomologist's Monthly Magazine 68: 193-204.
Pearman, J.V., 1934a. A new species of *Archipsocus* (Psocoptera).– Stylops 3: 112-113.
Pearman, J.V., 1934b. New and little-known African Psocoptera.– Stylops 3: 121-132.
Pearman, J.V., 1936a. The taxonomy of the Psocoptera: preliminary sketch.– Proceedings of the Royal Entomological Society of London, ser. B, 5: 58-62.
Pearman, J. V., 1942. Third note on Psocoptera from warehouses.– Entomologist's Monthly Magazine 78: 289-292.
Popescu, C., 1979. Natural selection in the industrial melanic psocid *Mesopsocus unipunctatus* (Müll.) (Insecta: Psocoptera) in northern England.– Heredity 42: 133-142.
Popescu, C., Broadhead, E. & Shorrocks, B., 1978. Industrial melanism in *Mesopsocus unipunctatus* (Müll.) (Psocoptera) in northern England.– Ecological Entomology 3: 209-219.
Ribaga, C., 1899. Descrizione di un nuovo genere e di una nuova specie di Psocidi trovato in Italia.– Rivista di Patologia Vegetale 8: 156-159.
Ribaga, C., 1904. Sul genere *Ectopsocus* MacLachl. e descrizione di una nuova varietà dell' *Ectopsocus briggsi* MacLachl.– Redia 1: 294-298.
Ribaga, C., 1908a. Un nuovo Copeognato dell' Isola di Giava.– Redia 5: 20-26.
Ribaga, C., 1908b. Copeognati eastraeuropei del Museo Civico di Storia Naturale di Genova.– Redia 5: 98-109.
Roesler, R., 1940c. Neue Copeognathen.– Arbeiten über morphologische und taxonomische Entomologie 7: 236-244.
Roesler, R., 1943. Über einige Copeognathengenera.– Stettiner Entomologische Zeitung 104: 1-14.
Roesler, R., 1944. Die Gattungen der Copeognathen.– Stettiner Entomologische Zeitung 105: 117-166.
Schmidt, E. R. & New, T. R. 2004. A systematic and phylogenetic revision of the family Elipsocidae (Insecta: Psocoptera), with the erection of two new families: Lesneiidae and Sabulopsocidae.– Invertebrate Systematics 18: 157-213.
Seeger, W., 1975. Funktionsmorphologie an Spezialbildungen der Fühlergeissel von Psocoptera und anderen Paraneoptera (Insecta: Psocodea); Psocodea als monophyletische Gruppe.– Zeitschrift für Morphologie der Tiere 81: 137-159.
Seeger, W., 1979. Spezialmerkmale an Eihüllen und Embryonen von Psocoptera im Vergleich zu anderen Paraneoptera (Insecta): Psocoptera als monophyletische Gruppe.– Stuttgarter Beiträge zur Naturkunde (ser. A) 329: 1-57.
Shipley, A. E., 1904. The orders of insects.– Zoologischer Anzeiger 27: 259-262.
Simpson, G. G., 1977. Too many lines: the limits of the Oriental and Australian zoogeographic regions.– Proceedings of the American Philosophical Society 121: 107-120.
Smithers, C. N., 1963a. *Psilopsocus mimulus* sp. n. (Psocoptera: Psilopsocidae), representing a family new to Australia.– Journal of the Entomological Society of Queensland 2: 56-59.
Smithers, C. N., 1964d. The Myopsocidae (Psocoptera) of Australia.– Proceedings of the Royal Entomological Society of London (B) 33: 133-138.
Smithers, C. N., 1965a. The Lepidopsocidae (Psocoptera) of Australia.– Journal of the Entomological Society of Queensland 4: 72-78.
Smithers, C. N., 1965c. *Caecilius soehardjani*, a new name for *Caecilius ferrugineus* Soehardjan and Hamann (Psocoptera: Caeciliidae).– Australian Zoologist 13: 136.

Smithers, C.N., 1965d. A bibliography of the Psocoptera (Insecta).– Australian Zoologist 13: 137–209.
Smithers, C.N., 1967a. A catalogue of the Psocoptera of the world.– Australian Zoologist 14: 1–145.
Smithers, C.N., 1967b. On the relationships of the Calopsocidae (Psocoptera).– Journal of the Australian Entomological Society 6: 61–64.
Smithers, C.N., 1972c. The classification and phylogeny of the Psocoptera.– Australian Museum Memoirs 14: 1–349.
Smithers, C.N., 1974c. Two new species and additional records of *Taeniostigma* Enderlein (Psocoptera: Stenopsocidae) from Australia.– Journal of the Australian Entomological Society 13: 211–217.
Smithers, C.N., 1974d. A new Australian species of *Stenopsocus* (Psocoptera, Stenopsocidae).– Australian Entomological Magazine 2: 21–23.
Smithers, C.N., 1975c. The names of Australian and New Zealand Myopsocidae (Psocoptera).– Australian Entomological Magazine 2: 76–78.
Smithers, C.N., 1976a. *Epipsocopsis mouldsi* sp. n. representing a family (Psocoptera: Epipsocidae) new to Australia.– Australian Entomological Magazine 3: 32–34.
Smithers, C.N., 1976b. The generic position of *Amphigerontia formosa* Banks (Psocoptera: Psocidae) and related species.– Australian Entomological Magazine 3: 67–73.
Smithers, C.N., 1977a. *Cycetes collessi* sp.n. (Psocoptera, Psocidae) representing a genus new to Australia.– Australian Entomological Magazine 3: 112–114.
Smithers, C.N., 1978a. A new genus of Myopsocidae (Psocoptera) from Queensland.– Journal of the Australian Entomological Society 17: 105–107.
Smithers, C.N., 1980b. The position of *Sigmatoneura* Enderlein (Psocoptera: Psocidae).– Australian Entomological Magazine 6: 93.
Smithers, C.N., 1981d. A new species of *Dirla* Navas (Psocoptera: Calopsocidae) with comments on the position of the genus.– Australian Entomological Magazine 8: 24–28.
Smithers, C.N., 1983a. A reappraisal of *Clematostigma* Enderlein with notes on related genera (Psocoptera: Psocidae).– Australian Entomological Magazine 9: 71–79.
Smithers, C.N., 1984b. The Psocoptera (Insecta) of South Australia.– Records of the South Australian Museum 18: 453–491.
Smithers, C.N., 1985a. Redefinition of *Copostigma* Enderlein (Psocoptera: Psocidae).– Australian Entomological Magazine 12: 61–62.
Smithers, C.N., 1985c. Name changes for some species of Myopsocidae (Psocoptera).– Psyche 92: 265–267.
Smithers, C.N., 1990a. Keys to the families and genera of Psocoptera (Arthropoda: Insecta).– Technical Reports of the Australian Museum 2: 1–82.
Smithers, C.N., 1991b. Psocoptera (psocids, booklice).– In: CSIRO (eds.). The Insects of Australia. Melbourne University Press, Melbourne: 412–420.
Smithers, C.N., 1994c. A redescription of *Anopistoscena specularifrons* Enderlein (Psocoptera: Hemipsocidae), confirming its position in the Hemipsocidae.– African Entomology 2: 25–29.
Smithers, C.N., 1995a. Final instar nymph of *Psilopsocus nebulosus* Mockford (Psocoptera: Psilopsocidae) redescribed and compared with two wood-boring species of the genus.– Beiträge zur Entomologie 45: 375–381.
Smithers, C.N., 1995b. *Psilopsocus mimulus* Smithers (Psocoptera: Psilopsocidae), the first known wood-boring psocopteran.– Journal of the Australian Entomological Society 34: 117–120.
Smithers, C.N., 1995d. Psocoptera (Insecta) of Christmas Island.– Invertebrate Taxonomy 9: 529–561.
Smithers, C.N., 1996a. Psocoptera.– In: Wells, A. (ed.). Zoological Catalogue of Australia 26. CSIRO, Australia: 1–79, 363–372.

Smithers, C. N., 2004 [2005]. A new genus for three species of Australian Myopsocidae (Insecta: Psocoptera). Thorntoniana. A commemorative volume for Ian W. B. Thornton (1926-2002).– Publicaciones Especiales del Instituto de Biologia, Universidad Nacional Autonoma de Mexico 20: 159-168.

Smithers, C. N. & Thornton, I. W. B., 1973. The Psilopsocidae (Psocoptera) of New Guinea.– Proceedings of the Linnean Society of New South Wales 98: 98-103.

Smithers, C. N. & Thornton, I. W. B., 1974b. The Myopsocidae (Psocoptera) of New Guinea and New Caledonia.– Transactions of the Royal Entomological Society of London 126: 91-127.

Smithers, C. N. & Thornton, I. W. B., 1975b. The first record of Stenopsocidae (Psocoptera) from New Guinea with descriptions of new species.– Proceedings of the Linnean Society of New South Wales 100: 156-166.

Smithers, C. N. & Thornton, I. W. B., 1977. A new genus and some new species of Epipsocidae (Psocoptera) from the Melanesian arc.– Proceedings of the Linnean Society of New South Wales 102: 60-75.

Smithers, C. N. & Thornton, I. W. B., 1979. Psilopsocidae and Myopsocidae (Insecta: Psocoptera) of the Bismarck Archipelago, Solomon Islands and New Hebrides.– Records of the Australian Museum 32: 513-545.

Smithers, C. N. & Thornton, I. W. B., 1981a. The Psocidae (Insecta: Psocoptera) of New Guinea, including a new coleopteriform genus from high on Mount Wilhelm.– Australian Journal of Zoology 29: 921-969.

Smithers, C. N. & Thornton, I. W. B., 1981b. The role of New Guinea in the evolution and biogeography of some families of psocopteran insects.– Monographia Biologicae 42: 621-638.

Smithers, C. N. & Thornton, I. W. B., 1990. Systematics and distribution of the Melanesian Psocidae (Psocoptera).– Invertebrate Taxonomy 3: 431-468.

Soehardjan, M., 1958. First contribution to a study of Copeognatha (Corrodentia) of the Indonesian archipelago.– Idea 11: 25-33.

Soehardjan, M. & Hamann, H. H. F., 1959. Second contribution to the Indonesian psocid fauna: notes on some recent collectings.– Idea 12: 1-13.

Spieksma, F. T. M. & Smits, C., 1975. Some ecological and biological aspects of the booklouse *Liposcelis bostrychophilus* Badonnel, 1931 (Psocoptera).– Netherlands Journal of Zoology 25: 219-230.

Thornton, I. W. B., 1959a. New species of *Peripsocus* Hagen, 1866 (Corrodentia: Peripsocidae) from Hong Kong Island, with further descriptions of *Peripsocus similis* Enderlein, 1903 and *Peripsocus quercicola* Enderlein, 1906.– Proceedings of the Royal Entomological Society of London (B) 28: 37-48.

Thornton, I. W. B., 1959b. A new genus of Philotarsidae (Corrodentia) and new species of this and related families from Hong Kong.– Transactions of the Royal Entomological Society of London 111: 331-349.

Thornton, I. W. B., 1960. New Psocidae and an aberant new myopsocid (Psocoptera) from Hong Kong.– Transactions of the Royal Entomological Society of London 112: 239-261.

Thornton, I. W. B., 1961a. The *Trichadenotecnum* group (Psocoptera: Psocidae) in Hong Kong, with descriptions of new species.– Transactions of the Royal Entomological Society of London 113: 1-24.

Thornton, I. W. B., 1961b. Comments on the geographical distribution of *Pseudocaecilius elutus* Enderlein (Psocoptera), and descriptions of related new species from Hong Kong.– Proceedings of the Royal Entomological Society of London (B) 30: 141-152.

Thornton, I. W. B., 1962b. Psocids (Psocoptera) from the Batu Caves, Malaya.– Pacific Insects 4: 441-455.

Thornton, I. W. B., 1962c. The Peripsocidae (Psocoptera) of Hong Kong.– Transactions of the Royal Entomological Society of London 114: 285-315.
Thornton, I. W. B., 1964. Air-borne Psocoptera trapped on ships and aircraft.– Pacific Insects 6: 285-291.
Thornton, I. W. B., 1980. Plate tectonics and the distribution of the insect family Philotarsidae (Order Psocoptera) in the southwest Pacific.– Palaeogeography, Palaeoclimatology, Palaeoecology 31: 251-266.
Thornton, I. W. B., 1981a. Psocoptera of the Fiji Islands.– Pacific Insects Monographs 37: 1-105.
Thornton, I. W. B., 1981f. The Psocoptera of the Hawaiian Islands. Parts I and II. Introduction and the nonendemic fauna.– Pacific Insects 23: 1-49.
Thornton, I. W. B., 1981g. The systematics, phylogeny and biogeography of the psocopteran family Philotarsidae.– Systematic Entomology 6: 413-452.
Thornton, I. W. B., 1984a. Psocoptera of the Hawaiian Islands. Part III. The endemic *Ptycta* complex (Psocidae): systematics, distribution and evolution.– International Journal of Entomology 26: 1-128.
Thornton, I. W. B., 1984b. Psocoptera and Wallace's Line: collections from the islands of Bali and Lombok.– Treubia 29: 83-177.
Thornton, I. W. B., 1984c. An unusual psocopteran from New Guinea and its relationships within the Philotarsidae.– International Journal of Entomology 26: 378-385.
Thornton, I. W. B., 1985a. The geographical and ecological distribution of arboreal Psocoptera.– Annual Review of Entomology 30: 175-196.
Thornton, I. W. B., 1985b. A preliminary survey of the psocopteran fauna of the Krakatau Islands.– In: Sastrapradja, D., Soemodihardjo, S., Soemartadipura, A. S., Soegiarto, A., Adisoemarto, S., Kusamadinata, K., Birowo, S., Riswan, S. & Lapian, A. B. (eds.). Proceedings of the Symposium on 100 years development of Krakatau and surroundings, Jakarta, 23-27 August 1983. Volume 1, Natural Sciences. L. I. P. I., Jakarta: 466-470.
Thornton, I. W. B. & Browning, J. A., 1992. Myopsocidae (Insecta: Psocoptera) from Java, including a discussion of the known Indonesian fauna.– Revue suisse de Zoologie 99: 343-367.
Thornton, I. W. B. & Harrell, J. C., 1965. Air-borne Psocoptera trapped on ships and aircraft. 2. Pacific ship trappings 1963-64.– Pacific Insects 7: 700-702.
Thornton, I. W. B. & New, T. R., 1977a. Philotarsidae (Psocoptera) of the Bismarck Archipelago.– Pacific Insects 17: 451-457.
Thornton, I. W. B. & Smithers, C. N., 1974. The Philotarsidae (Psocoptera) of New Caledonia.– Pacific Insects 16: 177-243.
Thornton, I. W. B. & Smithers, C. N., 1977. Philotarsidae (Psocoptera) of New Guinea.– Pacific Insects 17: 419-450.
Thornton, I. W. B. & Smithers, C. N., 1984. Systematics of the Calopsocidae, an Oriental and Melanesian family of Psocoptera.– Systematic Entomology 9: 183-244.
Thornton, I. W. B. & Wong, S. K., 1966. Some Psocoptera from West Bengal, India.– Transactions of the Royal Entomological Society of London 118: 1-21.
Thornton, I. W. B. & Wong, S. K., 1968. The peripsocid fauna (Psocoptera) of the Oriental Region and the Pacific.– Pacific Insects Monographs 19: 1-158.
Thornton, I. W. B., Lee, S. S. & Chui, W. D., 1972. Psocoptera.– Insects of Micronesia 8: 45-114.
Thornton, I. W. B., Wong, S. K. & Smithers, C. N., 1977. The Philotarsidae (Psocoptera) of New Zealand and the islands of the New Zealand plateau.– Pacific Insects 17: 197-228.
Thornton, I. W. B., New, T. R., McLaren, D. A., Sudarman, H. K. & Vaughan, P. J., 1988. Airborne arthropod fall-out on Anak Krakatau and a possible pre-vegetation pioneer community.– Philosophical Transactions of the Royal Society of London (series B) 322: 471-479.

Turner, B. D., 1973a. The distribution of webs and egg-batches of *Pseudocaecilius citricola* (Ashmead) (Insecta: Psocoptera: Pseudocaeciliidae) on the leaves of *Mangifera indica* L. and *Cordia collococca* L. in Jamaica, W. I.– Journal of Natural History 7: 561-565.
Turner, B. D., 1974b. The abdominal adhesive organs of *Caecilius equivocatus* Mockford (Caeciliidae, Psocoptera, Insecta).– Journal of Natural History 8: 427-431.
Turner, B. D., 1974c. The population dynamics of tropical arboreal Psocoptera (Insecta) on two species of conifers in the Blue Mountains, Jamaica.– Journal of Animal Ecology 43: 323-337.
Turner, B. D., 1975b. Energy flow in arboreal epiphytic communities. An empirical model of net primary productivity in the alga *Pleurococcus* on larch trees.– Oecologia 20: 179-188.
Turner, B. D., 1983a. Annual respiration and production estimates for collembolan and psocopteran epiphyte herbivores on larch trees in southern England.– Ecological Entomology 8: 213-228.
Turner, B. D., 1984a. Predation pressure on the arboreal epiphytic herbivores of larch trees in southern England.– Ecological Entomology 9: 91-100.
Turner, B. D., 1987. Forming a clearer view of *L. bostrychophilus*.– Environmental Health 95: 9-13.
Turner, B. D., 1994a. *Liposcelis bostrychophila* (Psocoptera: Liposcelididae), a stored food pest in the UK.– International Journal of Pest Management 40: 179-190.
Turner, B. D. & Broadhead, E. 1974. The diversity and distribution of psocid populations on *Mangifera indica* L. in Jamaica and their relationship to altitude and micro-epiphyte diversity.– Journal of Animal Ecology 43: 173-190.
Turner, B. D. & Maude-Roxby, H., 1988. Starvation survival of the stored product pest *Liposcelis bostrychophilus* Badonnel (Psocoptera; Liposcelidae).– Journal of Stored Products Research 24: 23-28.
van Steenis, C. G. G. J., 1950. The delimitation of Malesia and its main plant geographical divisions.– Flora Malesiana 1. 1: LXX-LXXXV.
Vaughan, P. J., Thornton, I. W. B. & New, T. R., 1989. The Psocoptera of the Krakatau Islands, Indonesia.– Treubia 30: 1-93.
Vaughan, P. J., Thornton, I. W. B. & New, T. R., 1991. Psocoptera from southern Sumatra and West Java, Indonesia: source faunas for colonisation of the Krakatau islands.– Treubia 30: 103-164.
Weidner, H., 1972. Copeognatha (Staubläuse).– Handbuch der Zoologie, Berlin. VI, 2. Hälfte, 2. Teil, Beitrag 16: 1-94.
Wolda, H. & Broadhead, E. 1985. Seasonality of Psocoptera in two tropical forests in Panama.– Journal of Animal Ecology 54: 519-530.
Wong, S. K. & Thornton, I. W. B., 1966. Chromosome numbers of some psocid genera.– Nature 211: 214-215.
Yoshizawa, K., 1995. Taxonomic study on the family Pseudocaeciliidae (Psocoptera: Psocomorpha) of Japan. 1. Revision of the genus *Pseudocaecilius*.– Japanese Journal of Entomology 63: 703-721.
Yoshizawa, K., 1996a. Taxonomic study on the family Pseudocaeciliidae (Psocoptera: Psocomorpha) of Japan. 2. Genera *Heterocaecilius* and *Phallocaecilius*.– Japanese Journal of Entomology 64: 3-16.
Yoshizawa, K., 1998a. A new genus, *Atrichadenotecnum*, of the Psocini (Psocoptera: Psocidae) and its systematic position.– Entomologica Scandinavica 29: 199-209.
Yoshizawa, K., 2000a. Redescription of *Mepleres suzukii* (Okamoto), with comments on synonymy among *Mepleres*, *Pseudoscottiella* and *Meniscopsocus* (Psocodea: 'Psocoptera': Pseudocaeciliidae).– Entomological Science 3: 669-674.

Yoshizawa, K., 2001a. A systematic revision of Japanese *Trichadenotecnum* Enderlein (Psocodea: 'Psocoptera': Psocidae: Ptyctini) with redefinition and subdivision of the genus.– Invertebrate Taxonomy 15: 159-204.

Yoshizawa, K., 2001b. Systematic study of Amphipsocidae in Japan (Psocodea: 'Psocoptera': Caeciliusetae), with comments on higher classification within the family.– Insecta Matsumurana (new series) 58: 1-25.

Yoshizawa, K., 2002. Phylogeny and higher classification of suborder Psocomorpha (Insecta: Psocodea: 'Psocoptera').– Zoological Journal of the Linnean Society 136: 371-400.

Yoshizawa, K. & Johnson, K.P., 2003. Phylogenetic position of Phthiraptera (insecta: Paraneoptera) and elevated rate of evolution in mitochondrial 12S and 16S rDNA.– Molecular Phylogenetics and Evolution 29: 102-114.

Yoshizawa, K. & Lienhard, C., 2004 [2005]. Systematics of *Trichadenotecnum* (Psocoptera: Psocidae) in Hong Kong. Thorntoniana. A commemorative volume for Ian W.B. Thornton (1926-2002).– Publicaciones Especiales del Instituto de Biología, Universidad Nacional Autonoma de Mexico 20: 121-149.

Yoshizawa, K. & Saigusa, T., 2001. Phylogenetic analysis of paraneopteran orders (Insecta: Neoptera) based on forewing base structure, with comments on monophyly of Auchenorrhyncha.– Systematic Entomology 26: 1-13.

Yoshizawa, K. & Saigusa, T., 2003. Reinterpretation of clypeus and maxilla in Psocoptera, and their significance in phylogeny of Paraneoptera (Insecta: Neoptera).– Acta Zoologica 84: 33-40.

Yoshizawa, K., Lienhard, C. & Idris, A.B., 2005. On the systematic position of *Podopterocus* Banks and *Dinopsocus* Banks, with a revised diagnosis of the genus *Sigmatoneura* Enderlein (Psocodea: 'Psocoptera': Psocidae).– Revue suisse de Zoologie 112: 831-848.

Index

This index lists all scientific names referred to in the text. The principal treatment of a taxon starts on the pages underlined; illustrations of a particular taxon are in **bold**.

N

O

P